Autophagy in
Microenviron

Sujit Kumar Bhutia
Editor

Autophagy in Tumor and Tumor Microenvironment

Springer

Editor
Sujit Kumar Bhutia
Department of Life Science
National Institute of Technology Rourkela
Rourkela, Odisha, India

ISBN 978-981-15-6932-6   ISBN 978-981-15-6930-2 (eBook)
https://doi.org/10.1007/978-981-15-6930-2

© Springer Nature Singapore Pte Ltd. 2020
This work is subject to copyright. All rights are reserved by the Publisher, whether the whole or part of the material is concerned, specifically the rights of translation, reprinting, reuse of illustrations, recitation, broadcasting, reproduction on microfilms or in any other physical way, and transmission or information storage and retrieval, electronic adaptation, computer software, or by similar or dissimilar methodology now known or hereafter developed.
The use of general descriptive names, registered names, trademarks, service marks, etc. in this publication does not imply, even in the absence of a specific statement, that such names are exempt from the relevant protective laws and regulations and therefore free for general use.
The publisher, the authors, and the editors are safe to assume that the advice and information in this book are believed to be true and accurate at the date of publication. Neither the publisher nor the authors or the editors give a warranty, expressed or implied, with respect to the material contained herein or for any errors or omissions that may have been made. The publisher remains neutral with regard to jurisdictional claims in published maps and institutional affiliations.

This Springer imprint is published by the registered company Springer Nature Singapore Pte Ltd.
The registered company address is: 152 Beach Road, #21-01/04 Gateway East, Singapore 189721, Singapore

Contents

1 **Role of Xenobiotic in Autophagy Inflection in Cell Death and Carcinogenesis** 1
   Durgesh Nandini Das and Prashanta Kumar Panda

2 **Autophagy: An Agonist and Antagonist with an Interlink of Apoptosis in Cancer** 35
   Vanishri Chandrashekhar Haragannavar, Roopa S. Rao,
   Kewal Kumar Mahapatra, Srimanta Patra, Bishnu Prasad Behera,
   Amruta Singh, Soumya Ranjan Mishra, Chandra Sekhar Bhol,
   Debasna Pritimanjari Panigrahi, Prakash Priyadarshi Praharaj,
   Sujit Kumar Bhutia, and Shankargouda Patil

3 **Cross-Talk Between DNA Damage and Autophagy and Its Implication in Cancer Therapy** 61
   Ganesh Pai Bellare, Pooja Gupta, Saikat Chakraborty,
   Mrityunjay Tyagi, and Birija Sankar Patro

4 **miRNAs and Its Regulatory Role on Autophagy in Tumor Microenvironment** 77
   Assirbad Behura, Abtar Mishra, Ashish Kumar, Lincoln Naik,
   Debraj Manna, and Rohan Dhiman

5 **Exploring the Metabolic Implications of Autophagy Modulation in Tumor Microenvironment** 103
   Subhadip Mukhopadhyay

6 **Mitophagy and Reverse Warburg Effect: Metabolic Compartmentalization of Tumor Microenvironment** 117
   Prajna Paramita Naik

7 **Mitochondrial Biogenesis, Mitophagy, and Mitophagic Cell Death in Cancer Regulation: A Comprehensive Review** 141
   Prakash Priyadarshi Praharaj, Bishnu Prasad Behera,
   Soumya Ranjan Mishra, Srimanta Patra, Kewal Kumar Mahapatra,
   Debasna Pritimanjari Panigrahi, Chandra Sekhar Bhol,
   and Sujit Kumar Bhutia

8	**Mechanical Stress-Induced Autophagy: A Key Player in Cancer Metastasis** 171
	Joyjyoti Das and Tapas Kumar Maiti

9	**The Interplay of Autophagy and the Immune System in the Tumor Microenvironment** 183
	Chandan Kanta Das, Bikash Chandra Jena, Ranabir Majumder, Himadri Tanaya Panda, and Mahitosh Mandal

10	**Relevance of Autophagy in Cancer Stem Cell and Therapeutic** 203
	Niharika Sinha

11	**The Autophagy Conundrum in Cancer Development, Progression and Therapeutics** 223
	Siddavaram Nagini, Palrasu Manikandan, and Rama Rao Malla

12	**Targeting Autophagy in Cancer: Therapeutic Implications** 249
	Dipanwita Das Mukherjee, Satabdi Datta Choudhury, and Gopal Chakrabarti

13	**Mechanistic Insights into Autophagosome–Lysosome Fusion in Cancer Therapeutics** 265
	Kewal Kumar Mahapatra, Soumya Ranjan Mishra, Bishnu Prasad Behera, Prakash Priyadarshi Praharaj, Debasna Pritimanjari Panigrahi, Chandra Sekhar Bhol, Srimanta Patra, and Sujit Kumar Bhutia

About the Editor

Sujit Kumar Bhutia is currently working as an Associate Professor at the Department of Life Science, National Institute of Technology in Rourkela, India. He received his doctorate in cell biology and cancer from the Indian Institute of Technology, Kharagpur, India in 2008, and subsequently pursued post-doctoral research on autophagy and cancer at Virginia Commonwealth University, USA. His research interests are focused on understanding the role of autophagy and autophagy-dependent cell death in cancer for the development of novel therapeutics.

He has published more than 80 research articles in international peer-reviewed journals and authored or co-authored numerous books and book chapters. He is also serving as a referee for several international journals. Prof. Bhutia has more than 10 years of teaching experience in the areas of cell biology, cell signaling, cancer biology, and cell death.

Role of Xenobiotic in Autophagy Inflection in Cell Death and Carcinogenesis

Durgesh Nandini Das and Prashanta Kumar Panda

Abstract

Macro-autophagy (herein referred to as autophagy) is considered a major degradation pathway for damaged organelles, aggregate-prone proteins, and pathogens. There is substantial evidence stating that dysfunctional autophagy is the cause of the manifestation of multifarious degenerative diseases and cancer. Xenobiotics (here, the known group I carcinogens), substances considered foreign to the human body, are associated with inciting multiple stresses such as the endoplasmic reticulum (ER) stress, mitochondrial stress, and dysfunctional lysosome. Furthermore, autophagy exhibits a dichotomous role in cancer, although a detailed description of the modulation of autophagy by the known important carcinogens is provided only by a limited number of reports. The pro-tumorigenic role of carcinogen-induced autophagy/mitophagy has been explored which maintains homeostasis in cancer. On the contrary, the association of carcinogens with the induction of autophagic cell death has been reported. In addition, certain xenobiotics for protecting cells through dampening of necrosis, inflammation, and maintenance of genome integrity have been proposed. So far, only a few

---

Both authors contributed equally to this work.

D. N. Das (✉)
Department of Life Science, National Institute of Technology Rourkela, Rourkela, Odisha, India

Department of Medicine, Texas Lung Injury Institute, University of Texas Health Science Centre at Tyler, Tyler, TX, USA
e-mail: dasd@wustl.edu

P. K. Panda
Department of Life Science, National Institute of Technology Rourkela, Rourkela, Odisha, India

Division of Medical Oncology, Department of Internal Medicine, Washington University School of Medicine, St. Louis, MO, USA
e-mail: prashantakpanda@wustl.edu

studies exploring the xenobiotic-associated autophagy modulation, both in vitro and in vivo, have been reported. The synergistic effect of environmental carcinogens in relation to autophagy has been explored, although quite little was discovered. Besides describing autophagy modulation by xenobiotics in the normal cells, there are reports illuminating how autophagy modulation could be utilized as an effective therapeutic approach for the impediment of carcinogenesis and to rescue cells from cytotoxicity. In addition, the application of chemopreventive compounds for autophagy modulation mitigating cellular toxicity and carcinogenesis have been described to achieve a safer and healthier human life.

Keywords

Carcinogenesis · Xenobiotic compounds (XCs) · Autophagy · Mitophagy · Group-1 carcinogen · Cytotoxicity

1.1 Introduction

Macro-autophagy (hereafter referred to as autophagy) is a process of lysosomal degradation in which intracellular cargo such as damaged organelles and aggregate-prone proteins are sequestered in the double-membrane vesicles known as autophagosomes, which subsequently fuse with the lysosome to form the autolysosome. The autolysosome is the organelle that ultimately degrades and recycles the autophagic cargo recycled (Levine and Klionsky 2004; Bhutia et al. 2013; Panda et al. 2015). Autophagy maintains homeostasis in a cell and saves the cell from various stressors such as amino acid starvation, genotoxic stress, hypoxia, and chemotherapeutics (Kimmelman and White 2017). Principally, there are three different types of autophagy—(a) microautophagy, (b) macroautophagy, and (c) chaperone-mediated autophagy (CMA) (Yim and Mizushima 2020). The dysregulation of autophagy is associated with the development of several diseases, including cancer. The role of autophagy in cancer is quite complicated while it plays a tumor-suppressive role during tumor initiation; it induces tumor promotion in the stages of tumor progression (Mathew and White 2011; Kimmelman and White 2017). The process of autophagy comprises five steps: (a) Nucleation of phagophore, (b) Expansion of phagophore, (c) Closure of phagophore to form autophagosome, (d) Fusion of the autophagosome with the lysosome, and (e) Degradation of the autophagic cargo (Galluzzi et al. 2015) (Fig. 1.1).

The International Agency for Research on Cancer (IARC) (Soto and Sonnenschein 2010) lists 107 agents, most of which are chemicals, as known human carcinogens (Group 1), 59 agents as probable human carcinogens (Group 2A), and 267 agents as possible human carcinogens (Group 2B). Most of the agents in Groups 2A and 2B are reported to be carcinogenic in animals, although there is no

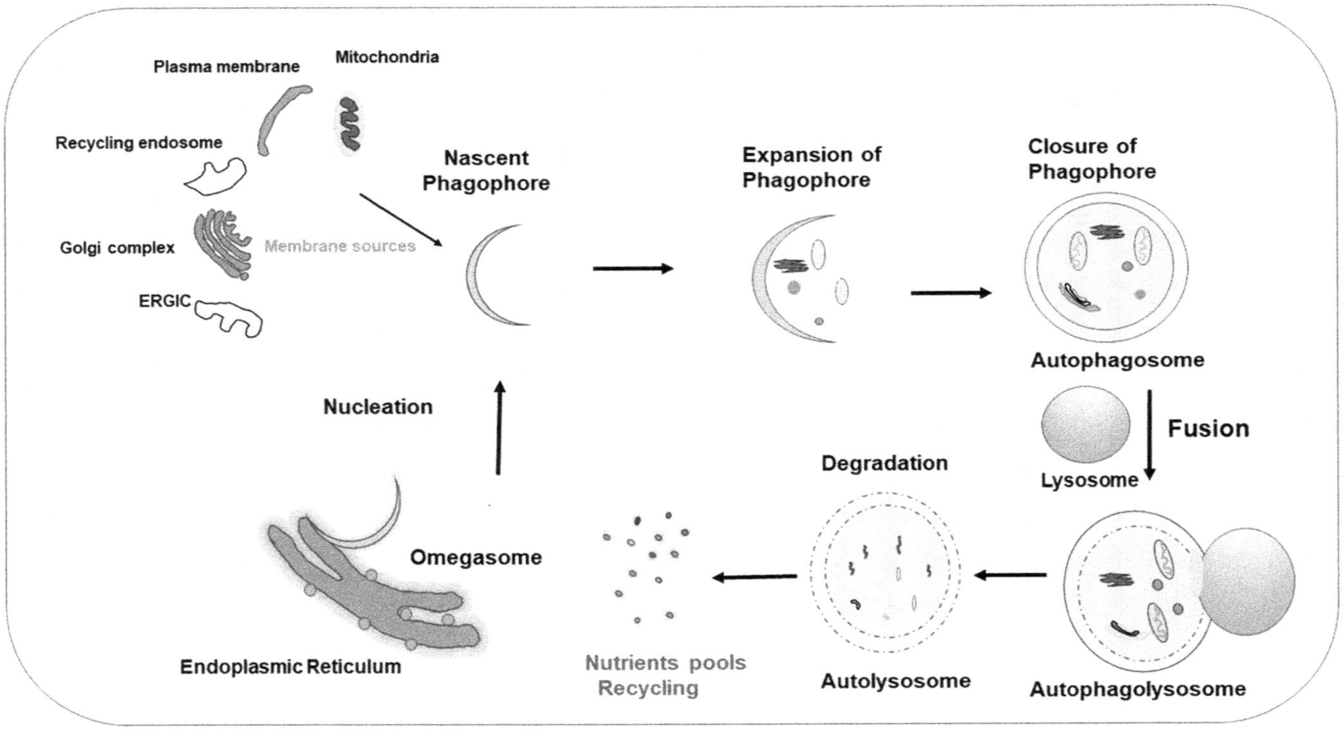

Fig. 1.1 A general schematic for mammalian autophagy process. The nucleation of phagophore begins de novo from the omegasome. However, several other membrane sources, ER–Golgi intermediate compartment (ERGIC), the Golgi, the plasma membrane, mitochondria, and the recycling endosomes, provide inputs for the elongation of phagophore or the isolation membrane. The expansion of phagophore is followed by complete sealing to form the autophagosome. The autophagosome then fuses with the lysosome to create autolysosomes, where the ultimate degradation of the autophagic cargo occurs. Subsequently, the degraded material is recycled back into the cytoplasm for the maintenance of homeostasis in the cell

definitive evidence regarding these being carcinogenic in humans as well. The present article focuses on certain Group 1 xenobiotic compounds (XCs), such as polycyclic aromatic hydrocarbons (PAHs), including Benzo[a]pyrene (B[a]P), 2,3,7,8-Tetrachlorodibenzodioxin (TCDD), dibenzofuran, and certain other inorganic compounds such as cadmium, arsenic, chromium, and nickel (Birkett et al. 2019).

Emerging evidence indicates that alteration in the autophagic pathway could be correlated to the onset of cytotoxicity resulting from chronic exposure to the aforestated XCs. These XCs contain several aryl hydrocarbon receptor (AhR) agonists, which upon activation lead to the induction of cytochrome P450 enzymes capable of converting procarcinogens into the carcinogens, which is a crucial event triggered in an individual for vulnerable to metastatic growth (Androutsopoulos et al. 2009; Hankinson 2016; Das et al. 2017a, b, c).

More importantly, the present article explores the consequences of autophagy modulation by XCs regarding cytotoxicity and carcinogenesis. In addition, autophagy, as well as autophagy-mediated cell death induced by known Group 1 carcinogens, are highlighted. Interestingly, the role of chemopreventive compounds in modulating autophagy and how these compounds could be utilized to rescue cells from toxicity and carcinogenesis as a complementary approach is discussed.

1.2 Aryl Hydrocarbon Receptor and Cytochrome P450 Regulates the Development of Carcinogenesis

The Aryl hydrocarbon receptor (AhR) is an intensively reported ligand-activated transcription factor that is adequately expressed in multiple organs and tissues. AhR contributes to the detoxification process for numerous xenobiotic substances and initiates phase I and phase II detoxification pathways. The XCs toxins that serve as the activators of AhR disrupt several cellular functions to extend the perception regarding the toxic and carcinogenic effects. The toxic compounds activate AhR, which may exhibit acute or chronic toxicity depending on the kind of toxin, its dose, and the health and age of the individual (Jaishankar et al. 2014; Arenas-Huertero et al. 2019). AhR plays an important role in xenobiotic-induced carcinogenesis. Several in vivo studies have demonstrated a substantial connection between the induction of aryl hydrocarbon hydroxylase activity and the carcinogenesis induced by XCs. Exposure to XCs is a major concern because once the XCs have entered the body, they can conveniently cross the cell membrane due to lipophilic in nature. In the cytoplasm, the XCs binds to the AhR, and the resulting system forms a complex with the chaperone proteins, namely, heat shock protein 90 (HSP90), co-chaperone protein X-associated protein 2 (XAP2), and p23 (Reyes et al. 1992; Tsai et al. 2015; Kudo et al. 2018). Binding of XCs to AhR indicates that the activation of the complex and its translocation to the nucleus has begun.

After forming a heterodimeric complex with the AhR nuclear translocator (ARNT), the complex again binds with the 5'-TNGCGTG-3' consensus sequence of the xenobiotic-responsive element (XRE) present in the promoter region of several genes, such as cytochrome P450 (CYP450), GST, UDP-GT, and quinine oxidoreductase (Gelboin 1980; Das et al. 2017a). Consequently, it induces the expression of various genes that are involved in XCs metabolism, including the CYP isoforms 1A1 and 1B1. Moreover, binding of XCs to AhR indicates the activation of a transcription factor that augments the expression of various genes, including those encoding the CYP450 enzymes, which metabolize XCs into mutagenic intermediates, ultimately leading to carcinogenesis (Fig. 1.2) (Das et al. 2017a, b, c). Interestingly, AhR is generally known to mediate cancer initiation via DNA damage, attributed to its role in the induction of CYP450 enzymes. The key findings regarding multiple cancer sites have elucidated that exposure to several of these persistent AhR ligands leads to an upsurge in cancer progression through the enhancement of tissue invasion and metastasis.

In AhR-stimulated human lung carcinoma A549 cells it has been demonstrated there is increased expression of E2F1 target genes such as RFC38 and PCNA, which are associated with cell cycle regulation (Watabe et al. 2010). Moreover, Ahr displays a central role for facilitating tumorigenesis, characterized by forming DNA adduct, reducing cell–cell adhesion, and increasing cellular proliferation in cigarette smoke-induced lung carcinogenesis (Tsay et al. 2013). It has been demonstrated that exposure to environmental carcinogen TCDD, activates AhR dependent pathway specifically by increasing expression of matrix metalloproteinases (MMPs), which are involved with increased invasive potential for generating melanoma tumorigenesis. (Villano et al. 2006). TCDD has also been demonstrated to augment MMP10 expression in keratinocytes (De Abrew et al. 2014). The carcinogenicity prospective of XCs is associated with their ability to bind to the DNA, thereby enhancing DNA cross-linking, leading to a series of disrupting effects, which may ultimately result in tumor initiation. These XCs increase cellular toxicity by regulating the generation of reactive oxygen species (ROS), which mediate apoptosis. Similarly, AhR-dependent tumor promoters may serve as significant tumorigenic agents as they possess the capability to enhance the repair of any DNA damage and the development of the initiated cells ultimately driving tumor progression (Dietrich and Kaina 2010).

Cellular toxicity results due to XCs, disturb the homeostasis in the cell by modulating autophagy, which results in unusual proliferation and leading to carcinogenesis. A previous study demonstrated that particulate matter stimulated AhR regulates autophagy in keratinocytes (Jang et al. 2019). Conversely, AhR activation by TCDD led to repressed autophagy in HaCaT cells and normal human epidermal keratinocytes (NHEKs) (Kim et al. 2020). The cellular mechanisms contributing to the manifestation of toxicities are examined by comparing a series of events that begin with exposure, involve a multitude of interactions between the invading toxicant and the host, and culminate in a toxic effect.

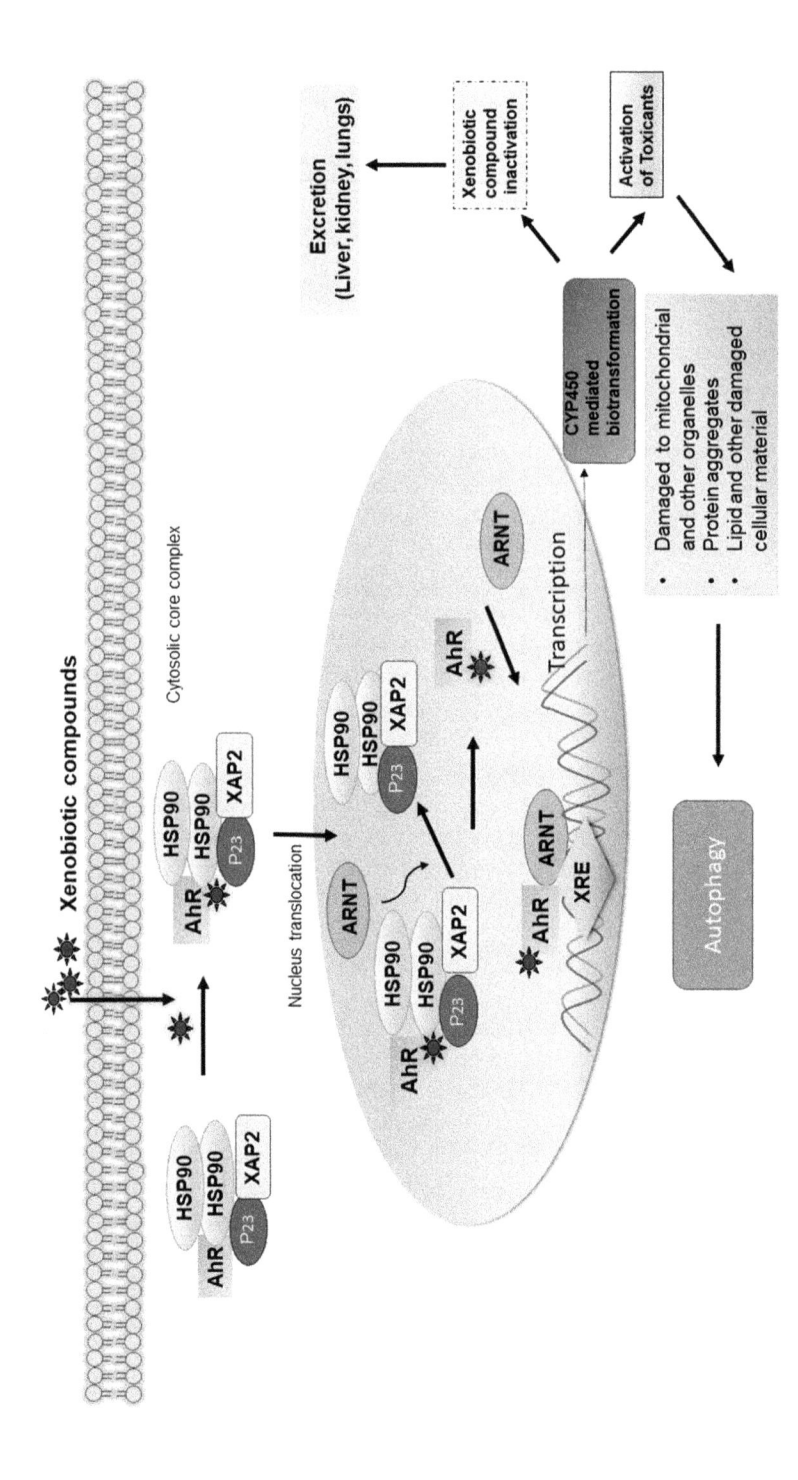

Fig. 1.2 Molecular representation of the biotransformation of a xenobiotic compounds. The schematic diagram illustrates the molecular events that occur following the entry of an aryl hydrocarbon receptor (AhR) ligand-activated transcription regulator for binding to xenobiotic compounds. In the absence of a ligand, the AhR exists within the cytosol as a complex with the dimer of the molecular chaperone named heat shock protein 90 (HSP90), the HSP90-interacting

protein p23, and the co-chaperone protein X-associated protein 2 (XAP2). Molecular chaperone HSP90 is abundant in eukaryotic cells and performs the function of regulating more than 300 target substrates. HSP90 also regulates AhR to maintain the stability of the AhR complex in a ligand-free state within the cytoplasm. In the presence of a ligand, the AhR forms a heterodimer with the aryl hydrocarbon receptor nuclear translocator (Arnt) and translocates into the nucleus, where it binds with the xenobiotic responsible element (XRE) to act as a transcription factor that induces the toxicant-metabolizing enzyme cytochrome P450. The activated xenobiotic compound causes damage to cellular components through, for example, protein aggregates, mitochondrial damage, or other stress

1.3 Cellular Mechanisms Underlying Xenobiotic Compound-Induced Toxicity and Carcinogenesis

Xenobiotic metabolic enzymes are divided into phases I and II that are required for metabolizing of xenobiotics compounds including environmental carcinogens, drugs, and pesticides. Phase I xenobiotic-metabolizing enzymes like Cytochrome P450 (CYPs) are linked with the biotransformation of environmental pollutants and associated with the development of cancer. On the contrary phase II xenobiotic-metabolizing enzymes are linked to biotransformation of xenobiotics to more excretable form while the inability to detoxify ends with the formation of carcinogens which lead to cancer (Nebert and Dalton 2006; Jancova et al. 2010).

Autophagy is stimulated in response to distinct cellular stresses caused by XCs, such as the ER stress, oxidative stress, lysosomal damage, and DNA damage (Lafleur et al. 2013; El-Demerdash et al. 2018; Ashoor et al. 2013; Kvitko et al. 2012) (Fig. 1.3). Interestingly, autophagy operates as a cytoprotective mechanism, degrading the damaged cellular proteins and organelles, which could be toxic to the cell, thereby restoring cellular equilibrium in the cell (Ogata et al. 2006; Moreau et al. 2010). It is noteworthy that autophagy is a catabolic pathway activated in response to various cellular stressors, such as damaged organelles, ER stress, ROS, DNA damage, and the accumulation of misfolded or unfolded proteins. The high levels of ROS in the early stage of XCs related carcinogenesis or cell transformation are oncogenic and cause DNA damage, inhibition of DNA repair, and alterations in normal signal transduction, ultimately leading to malignant transformation.

1.3.1 Endoplasmic Reticulum (ER) Stress Regulates Xenobiotic Compounds Associated Toxicity and Carcinogenesis

ER is considered as an important organelle for containing enzymes needed for xenobiotic biotransformation. The endoplasmic reticulum plays an important role in protein folding and when the load of the unfolded protein dominates it exerts ER stress. Moreover, ER maintains homeostasis in the cell by protein folding (Xu et al. 2005). Modulation of the ER stress signaling pathways is an important concern for protecting against the cellular damage induced by xenotoxicants.

In this context, XCs, such as cadmium, induce the expression of Grp78 through the phosphorylation of eIF2 alpha, which increased the translational activity of ATF4 resulting in ER stress response, which displays a protective role against cadmium-induced cytotoxicity (Liu et al. 2006). In addition, ER stress and change in calcium homeostasis are associated with cadmium triggered apoptosis (Biagioli et al. 2008). Furthermore, cadmium induces ER stress and autophagy in proximal convoluted tubule cells (Chargui et al. 2011). Moreover, cadmium-tempted cytotoxicity by accelerating ER stress and autophagy in retinal pigment epithelial cells (Zhang et al. 2019a).

According to a previous report, cigarette smoke induces protracted ER stress and autophagic cell death in human umbilical vein endothelial cells (Csordas et al. 2011).

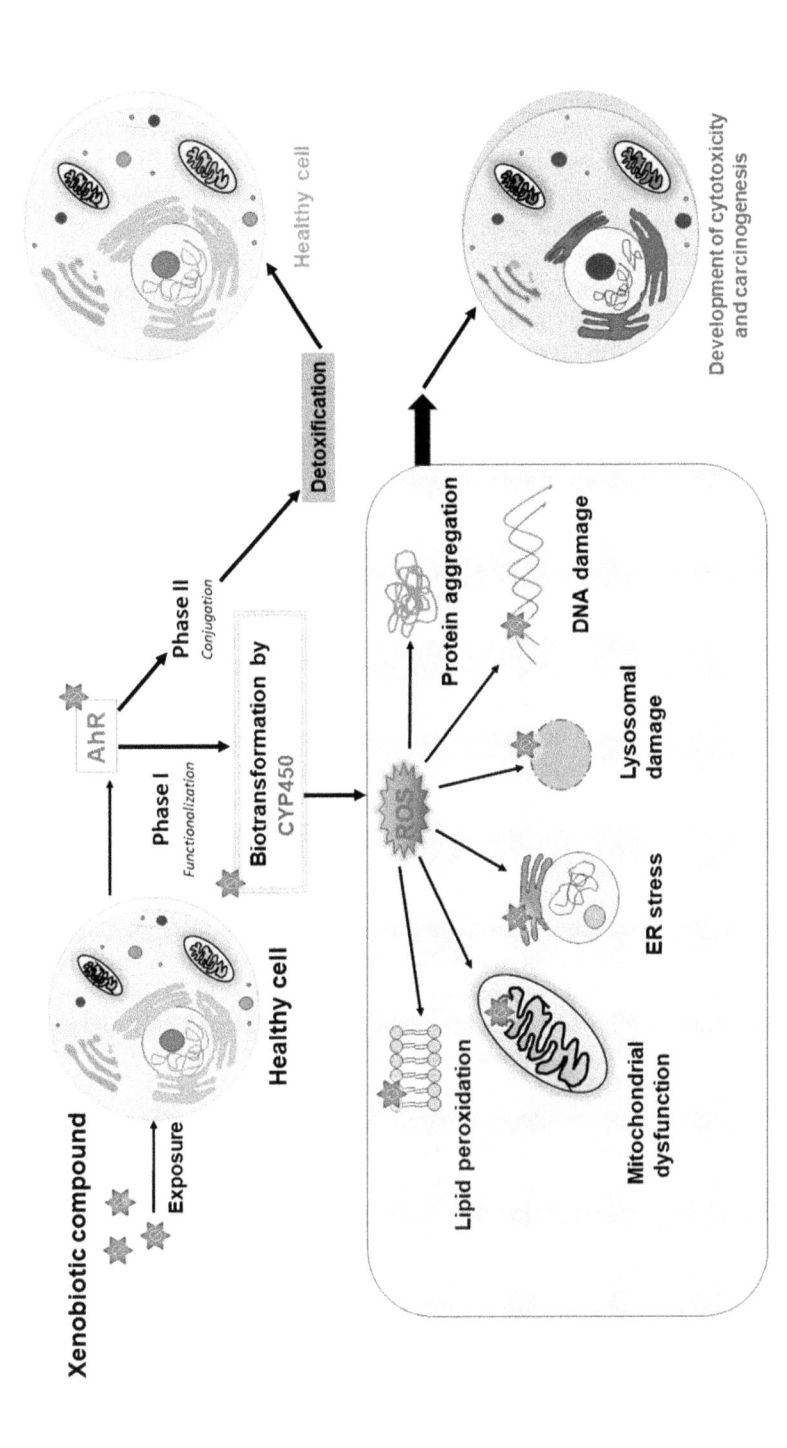

Fig. 1.3 The metabolism Xenobiotic compounds (XCs) regulated by xenobiotic metabolizing enzymes (XMEs). There are two types of XMEs which are classified as phase I (functionalization) and phase II (conjugation) reaction. The CYP450 enzymes play a dominant role in phase I Reaction for the biotransformation of most of the xenobiotic compounds. The complex of the xenobiotic compound with AhRdimerizes with ARNT and recognizes the XRE sequence, which transcribes the enzymes that modulate different cellular processes such as lipid peroxidation, mitochondrial damage, ER stress, lysosomal damage, DNA damage, protein aggregate leading to carcinogenesis. However, some of the XCs bypass the phase I enzymes and follows phase II reaction for detoxification and excreted. However, it is not rigid for the exclusive division of phase I and phase II reactions which need more precise study

Similarly, the elevation of autophagy by arsenic is associated with ER stress, initiation of UPR, and the deposition of protein aggregates in human lymphoblastoid cells (Bolt et al. 2012). In addition, TCDD is reported to induce ER stress in PC12 cells via the PERK-eIF2α signaling pathway (Duan et al. 2014a). However, it has been elucidated that lead compound-induced toxicity stimulates the mTORC1 pathway of autophagy in cardio fibroblasts to ensure survival under ER stress conditions (Sui et al. 2015). Furthermore, B[a]P was reported to induce cell cycle arrest and apoptosis in human choriocarcinoma cancer cells via the ROS-induced ER stress pathway (Kim et al. 2017). Recent findings revealed that the mutual communication between autophagy and ER stress in response to cigarette smoke extract (CSE) exposure stimulates apoptosis in human bronchial epithelial cells (He et al. 2019).

1.3.2 Mitochondrial Dysfunction and Xenobiotic Compounds Triggered Toxicity and Carcinogenesis

In recent decades, there has been an increase in the reports describing the toxic effects of pollutants on the mitochondria. Mitochondria are the major locations for energy production and the execution of oxidative reactions within the cells (Hamanaka and Chandel 2010). Xenotoxicants induce mitochondrial impairment via multiple mechanisms; therefore, several methods are required to evaluate mitotoxicity. Previous studies have reported that mitochondrial dysfunctions lead to increased levels of ROS, mainly to activate autophagy. Moreover, the majority of the ROS are produced in mitochondria. Another factor influencing the generation of mtROS is calcium signaling, in which calcium ions are transferred from the ER to the mitochondria "quasi-synaptically," that is, through closely placed mitochondria-associated ER membranes (Marchi et al. 2017). Calcium encourages ATP synthesis by exhilarating ATP synthase and the enzymes involved in the tricarboxylic acid cycle (Rizzuto et al. 2000), which suggest increased mitochondrial metabolic rate, oxygen consumption, and mitochondrial ROS generation. Calcium accrual in mitochondria results in augmented mitochondrial ROS (Hansson et al. 2008).

B[a]P induces abnormal mitochondria and cellular demise in Hep3B cells (Jiang et al. 2011). Furthermore, PM2.5 stimulates oxidative stress, which triggers the autophagy pathway in A549 human lung epithelial cells (Deng et al. 2013). Cadmium-based quantum dots increase the intracellular ROS levels, affect mitochondrial function, and induce autophagy, leading to apoptosis in mouse renal adenocarcinoma cells (Luo et al. 2013). Studies evaluating the toxicity of zinc oxide nanoparticles (NPs) revealed that these NPs induced cell death in normal skin cells through autophagic vacuole accumulation and damage to mitochondria via ROS production (Yu et al. 2013).

Cigarette smoke exposure induces the stimulation of autophagy and dysregulation of mitochondrial repair machinery resulting in cell death in granulosa cells (Gannon et al. 2013). In accordance with this, another report suggests persistent exposure to cigarette smoke alters mitochondrial structure and function in airway epithelial cells leading to COPD pathogenesis (Hoffmann et al. 2013).

Mitochondrial targeting of CYP1B1 and its role in PAH-induced mitochondrial dysfunction has also been elucidated previously (Bansal et al. 2014). Cadmium also activates ROS induced PINK1/Parkin dependent mitophagy in mice kidneys (Wei et al. 2014). Silica nanoparticles exposure causes ROS-triggered autophagy in MRC-5 cells, which could be a mechanism for cell survival (Petrache Voicu et al. 2015). Similarly, exposure to amorphous silica nanoparticles induces vascular endothelial cell injury following both apoptosis and autophagy via ROS-facilitated MAPK/Bcl-2 and PI3K/Akt/mTOR signaling axis (Guo et al. 2016). Interestingly, TCDD-induced toxicity mediated by mitoAhR localized to the intermembrane space (IMS) influences mitochondrial dysfunction (Hwang et al. 2016). Exposure to particulate matter (PM) induces autophagy in macrophages through the oxidative stress intermediated PI3K/Akt/mTOR signaling pathway (Su et al. 2017). Our group has recently deciphered that the exclusion of dysfunctional mitochondria through mitophagy represses B[a]P-triggered apoptosis in HaCaT cells (Das et al. 2017c). Furthermore, the use of electronic cigarettes induces mitochondrial stress in neural stem cells (Zahedi et al. 2019).

1.3.3 Lysosomal Disruption and Xenobiotic Compounds Prompted Toxicity and Carcinogenesis

Lysosomes play a central role in cellular catabolism, trafficking, and processing of foreign particles. Lysosomes facilitate detoxification and cell survival through the storage and degradation of genotoxic materials. Lysosome pathology may imply cytotoxicity, conceivably leading to cell death, and should, therefore, be considered adverse for cellular injury and dysfunction. Mechanistic investigations may involve the evaluation of cell or tissue-specific clearance pathways and mitochondrial toxicity. Similarly, an impaired lysosomal function may have an impact on autophagy and ultimately lead to an increase in oxidative stress, mitochondrial dysfunction, inflammation, and cell death (Boya et al. 2005; Martini-Stoica et al. 2016).

Moreover, defects in lysosomal capacity result from a modification in the lumen pH and/or altering in lysosomal membrane permeabilization which interrupts the autophagosome and lysosomal fusion. Prevalent variations in membrane permeability may cause acidified cytosol and cellular necrosis (Martini-Stoica et al. 2016). Agricultural insecticide lindane disrupts the maturation of an autophagosome into an autolysosome following aberrant activation of the ERK pathway found in several types of cancer (Corelle et al. 2006). Ji et al. observed that graphene oxide quantum dots blocked the autophagic flux by decreasing the activity of cathepsin B and obstructing the lysosome proteolytic potential in GC-2 and TM4 cells (Ji et al. 2016). Similarly, in hepatocytes silica nanoparticles induced dysfunctional autophagy through lysosomal impairment leading to inhibition of autophagosome and lysosome fusion (Wang et al. 2017a). Likewise, lead disrupts autophagic flux by impeding the formation and activity of lysosomes in the neural cells

(Gu et al. 2019). Furthermore, arsenic nanoparticles induce apoptosis and impairment of mitochondria and lysosomes in isolated rat hepatocytes (Jahangirnejad et al. 2020).

1.3.4 Induction of DNA Damage in Response to Xenobiotic Compounds

Autophagy and DNA damage response (DDR) are two important biological processes that are crucial for cellular and organismal homeostasis. DNA damage activates autophagy, while autophagy is essential for several functional consequences of DDR signaling, including senescence, cell death, the repair of DNA lesions, and cytokine release. DNA damage is the initial crucial step during the process of carcinogenesis. Chemical carcinogens are capable of causing the formation of carcinogen DNA adducts or encouraging other modifications to the DNA, such as oxidative damage and amendments to the DNA ultrastructure (e.g., DNA strand breakage, strand cross-linking, chromosomal rearrangements, and deletions). Prolonged exposure to low levels of arsenic or cadmium leads to cell transformation in the target tissues. Although the mechanism of this cell transformation is not completely understood yet, it is supposed that defective autophagy leading to the accumulation of genomic mutations and epigenetic alterations is a contributor (Mathew et al. 2007). Cigarette smoke induces oxidative stress and DNA damage, and it is more severe as a carcinogen in mice exposed to the chemical from birth (Micale et al. 2013). In A549 cells particulate matter 2.5 (PM2.5) enhanced autophagy, elevated oxidative stress, and activated the tumor necrosis factor-alpha (TNF-α) causing cytotoxicity (Deng et al. 2014). Similarly, particulate matter 10 (PM10) exposure resulted in an elevation in the ROS levels, inflammatory cytokines, DNA damage, and autophagy in human lung cells (de Oliveira Alves et al. 2017). PM2.5 induced oxidative stress via ROS generation, which led to DNA damage, lipid peroxidation, and protein carbonylation; consequently tempted ER stress, depolarized mitochondria, and autophagy, ultimately causing apoptosis in both in vitro and in vivo (Piao et al. 2018). Similarly, cadmium exerts toxic molecular effects and consequently increases DNA strand breaks, elevates the ER stress, increases ROS production, and disturbs the calcium homeostasis. Different signaling pathways such as calcium-ERK and PERK-elF2α have been implicated in cadmium-activated autophagy (Messner et al. 2016). Arsenic induces the production of ROS/RNS, which may generate a mechanism for the disruptions of DNA repair (Tam et al. 2020).

1.4 Autophagy Plays a Dual Role in Cellular Stress Response: Cell Survival or Cell Death

Increasing evidence has been indicating that several xenobiotic compounds modulate autophagy. Since autophagy plays a dual role, there is an incessant debate on whether autophagy acts as a cell death mechanism or conversely as a cytoprotecting one in the presence of XCs.

1.4.1 Triggering Autophagy/Mitophagy Rescues from Xenobiotic Compounds Triggered Cytotoxicity

There is extensive evidence that autophagy protects against XCs-induced cytotoxicity. Defective mitophagy leads to cigarette smoke-induced lung cellular senescence in chronic airway diseases (Ahmad et al. 2015). For instance, TCDD exposure induced protective autophagy mechanism to ameliorate ROS-induced cytotoxic effects in human SH-SY5Y neuronal cells (Zhao et al. 2016). A study by our research group revealed that B[a]P-induced mitophagy in HaCaT cells as a cytoprotective mechanism to resist cell death (Das et al. 2017c). Moreover, cadmium-initiated autophagy in rat renal mesangial cells has been reported to assist in rescuing against apoptosis- and necrosis-mediated cell death (Fig. 1.4) (Fujishiro et al. 2018).

Bisphenol A (BPA) is a chemical used commonly in the production of polycarbonate plastics and epoxy resins. BPA enters the human body via different routes such as food and drinking water. The current literature reports that autophagy inhibition because of the disruption of autophagosome–lysosome fusion is the main cause underlying the deposition of toxic lipids in the liver. Furthermore, facilitating autophagy by using mTOR inhibitor Torin2 is reported to increase the degradation of toxic lipids, suggesting that autophagy could be used for therapeutic benefit to reduce toxic lipid deposition in the liver (Song et al. 2019).

Recently, it was established that heme-induced toxicity was enhanced upon the inhibition of autophagy in H9c2 cardiomyoblast cells, further corroborating that the differential role of autophagy inhibition depends on the cellular context, dose, and time (Gyongyosi et al. 2019).

Furthermore, it was observed that cadmium triggered cytotoxicity in mouse liver cells which is liknked with the disruption of autophagic flux due to inhibition of autophagosome-lysosome fusion (Zou et al. 2020) (Fig. 1.4).

1.4.2 Autophagic Cell Death Induced by Xenobiotic Compounds

Morphologically, autophagic cell death is characterized by huge autophagic vacuolization of the cytoplasm in deficiency of chromatin condensation (Kroemer et al. 2009). In fact, there is a paucity of established reports where autophagy inhibition completely inhibited cell death induced by xenobiotics. Delayed cell

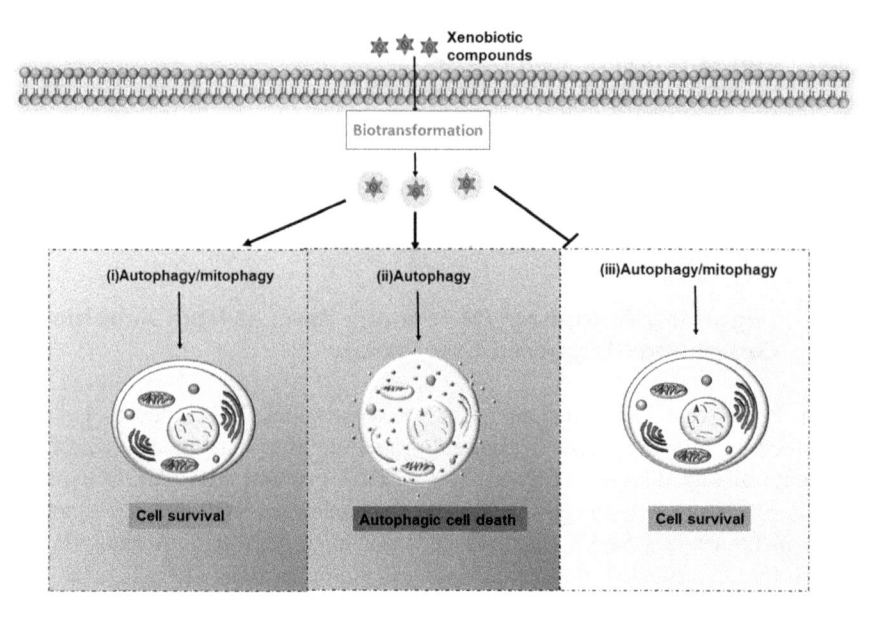

Fig. 1.4 Toxicity and autophagy modulation by xenobiotic compounds. (i) Benzo[a]pyrene and TCDD protect cells by inducing autophagy, whereas bisphenol A, cadmium, cigarette smoke, and heme induced toxicity could be alleviated by stimulating autophagy. (ii) Cigarette smoke extract (CSE) and TCDD-triggered autophagic cell death. (iii) Mitigating autophagy protect cells from cigarette smoke and particulate matter triggered cytotoxicity

death was demonstrated in 3-methyladenine (3-MA) treated or ATG5 knockdown human umbilical vein endothelial cells (HUVECs) which were exposed with cigarette smoke extract. On the contrary, cell death was unaltered after treating with apoptosis inhibitor BCL-XL suggesting manifestation of autophagic cell death induced by CSE is independent of apoptosis (Csordas et al. 2011). Furthermore, TCDD activates cell death through the induction of autophagy in bovine kidney cells (MDBK cells) (Fiorito et al. 2011). Similarly, nuclear receptor 77 (Nur77) was found to promote cigarette smoke-induced autophagic cell death by increasing the dissociation of B-cell lymphoma 2 (Bcl2) from Beclin-1 in lung cells (Qin et al. 2019) (Fig. 1.4).

1.4.3 Mitigating Autophagy/Mitophagy Protects Xenobiotic Compounds Induced Cytotoxicity

In addition to evaluating the effects of autophagy by inducing autophagy, it is equally important to analyze in what way autophagy inhibition regulates the XCs-induced cytotoxicity. Previous studies have reported substantially reduced levels of apoptosis in the lungs of $LC3B^{-/-}$ mice compared to wild type mice, and enhanced resistance to emphysema upon cigarette smoke exposure (Chen et al. 2010). The treatment with autophagy inhibitors, like 3-methyladenine (3-MA) or spautin-1, reduced the airway injury in particulate matter (PM)-treated mice (Xu et al. 2017). In addition, mice with knocked-down autophagy-related gene Beclin1 or LC3B exhibited reduced airway inflammation and mucus hypersecretion in response to PM exposure (Chen et al. 2019a). Cigarette smoke-induced Nix/BNIP3L-dependent mitophagy triggers airway epithelial cell and mitochondria injury and causes COPD pathogenesis (Zhang et al. 2019b). This result further corroborates that the use of autophagy inhibitors could serve as a therapeutic strategy for inhibiting PM-induced airway inflammation (Fig. 1.4).

1.5 Molecular and Cellular Signaling Responsible for Causing Carcinogenesis upon Exposure to Xenobiotic Compounds

Cancer initiation, promotion, and progression are exceptionally complex processes. XCs induced carcinogenesis may occur via multiple mechanisms (Barrett 1993; Patterson et al. 2018). Evidence suggests that chemical toxicants may operate through genotoxic, cytotoxic, as well as epigenetics pathways, which further complicates the pursuit of alleviating chemical toxicant-associated diseases and cancers. It is also suggested that XCs may induce carcinogenic effects through the disruption of important signal transduction pathways.

1.5.1 PI3K/Akt/mTOR Signaling Pathway

Phosphatidylinositol 3-kinase (PI3K), protein kinase B(Akt), and mammalian target of rapamycin (mTOR), the components of PI3K/Akt/mTOR signaling pathway, play important functions in a cell, such as pathologic changes, cellular physiology, and cell survival. Therefore, disruptions of this pathway result in different types of cancer (Levine 2007; Bartholomeusz and Gonzalez-Angulo 2012; Kandoth et al. 2013; Tai et al. 2017). Chemical toxicants, such as all the members of Group I carcinogens, are capable of inducing malignant cell transformation via PI3K/Akt/mTOR pathway. PI3K/Akt and mTOR signaling pathways are crucial to several aspects of cellular growth and survival in normal physiological conditions as well as during carcinogenesis.

PI3K/Akt/mTOR pathway plays a critical role in multiple cellular functions and is a major regulator of autophagy (McAuliffe et al. 2010). Various growth factor receptors and oncogenes activate PI3K. In fact, elevation in PI3K signaling is regarded as a distinct marker of cancer (Fruman et al. 2017). Members of protein kinase B (Akt)-serine/threonine kinase family mainly exist in three isoforms (Akt1, Akt2, and Akt3) and are common downstream effectors of the PI3K signaling pathway (Fresno Vara et al. 2004). Akt is the master regulator of tumor cell invasion, migration, and metastasis. Current evidence suggests that mTOR is associated with a myriad of functions including lipid generation, nucleotide precursors biosynthesis, metabolic alteration, and metastasis (Yecies and Manning 2011; Ben-Sahra et al. 2013; Valvezan et al. 2017).

Tobacco smoke (TS) is reported to induce lung tumorigenesis through the upregulation of the Akt/mTOR pathway (Memmott and Dennis 2010). Slug induced by B[a]P is involved in the regulation of the invasive properties of fibroblast-like synoviocytes (FLS) in rheumatoid arthritis following PI3K/Akt/mTOR pathway (Lee et al. 2013). Similarly, silica nanoparticles are reported to suppress phosphorylated PI3K, Akt, and mTOR in endothelial cells in a dose-dependent manner (Duan et al. 2014b). Roy et al. reported that zinc oxide nanoparticles induced apoptosis through enhancement of autophagy via PI3K/Akt/mTOR inhibition. The levels of phosphorylated PI3K, Akt, and mTOR were significantly decreased in macrophages upon exposure to zinc oxide nanoparticles (Roy et al. 2014).

Furthermore, PM2.5 stimulates autophagy in human bronchial epithelial cells through suppression of the PI3K/Akt/mTOR pathway (Liu et al. 2015a). In addition, PM2.5 exposure induces autophagy in lung macrophages through the oxidative stress-mediated PI3K/Akt/mTOR pathway (Su et al. 2017). Moreover, inactivating mTOR augments autophagy-mediated epithelial injury in airway inflammation caused by particulate matter (Wu et al. 2020a).

Wang et al. reported that arsenic disulfide attenuates the Akt/mTOR signaling pathway, thereby prompting both autophagy and apoptosis in osteosarcoma (Wang et al. 2017b). B[a]P, a known carcinogen, induces pyroptotic and autophagic cell death in HL-7702 human normal liver cells through the inhibition of the PI3K/Akt signaling pathway (Li et al. 2019a).

1.5.2 MAPK/ERK Signaling Pathway

Mitogen-activated protein kinases (MAPKs) involve extracellular signaling-regulated kinase. Sustained activation of the MAPK/ERK pathway by carcinogens causes a selective alteration in autophagy at the maturation step, resulting in the giant defective autolysosomes accumulation (Corelle et al. 2006). Studies suggest that the activation of extracellular signal-regulated kinases (ERK) could be a contributor to the autophagic effects and promote cell survival (Ogier-Denis et al. 2000; Cagnol and Chambard 2010). B[a]P exposure to HepG2 cells is reported to induce p53-dependent cell death, under the regulation of p38 MAPK and ERK pathway (Lin et al. 2008). Similarly, arsenic is reported to stimulate cell proliferation through enhanced ROS generation, ERK signaling, and Cyclin A expression in HaCaT and Int407 cells (Chowdhury et al. 2010). Furthermore, iron oxide nanoparticles are reported to induce autophagy in RAW 264.7 macrophage in a dose-dependent manner together with phosphorylated ERK (Park et al. 2014). Copper oxide nanoparticle-induced cytotoxicity in human keratinocytes and mouse embryonic fibroblasts mediated via p53 and ERK activation (Luo et al. 2014). In addition, ERK activation plays an important role in enhancing the radiosensitivity of silver nanoparticles; while the inhibition of ERK reduces autophagy, the ERK levels triggered by silver nanoparticles could reduce apoptosis in glioma cells (Wu et al. 2015). Rinna et al. explored the effects of silver nanoparticles on MAPK activation and confirmed the role of ROS in DNA damage during silver nanoparticles-elevated toxicity in human embryonic epithelial cells (Rinna et al. 2015). In lung cancer cells cadmium induces cell migration and invasion through the activation of the ERK pathway (Zhai et al. 2019).

1.5.3 Hypoxia-Inducible Factor (HIF)

Hypoxia-inducible factor-1 is a heterodimer encompassing α and β subunits. It is a transcription factor that mediates the adaptive mechanism to hypoxia. Hypoxia-inducible factor-1 is regulated mainly by oxygen-dependent changes and could be responsible for regulating autophagy and other hypoxia-related responses (Bruick and McKnight 2001). Studies have reported that nicotine encourages the accumulation of hypoxia-inducible factor-1α protein and vascular endothelial growth factor (VEGF) expression in human lung cancer cells via nicotinic acetylcholine receptors (Zhang et al. 2007). In addition, in human non-small cell lung cancer cells mitochondrial reactive oxygen species facilitate nicotine in elevating the expression of hypoxia-inducible factor-1α (Guo et al. 2012). TCDD is reported to induce hypoxia-inducible factor-1α pathway, oxidative stress, and metabolic stress, contributing to trophoblastic toxicity (Liao et al. 2014). Zinc oxide nanoparticles are reported to enhance ROS generation, apoptosis, autophagy, and hypoxia-inducible factor-1α signaling pathway in HEK-293 cells and mouse kidney tissues (Lin et al. 2016). Hypoxia-inducible factor-1α inhibits the mitochondria-mediated apoptosis induced by silver nanoparticles in human lung cancer cells through the regulation of autophagic flux via ATG5, LC3-II, and p62 regulation (Jeong et al. 2016).

1.5.4 NF-κB Signaling Pathway

Nuclear factor kappa-light-chain-enhancer of activated B cells (NF-κB) is an inducible transcription factor regulated by signal activation cascades. NF-κB modulates the expression of several genes involved in diverse cellular processes such as cell proliferation and apoptosis and the stress responses to a variety of noxious stimuli, thereby promoting carcinogenesis and cancer progression. It has been reported that PAHs exposure activates the NF-κB transcription factor in the hepatoma cell line (Volkov and Kobliakov 2011). In addition, B[a]P stimulates oxidative stress and endothelial progenitor cell dysfunction by activating the NF-κB pathway (Ji et al. 2013). Moreover, it has been investigated activation of the NF-κB pathway takes place during chronic exposure of B[a]P in hepatocellular carcinoma (Ba et al. 2015). Furthermore, cadmium induces nephrotoxicity through an elevation in the levels of ROS involved in NF-κB -mediated apoptosis (Ansari et al. 2017), while PM2.5 induces apoptosis by upregulating NF-κB signaling in Chinese hamster ovary cells (Peng et al. 2017). Similarly, arsenic induces apoptosis in p53-proficient (p53+/+)

and p53-deficient (p53−/−) cells via differential alteration of the NF-κB pathway (Yin and Yu 2018). Intriguingly, cigarette smoke encourages HMGB1 translocation and release, contributing to migration and NF-κB stimulation through the induction of autophagy in lung macrophages (Le et al. 2020).

1.5.5 p53 Signaling Pathway

The "cellular gatekeeper" p53 acts through transcription-dependent as well as transcription-independent mechanisms which transmits a type of stress-inducing signals for various antiproliferative cellular responses (Zilfou and Lowe 2009). A previous study reported that B[a]P-induced toxicity related to DNA damage and p53 modulation in HepG2 cells (Park et al. 2006). p53 regulates autophagy in an ambiguous manner such as p53 stimulated autophagy leads to cell death or shows protective response depending upon the different cellular contexts (Crighton et al. 2007; Amaravadi et al. 2007). Moreover, B[a]P-induced DNA damage instigates p53-independent necroptotic cell death via a Bax/Bcl-2-dependent mitochondrial pathway in human non-small cell lung carcinoma cell line (Jiang et al. 2013). Besides, B[a]P 7,8-diol-9,10-epoxide promotes p53-independent necrosis following the mitochondria-linked pathway involving Bak and Bax activation (Zhang et al. 2015). In addition, TCDD was reported to induce cell death through autophagy in bovine cells with decreased Mdm2 and increased p53 levels (Fiorito et al. 2011). A study by our research group deciphered that TCDD instigates p53-regulated apoptosis through the activation of cytochrome P450/aryl hydrocarbon receptors in the HaCaT cell line (Das et al. 2017b).

1.6 Xenobiotic Compounds (Group I Carcinogens) Induce Carcinogenesis In Vitro and In Vivo

Carcinogenic compounds may cause cancer either by directly inducing DNA damage or through indirect cellular or physiological effects. Disruptive XCs may contribute to multiple stages of tumor development by influencing the tumor microenvironment. The tumor microenvironment involves intricate interactions among the blood vessels that supply nutrient pool to tumor cells (Casey et al. 2015).

Fibroblast growth factor 9 (FGF9) that plays a substantial role in B[a]P-induced lung adenocarcinoma CL5 cell invasion as well as the progression of human lung adenocarcinoma (Ueng et al. 2010). It has been demonstrated that B[a]P upsurges breast cancer cell migration and invasion through upregulation of the ROS-stimulated ERK pathway and promotes the activation of matrix metalloproteinase-9 (Guo et al. 2015). B[a]P was demonstrated to promote A549 cell migration, invasion, and EMT through the up-regulation of linc00673 expression in an AhR-dependent manner (Wu et al. 2020b). Moreover, the study of the effects of B[a]P on cancer metastasis and progression reported the NF-κB pathway as a potential target. Increased aggressiveness of B[a]P-triggered squamous carcinomas were observed in PACE4 overexpressed transgenic mice (Bassi et al. 2015). Furthermore, it was also found B[a]P activates the ERK pathway, as well as its downstream partner phosphorylated checkpoint kinase-1 (Chk1), is involved with cellular proliferation in human lung cancer cells (Wang et al. 2015). B[a]P promotes migration, invasion, and metastasis in lung adenocarcinoma cells through the upregulation of the TG-interacting factor (Yang et al. 2018). Moreover, the p38 MAPK pathway is reported to be intricately involved in B[a]P-induced migration and invasion in hepatoma cells (Wang et al. 2019a). Similarly, AhR mediates cell proliferation enhanced by B[a]P in human lung cancer 3D spheroids (Jimma et al. 2019).

The carcinogen nitrosamine 4-(methylnitrosamino)-1-butanone (NNK) found in the cigarette smoke that induces migration and invasion via the activation of a c-Src/PKCι/FAK loop, which may promote the development of human lung cancer (Shen et al. 2012). Electronic-cigarette smoke is reported to induce lung adenocarcinoma and urothelial hyperplasia in FVB/N mice (Tang et al. 2019). Equally increasing evidence suggests that CSE is also found to modulate the expression of Claudin-1, E-Cadherin, and miR-21, which might be associated with increased migration of cancer cells (Dino et al. 2019).

Cadmium is classified as a Group 1 carcinogen and has been demonstrated to be directly associated with tumors of the lung, breast, and prostate (Person et al. 2013; Divekar et al. 2019; Zimta et al. 2019). Similarly, arsenic and cadmium exhibit estrogen-like activity that contributes to the risk of developing mammary tumorigenesis (Divekar et al. 2019). The interaction between Atg4B and Bcl-2 plays an important role in cadmium-induced cross-talk between apoptosis and autophagy through the disassociation of Bcl-2 from Beclin1 in A549 cells (Li et al. 2019b).

Substantial report suggests that TCDD exposure causes disruption of mitochondria changing the mitochondrial membrane potential ($\Delta\Psi$m) and engrosses with mitochondria to nucleus stress signaling (Biswas et al. 2008). TCDD also promotes lung tumors through attenuation of apoptosis via Akt and ERK1/2 signaling pathways activation in female A/J mice (Chen et al. 2014). Moreover, Vk*Myc mouse exposure to TCDD provokes splenomegaly, blood cell abnormalities, and plasma cell carcinoma resembling multiple myeloma (Wang et al. 2019b).

Assessment of the carcinogenic effect of TCDD in vivo using mouse embryonic stem cells revealed the formation of teratoma (Yang et al. 2019). Activated macrophages were reported to be crucial during acute PM2.5-persuaded angiogenesis in lung cancer in a mouse model (Li et al. 2020).

1.7 Autophagy Modulation Induced by Xenobiotic Compounds Regulates Carcinogenesis

Accumulating evidence indicates that autophagy modulation could serve as an effective therapeutic strategy for combating cancer. Studies concerning XCs and autophagy modulation in relation to carcinogenesis are gaining increasing interest from a therapeutic perspective. Evidence suggests that cigarette smoke-induced autophagy in head and neck squamous cell carcinoma (HNSCC) cells and oral keratinocytes (OKF6/TERT2) cells result in the upregulation of ΔNp63α protein expression and a consequent increase in the NOS2 expression. Conversely, downregulation of ΔNp63α, IRF6, or NOS2 mitigates the autophagic process, which further suggests a relationship between smoke-induced autophagy and ΔNp63α/IRF6/NOS2 signaling and corroborates that modulation of ΔNp63α/IRF6/NOS2 signaling and consequently autophagy could serve as an effective therapeutic strategy against cigarette smoke-induced carcinogenesis (Ratovitski 2011). It is also known that cigarette smoke exposure induces cancer-associated fibroblast phenotype through the induction of autophagy, mitophagy, and DNA damage. Moreover, stromal fibroblasts secrete lactate and ketone bodies as a stress response strategy by fueling the oxidative mitochondrial metabolism (OXPHOS) in neighboring epithelial cells (Salem et al. 2013). On the contrary, nicotine, which is one of the active components in cigarette smoke, exhibits protective effects in ulcerative colitis (UC) patients through autophagy induction (Pelissier-Rota et al. 2015). CSE and B[a]P diol epoxide (BPDE) were observed to be involved in the transformation of human bronchial epithelial cells (HBECs) through the upregulation of vasorin expression. Vasorin-induced lung carcinogenesis was enhanced upon inhibition of autophagy-mediated apoptosis in the cigarette smoke treated cells (Chen et al. 2019b).

The study on the mechanism underlying metal-induced carcinogenesis is receiving increasing interest as metals are considered pollutants for several organisms. Only a few reports have investigated the role of autophagy in metal-induced carcinogenesis. Cadmium, which is considered one of the most hazardous materials, affects different organs and organisms. Increasing evidence suggests that cadmium-induced carcinogenesis is associated with the inhibition of autophagy, indicated by the accumulation of autophagosomes and the adaptor protein p62 (Wang et al. 2018; Ashrafizadeh et al. 2019). Inhibition of autophagic flux in cadmium-exposed oral squamous cell carcinoma (OSCC) CAL27 cells results in reduced migration and invasion, suggesting that autophagy plays a crucial role in cadmium-exposed carcinogenesis (Fan et al. 2019). In addition, Psoralidin (Pso), a nontoxic natural compound, inhibits autophagic flux and induces apoptosis in prostate cancer, demonstrating that autophagy inhibition could serve as an effective therapy in cadmium-induced carcinogenesis (Pal et al. 2017).

Besides autophagy modulation in response to exposure to several known carcinogens, some reports demonstrate selective autophagy, such as mitophagy, playing a master role in carcinogenesis (Chang et al. 2017). It is well established that the presence of damaged mitochondria is associated with the initiation of both central and peripheral COPD-associated non-small cell lung cancer (NSCLC). Excessive dysfunctional mitochondria, which lead to oxidative stress, are observed in COPD patients, and evidence suggests that increased oxidative stress leads to carcinogenesis (Ryter et al. 2018; Ng Kee Kwong et al. 2017). Moreover, COPD and NSCLC are connected through a common mechanistic linkage (Houghton 2013; Ng Kee Kwong et al. 2017). Accumulating evidence display that aberrant lung function upsurges the occurrence of lung cancer pathogenesis (Mannino et al. 2003; Purdue et al. 2007). More importantly, it has been demonstrated that excessive mitophagy is the cause of COPD pathogenesis (Mizumura et al. 2014). Surprisingly, it is important to know the role of autophagy/mitophagy which plays a crucial role in pulmonary diseases (Aggarwal et al. 2016). Furthermore, more mechanistic demonstration of autophagy/mitophagy is required when connecting COPD associated carcinogenesis. There are also reports describing autophagy effect in response to environmental carcinogens such as nickel, which is a Group 1 carcinogen. According to one study, nickel-induced carcinogenesis occurs in human bronchial epithelial cells via a novel SQSTM1 regulatory network (Huang et al. 2016). Another report demonstrated nickel-induced carcinogenesis in human bronchial epithelial cells via increased hexokinase 2 (HK2) expression (Kang et al. 2017). Furthermore, particulate matter (PM) with an aerodynamic diameter of less than 2.5 μm (PM2.5) induces oxidative stress-mediated autophagy in A549 cells. In addition, PM2.5 is considered a novel player for epithelial-to-mesenchymal transition, which contributes to several malignant characteristics observed in cancer (Xu et al. 2019).

The combination of two persistent organic pollutants TCDD and endosulfan disturbs mitochondrial homeostasis and ultimately leads to cell death through the induction of mitochondrial apoptosis associated with an early onset of autophagy (Rainey et al. 2017). Although a few reports regard autophagy as a biomarker for metal-induced toxicity (Di Gioacchino et al. 2008), several studies have established that metal exposure induces autophagic cell death in several cancers. Arsenic, an important carcinogen, is reported to induce autophagy or autophagic cell death, depending on the cellular context. It has been demonstrated that increased SnoN facilitates Beclin-1-independent protective autophagy against arsenic trioxide (As_2O_3)-induced cell death in ovarian carcinoma cells (Smith et al. 2010). Arsenic sulfide (As_2S_2)-induced cell death is promoted upon treatment with autophagy inhibitor 3-MA, suggesting the protective role of autophagy in human osteosarcoma cells (Wang et al. 2017b). According to a report, reduced autophagic flux due to disrupted autophagosome-lysosome fusion was observed in human keratinocytes in response to acute arsenic exposure, which resulted in skin carcinogenesis (Wu et al. 2019). This report delineated that the p62/Nrf2 feedback loop regulates arsenic-induced carcinogenesis. Recent reports suggest that arsenic trioxide is also involved in autophagic degradation in several cancers such as glioma, non-small cell lung cancer (NSCLC), and myeloid leukemia (Kanzawa et al. 2005; Mao et al. 2018; Liu et al. 2020).

1.8 Autophagy Induced by the Synergistic Effect of Cigarette Smoke and PM2.5 Regulates Carcinogenesis

Several reports confirm the association of cigarette smoke exposure and carcinogenesis. However, only a few reports describe the relationship between combinatorial exposure-induced autophagy and carcinogenesis. For instance, combinatorial exposure to cigarette smoke and PM2.5 was reported to increase the levels of autophagy proteins (ATG5, Beclin1, and LC3-II), demonstrating the association of autophagy induction with lung cancer progression. Furthermore, autophagy involved with cell invasion, migration, and EMT was observed in response to combinatorial treatment with cigarette smoke and PM2.5, as evidenced by the siRNA study of the autophagy gene Atg5. Together, these findings suggest that autophagy inhibition could be applied in a therapeutic intervention (Fig. 1.5) (Lin et al. 2018). The mechanisms underlying Group 1 carcinogen-induced autophagy modulation and carcinogenesis remain unexplored to date, and further mechanistic investigations on the role of autophagy and selective autophagy are required for precise therapeutic intervention.

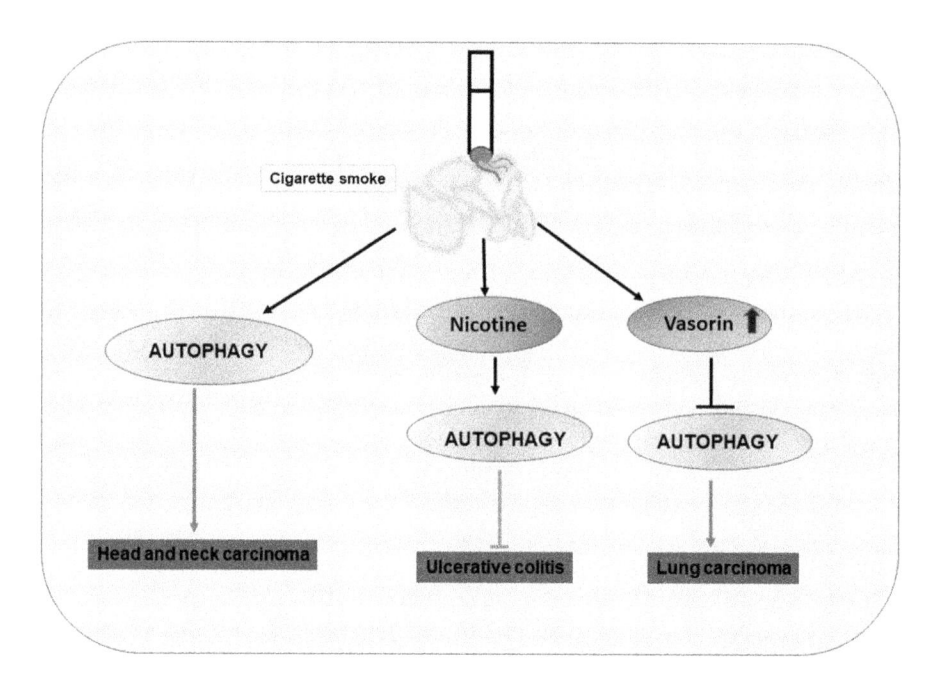

Fig. 1.5 Carcinogenesis and autophagy modulation by xenobiotic compounds. Xenobiotic compounds for example cigarette smoke regulate carcinogenesis via context-dependent regulation of autophagy. Cigarette smoke exposure leads to Head and Neck carcinoma via autophagy induction, while nicotine, one of the active components of cigarette smoke, rescues the cells from ulcerative colitis through autophagy induction. Vasorin upregulation in response to cigarette smoke exposure associated with the inhibition of autophagy which stimulates lung carcinogenesis

1.9 Impact of Chemopreventive Compounds on Xenobiotic Compounds-Induced Toxicity and Carcinogenesis

The chemopreventive compounds exhibited beneficial effects for the prevention of the inhibition of XCs induced cytotoxicity and carcinogenesis. Most of the existing studies on chemopreventive compounds capable of suppressing xenobiotic-induced toxicity have been investigated. For example, capsaicin was reported to modulate the pulmonary antioxidant defense system in B[a]P-induced lung cancer in Swiss albino mice. Moreover, capsaicin also mitigated lysosomal damage in B[a]P-induced lung cancer proliferation (Anandakumar et al. 2008). Another compound baicalein was reported to mitigate the levels of lysosomal enzymes and xenobiotic-metabolizing enzymes in B[a]P-induced lung carcinogenesis in Swiss albino mice (Naveenkumar et al. 2014). Fascinatingly, combination therapy of curcumin and resveratrol was reported to modulate drug-metabolizing enzymes as well as antioxidant indices during lung carcinogenesis in mice (Liu et al. 2015b). Eicosapentaenoic acid inhibits TCDD-induced upstream events of MAPK phosphorylation, the increase in the $[Ca^{2+}]_i$ levels, and the cell surface changes in the microvilli of HepG2 cells (Palanisamy et al. 2015). Likewise, S-Allylcysteine acts as an inhibitor of B[a]P-induced precancerous carcinogenesis in human lung cells by inhibiting the activation of NF-κB (Wang et al. 2019c).

Remarkably, B[a]P-trigger apoptosis was rescued by *Bacopa monnieri* treatment, which provided cytoprotection through Beclin-1-dependent autophagy (Das et al. 2016). Similarly, antidiabetic drug metformin could suppress nickel-induced autophagy and apoptosis by alleviating hexokinase-2 expression and activating lipocalin-2 expression in lung cancer (Kang et al. 2017). Nowadays the precipitous interest has been focused on citrus peel polymethoxy flavones as it prevents B[a]P/dextran sodium sulfate-induced colorectal carcinogenesis by modulating xenobiotic metabolism and ameliorating autophagic defect in ICR mice (Wu et al. 2018). Sulforaphane, a natural dietary compound generally found in the cruciferous vegetables such as broccoli, prevents cadmium-induced carcinogenesis by restoring autophagy, diminishing Nrf2, and reducing apoptosis resistance (Wang et al. 2018). Recently another group investigated natural flavonoid iso-orientin, attenuates benzo[a]pyrene-induced liver injury in vitro, and in vivo through the inhibition of autophagy and pyroptosis (Xueyi et al. 2019).

1.10 Conclusion

The present article discusses the role of the important Group 1 carcinogens in modulating autophagy. Although it is well established that autophagy plays an important role in cancer cell maintenance, increasing evidence has been suggesting that autophagy inhibition or dysfunctional autophagy is also associated with the induction of carcinogenesis. Several studies have reported the association of carcinogens with cancer development, the literature concerning the regulation of carcinogenesis by XCs-induced autophagy is scanty. Moreover, little research has been conducted to decipher the roles of selective autophagy, such as mitophagy, ER-phagy, pexophagy, lysophagy, and ciliophagy, for the regulation of carcinogenesis induced by Group 1 carcinogens; this area of research requires exploration for precise dissection of autophagy in search of better therapeutics. The study of XCs-induced toxicity and carcinogenesis is required for planning and implementing better therapeutic strategies in the future. The study of the combinatorial effect of important carcinogens in relation to autophagy represents another research area to be explored in the future to discover possible therapeutic benefits. Moreover, it is imperative to understand the mechanism underlying the carcinogenesis regulated by XCs stimulated autophagy. It is vital to search for potent autophagy inhibitors, only a few of which are reported in the literature so far. The currently available data regarding metal-induced carcinogenesis has also been discussed, with special emphasis on autophagy. The current research on XCs induced autophagy and metabolism alteration in relation to carcinogenesis is at a stage of infancy which should be addressed. In summary, delineating the complicated interrelationship between xenobiotics and autophagy modulation will attract autophagy scientists for investigating autophagy intonation could be effective therapeutics in the case of cytotoxicity and carcinogenesis.

Acknowledgments Authors thankful to the funding agencies for supporting the research in the lab.

References

Aggarwal S, Mannam P, Zhang J (2016) Differential regulation of autophagy and mitophagy in pulmonary diseases. Am J Physiol Lung Cell Mol Physiol 311:L433–L452

Ahmad T, Sundar IK, Lerner CA, Gerloff J, Tormos AM, Yao H, Rahman I (2015) Impaired mitophagy leads to cigarette smoke stress-induced cellular senescence: implications for chronic obstructive pulmonary disease. FASEB J 29:2912–2929

Amaravadi RK, Yu D, Lum JJ, Bui T, Christophorou MA, Evan GI, Thomas-Tikhonenko A, Thompson CB (2007) Autophagy inhibition enhances therapy-induced apoptosis in a Myc-induced model of lymphoma. J Clin Invest 117:326–336

Anandakumar P, Kamaraj S, Jagan S, Ramakrishnan G, Vinodhkumar R, Devaki T (2008) Capsaicin modulates pulmonary antioxidant defense system during benzo(a)pyrene-induced lung cancer in Swiss albino mice. Phytother Res 22:529–533

Androutsopoulos VP, Tsatsakis AM, Spandidos DA (2009) Cytochrome P450 CYP1A1: wider roles in cancer progression and prevention. BMC Cancer 9:187

Ansari MA, Raish M, Ahmad A, Alkharfy KM, Ahmad SF, Attia SM, Alsaad AMS, Bakheet SA (2017) Sinapic acid ameliorate cadmium-induced nephrotoxicity: in vivo possible involvement of oxidative stress, apoptosis, and inflammation via NF-κB downregulation. Environ Toxicol Pharmacol 51:100–107

Arenas-Huertero F, Zaragoza-Ojeda M, Sánchez-Alarcón J, Milić M, Šegvić Klarić M, Montiel-González JM, Valencia-Quintana R (2019) Involvement of Ahr pathway in toxicity of aflatoxins and other mycotoxins. Front Microbiol 10:2347

Ashoor R, Yafawi R, Jessen B, Lu S (2013) The contribution of lysosomotropism to autophagy perturbation. PLoS One 8:e82481

Ashrafizadeh M, Ahmadi Z, Farkhondeh T, Samarghandian S (2019) Back to nucleus: combating with cadmium toxicity using Nrf2 signaling pathway as a promising therapeutic target. Biol Trace Elem Res. https://doi.org/10.1007/s12011-019-01980-4

Ba Q, Li J, Huang C, Qiu H, Li J, Chu R, Zhang W, Xie D, Wu Y, Wang H (2015) Effects of benzo[a]pyrene exposure on human hepatocellular carcinoma cell angiogenesis, metastasis, and NF-κB signaling. Environ Health Perspect 123:246–254

Bansal S, Leu AN, Gonzalez FJ, Guengerich FP, Chowdhury AR, Anandatheerthavarada HK, Avadhani NG (2014) Mitochondrial targeting of cytochrome P450 (CYP) 1B1 and its role in polycyclic aromatic hydrocarbon-induced mitochondrial dysfunction. J Biol Chem 289:9936–9951

Barrett JC (1993) Mechanisms of multistep carcinogenesis and carcinogen risk assessment. Environ Health Perspect 100:9–20

Bartholomeusz C, Gonzalez-Angulo AM (2012) Targeting the PI3K signaling pathway in cancer therapy. Expert Opin Ther Targets 16:121–130

Bassi DE, Cenna J, Zhang J, Cukierman E, Klein-Szanto AJ (2015) Enhanced aggressiveness of benzopyrene-induced squamous carcinomas in transgenic mice overexpressing the proprotein convertase PACE4 (PCSK6). Mol Carcinog 54:1122–1131

Ben-Sahra I, Howell JJ, Asara JM, Manning BD (2013) Stimulation of de novo pyrimidine synthesis by growth signaling through mTOR and S6K1. Science 339:1323–1328

Bhutia SK, Mukhopadhyay S, Sinha N, Das DN, Panda PK, Patra SK, Maiti TK, Mandal M, Dent P, Wang XY, Das SK, Sarkar D, Fisher PB (2013) Autophagy: cancer's friend or foe? Adv Cancer Res 118:61–95

Biagioli M, Pifferi S, Ragghianti M, Bucci S, Rizzuto R, Pinton P (2008) Endoplasmic reticulum stress and alteration in calcium homeostasis are involved in cadmium-induced apoptosis. Cell Calcium 43:184–195

Birkett N, Al-Zoughool M, Bird M, Baan RA, Zielinski J, Krewski D (2019) Overview of biological mechanisms of human carcinogens. J Toxicol Environ Health B Crit Rev 22:288–359

Biswas G, Srinivasan S, Anandatheerthavarada HK, Avadhani NG (2008) Dioxin-mediated tumor progression through activation of mitochondria-to-nucleus stress signaling. Proc Natl Acad Sci U S A 105:186–191

Bolt AM, Zhao F, Pacheco S, Klimecki WT (2012) Arsenite-induced autophagy is associated with proteotoxicity in human lymphoblastoid cells. Toxicol Appl Pharmacol 264:255–261

Boya P, González-Polo RA, Casares N, Perfettini JL, Dessen P, Larochette N, Métivier D, Meley D, Souquere S, Yoshimori T, Pierron G, Codogno P, Kroemer G (2005) Inhibition of macroautophagy triggers apoptosis. Mol Cell Biol 25:1025–1040

Bruick RK, McKnight SL (2001) A conserved family of prolyl-4-hydroxylases that modify HIF. Science 294:1337–1340

Cagnol S, Chambard JC (2010) ERK and cell death: mechanisms of ERK-induced cell death—apoptosis, autophagy and senescence. FEBS J 277:2–21

Casey SC, Vaccari M, Al-Mulla F, Al-Temaimi R, Amedei A, Barcellos-Hoff MH, Brown DG, Chapellier M, Christopher J, Curran CS, Forte S, Hamid RA, Heneberg P, Koch DC, Krishnakumar PK, Laconi E, Maguer-Satta V, Marongiu F, Memeo L, Mondello C, Raju J, Roman J, Roy R, Ryan EP, Ryeom S, Salem HK, Scovassi AI, Singh N, Soucek L, Vermeulen L, Whitfield JR, Woodrick J, Colacci A, Bisson WH, Felsher DW (2015) The effect of environmental chemicals on the tumor microenvironment. Carcinogenesis 1:160–183

Chang JY, Yi HS, Kim HW, Shong M (2017) Dysregulation of mitophagy in carcinogenesis and tumor progression. Biochim Biophys Acta Bioenerg 1858:633–640

Chargui A, Zekri S, Jacquillet G, Rubera I, Ilie M, Belaid A, Duranton C, Tauc M, Hofman P, Poujeol P, El May MV, Mograb B (2011) Cadmium-induced autophagy in rat kidney: an early biomarker of subtoxic exposure. Toxicol Sci 121:31–42

Chen ZH, Lam HC, Jin Y, Kim HP, Cao J, Lee SJ, Ifedigbo E, Parameswaran H, Ryter SW, Choi AM (2010) Autophagy protein microtubule-associated protein 1 light chain-3B (LC3B) activates extrinsic apoptosis during cigarette smoke-induced emphysema. Proc Natl Acad Sci U S A 107:18880–18885

Chen RJ, Siao SH, Hsu CH, Chang CY, Chang LW, Wu CH, Lin P, Wang YJ (2014) TCDD promotes lung tumors via attenuation of apoptosis through activation of the Akt and ERK1/2 signaling pathways. PLoS One 9:e99586

Chen ZH, Wu YF, Wang PL, Wu YP, Li ZY, Zhao Y, Zhou JS, Zhu C, Cao C, Mao YY, Xu F, Wang BB, Cormier SA, Ying SM, Li W, Shen HH (2019a) Autophagy is essential for ultrafine particle-induced inflammation and mucus hyperproduction in airway epithelium. Autophagy 12:297–311

Chen W, Wang Q, Xu X, Saxton B, Tessema M, Leng S, Choksi S, Belinsky SA, Liu ZG, Lin Y (2019b) Vasorin/ATIA promotes cigarette smoke-induced transformation of human bronchial epithelial cells by suppressing autophagy-mediated apoptosis. Transl Oncol 13:32–41

Chowdhury R, Chatterjee R, Giri AK, Mandal C, Chaudhuri K (2010) Arsenic-induced cell proliferation is associated with enhanced ROS generation, Erk signaling and Cyclin A expression. Toxicol Lett 198:263–271

Corelle E, Nebout M, Bekri S, Gauthier N, Hofman P, Poujeol P, Fénichel P, Mograbi B (2006) Disruption of autophagy at the maturation step by the carcinogen lindane is associated with the sustained mitogen-activated protein kinase/extracellular signal-regulated kinase activity. Cancer Res 66:6861–6869

Crighton D, O'Prey J, Bell HS, Ryan KM (2007) p73 regulates DRAM-independent autophagy that does not contribute to programmed cell death. Cell Death Differ 14:1071–1079

Csordas A, Kreutmayer S, Ploner C, Braun PR, Karlas A, Backovic A, Wick G, Bernhard D (2011) Cigarette smoke extract induces prolonged endoplasmic reticulum stress and autophagic cell death in human umbilical vein endothelial cells. Cardiovasc Res 92:141–148. https://doi.org/10.1093/cvr/cvr165

Das DN, Naik PP, Nayak A, Panda PK, Mukhopadhyay S, Sinha N, Bhutia SK (2016) Bacopa monnieri-induced protective autophagy inhibits benzo[a]pyrene-mediated apoptosis. Phytother Res 30:1794–1801

Das DN, Panda PK, Naik PP, Mukhopadhyay S, Sinha N, Bhutia SK (2017a) Phytotherapeutic approach: a new hope for polycyclic aromatic hydrocarbons induced cellular disorders, autophagic and apoptotic cell death. Toxicol Mech Methods 27:1–17

Das DN, Panda PK, Sinha N, Mukhopadhyay S, Naik PP, Bhutia SK (2017b) DNA damage by 2,3,7,8-tetrachlorodibenzo-p-dioxin-induced p53-mediated apoptosis through activation of cytochrome P450/aryl hydrocarbon receptor. Environ Toxicol Pharmacol 55:175–185

Das DN, Naik PP, Mukhopadhyay S, Panda PK, Sinha N, Meher BR, Bhutia SK (2017c) Elimination of dysfunctional mitochondria through mitophagy suppresses benzo[a]pyrene-induced apoptosis. Free Radic Biol Med 112:452–463

De Abrew KN, Thomas-Virnig CL, Rasmussen CA, Bolterstein EA, Schlosser SJ, Allen-Hoffmann BL (2014) TCDD induces dermal accumulation of keratinocyte-derived matrix metalloproteinase-10 in an organotypic model of human skin. Toxicol Appl Pharmacol 276:171–178

de Oliveira Alves N, Vessoni AT, Quinet A, Fortunato RS, Kajitani GS, Peixoto MS, Hacon SS, Artaxo P, Saldiva P, Menck CFM, Batistuzzo de Medeiros SR (2017) Biomass burning in the Amazon region causes DNA damage and cell death in human lung cells. Sci Rep 7:10937

Deng X, Zhang F, Rui W, Long F, Wang L, Feng Z, Chen D, Ding W (2013) PM2.5-induced oxidative stress triggers autophagy in human lung epithelial A549 cells. Toxicol In Vitro 27:1762–1770

Deng X, Zhang F, Wang L, Rui W, Long F, Zhao Y, Chen D, Ding W (2014) Airborne fine particulate matter induces multiple cell death pathways in human lung epithelial cells. Apoptosis 19:1099–1112

Di Gioacchino M, Petrarca C, Perrone A, Farina M, Sabbioni E, Hartung T, Martino S, Esposito DL, Lotti LV, Mariani-Costantini R (2008) Autophagy as an ultrastructural marker of heavy metal toxicity in human cord blood hematopoietic stem cells. Sci Total Environ 392:50–58

Dietrich C, Kaina B (2010) The aryl hydrocarbon receptor (AhR) in the regulation of cell-cell contact and tumor growth. Carcinogenesis 31:1319–1328

Dino P, D'Anna C, Sangiorgi C, Di Sano C, Di Vincenzo S, Ferraro M, Pace E (2019) Cigarette smoke extract modulates E-Cadherin, Claudin-1 and miR-21 and promotescancer invasiveness in human colorectal adenocarcinoma cells. Toxicol Lett 317:102–109

Divekar SD, Li HH, Parodi DA, Ghafouri TB, Chen R, Cyrus K, Foxworth AE, Fornace AJ, Byrne C, Martin MB (2019) Arsenite and cadmium promote the development of mammary tumors. Carcinogenesis. https://doi.org/10.1093/carcin/bgz176

Duan Z, Zhao J, Fan X, Tang C, Liang L, Nie X, Liu J, Wu Q, Xu G (2014a) The PERK-eIF2α signaling pathway is involved in TCDD-induced ER stress in PC12 cells. Neurotoxicology 44:149–159

Duan J, Yu Y, Yu Y, Li WJ, Geng W, Jiang L, Li Q, Zhou X, Sun Z (2014b) Silica nanoparticles induce autophagy and endothelial dysfunction via the PI3K/Akt/mTOR signaling pathway. Int J Nanomedicine 9:5131–5141

El-Demerdash FM, Tousson EM, Kurzepa J, Habib SL (2018) Xenobiotics, oxidative stress, and antioxidants. Oxidative Med Cell Longev 2018:9758951

Fan T, Chen Y, He Z, Wang Q, Yang X, Ren Z, Zhang S (2019) Inhibition of ROS/NUPR1-dependent autophagy antagonises repeated cadmium exposure-induced oral squamous cell carcinoma cell migration and invasion. Toxicol Lett 314:142–152

Fiorito F, Ciarcia R, Granato GE, Marfe G, Iovane V, Florio S, De Martino L, Pagnini U (2011) 2,3,7,8-tetrachlorodibenzo-p-dioxin induced autophagy in a bovine kidney cell line. Toxicology 290:258–270

Fresno Vara JA, Casado E, de Castro J, Cejas P, Belda-Iniesta C, Gonzalez-Baron M (2004) PI3K/Akt signalling pathway and cancer. Cancer Treat Rev 30:193–204

Fruman DA, Chiu H, Hopkins BD, Bagrodia S, Cantley LC, Abraham RT (2017) The PI3K pathway in human disease. Cell 170:605–635

Fujishiro H, Liu Y, Ahmadi B, Templeton DM (2018) Protective effect of cadmium-induced autophagy in rat renal mesangial cells. Arch Toxicol 92:619–631

Galluzzi L, Pietrocola F, Bravo-San Pedro JM, Amaravadi RK, Baehrecke EH, Cecconi F, Codogno P, Debnath J, Gewirtz DA, Karantza V, Kimmelman A, Kumar S, Levine B, Maiuri MC, Martin SJ, Penninger J, Piacentini M, Rubinsztein DC, Simon HU, Simonsen A, Thorburn AM, Velasco G, Ryan KM, Kroemer G (2015) Autophagy in malignant transformation and cancer progression. EMBO J 34:856–880

Gannon AM, Stämpfli MR, Foster WG (2013) Cigarette smoke exposure elicits increased autophagy and dysregulation of mitochondrial dynamics in murine granulosa cells. Biol Reprod 88:63

Gelboin HV (1980) Benzo[alpha]pyrene metabolism, activation and carcinogenesis: role and regulation of mixed-function oxidases and related enzymes. Physiol Rev 60:1107–1166

Gu X, Han M, Du Y, Wu Y, Xu Y, Zhou X, Ye D, Wang HL (2019) Pb disrupts autophagic flux through inhibiting the formation and activity of lysosomes in neural cells. Toxicol In Vitro 55:43–50

Guo L, Li L, Wang W, Pan Z, Zhou Q, Wu Z (2012) Mitochondrial reactive oxygen species mediates nicotine-induced hypoxia-inducible factor-1α expression in human non-small cell lung cancer cells. Biochim Biophys Acta 822:852–861

Guo J, Xu Y, Ji W, Song L, Dai C, Zhan L (2015) Effects of exposure to benzo[a]pyrene on metastasis of breast cancer are mediated through ROS-ERK-MMP9 axis signaling. Toxicol Lett 234:201–210

Guo C, Yang M, Jing L et al (2016) Amorphous silica nanoparticles trigger vascular endothelial cell injury through apoptosis and autophagy via reactive oxygen species-mediated MAPK/Bcl-2 and PI3K/Akt/mTOR signaling. Int J Nanomedicine 11:5257–5276

Gyongyosi A, Szoke K, Fenyvesi F, Fejes Z, Debreceni IB, Nagy B Jr, Tosaki A, Lekli I (2019) Inhibited autophagy may contribute to heme toxicity in cardiomyoblast cells. Biochem Biophys Res Commun 511:732–738

Hamanaka RB, Chandel NS (2010) Mitochondrial reactive oxygen species regulate cellular signaling and dictate biological outcomes. Trends Biochem Sci 35:505–513

Hankinson O (2016) The role of AHR-inducible cytochrome P450s in metabolism of polyunsaturated fatty acids. Drug Metab Rev 48:342–350

Hansson MJ, Månsson R, Morota S, Uchino H, Kallur T, Sumi T, Ishii N, Shimazu M, Keep MF, Jegorov A, Elmér E (2008) Calcium-induced generation of reactive oxygen species in brain mitochondria is mediated by permeability transition. Free Radic Biol Med 45:284–294

He B, Chen Q, Zhou D, Wang L, Liu Z (2019) Role of reciprocal interaction between autophagy and endoplasmic reticulum stress in apoptosis of human bronchial epithelial cells induced by cigarette smoke extract. IUBMB Life 71:66–80

Hoffmann RF, Zarrintan S, Brandenburg SM, Kol A, de Bruin HG, Jafari S, Dijk F, Kalicharan D, Kelders M, Gosker HR, Ten Hacken NH, van der Want JJ, van Oosterhout AJ, Heijink IH (2013) Prolonged cigarette smoke exposure alters mitochondrial structure and function in airway epithelial cells. Respir Res 14:97

Houghton AM (2013) Mechanistic links between COPD and lung cancer. Nat Rev Cancer 13:233–245

Huang H, Zhu J, Li Y, Zhang L, Gu J, Xie Q, Jin H, Che X, Li J, Huang C, Chen LC, Lyu J, Gao J, Huang C (2016) Upregulation of SQSTM1/p62 contributes to nickel-induced malignant transformation of human bronchial epithelial cells. Autophagy 12:1687–1703

Hwang HJ, Dornbos P, Steidemann M, Dunivin TK, Rizzo M, LaPres JJ (2016) Mitochondrial-targeted aryl hydrocarbon receptor and the impact of 2,3,7,8-tetrachlorodibenzo-p-dioxin on cellular respiration and the mitochondrial proteome. Toxicol Appl Pharmacol 304:121–132

Jahangirnejad R, Goudarzi M, Kalantari H, Najafzadeh H, Rezaei M (2020) Subcellular organelle toxicity caused by arsenic nanoparticles in isolated rat hepatocytes. Int J Occup Environ Med 11:41–52

Jaishankar M, Tseten T, Anbalagan N, Mathew BB, Beeregowda KN (2014) Toxicity, mechanism and health effects of some heavy metals. Interdiscip Toxicol 7:60–72

Jancova P, Anzenbacher P, Anzenbacherova E (2010) Phase II drug metabolizing enzymes. Biomed Pap Med Fac Univ Palacky Olomouc Czech Repub 154:103–116

Jang HS, Lee JE, Myung CH, Park JI, Jo CS, Hwang JS (2019) Particulate matter-induced aryl hydrocarbon receptor regulates autophagy in keratinocytes. Biomol Ther (Seoul) 11:570–576

Jeong JK, Gurunathan S, Kang MH et al (2016) Hypoxia-mediated autophagic flux inhibits silver nanoparticle-triggered apoptosis in human lung cancer cells. Sci Rep 6:21688

Ji K, Xing C, Jiang F, Wang X, Guo H, Nan J, Qian L, Yang P, Lin J, Li M, Li J, Liao L, Tang J (2013) Benzo[a]pyrene induces oxidative stress and endothelial progenitor cell dysfunction via the activation of the NF-κB pathway. Int J Mol Med 31:922–930

Ji X, Xu B, Yao M, Mao Z, Zhang Y, Xu G, Tang Q, Wang X, Xia Y (2016) Graphene oxide quantum dots disrupt autophagic flux by inhibiting lysosome activity in GC-2 and TM4 cell lines. Toxicology 374:10–17

Jiang Y, Zhou X, Chen X, Yang G, Wang Q, Rao K, Xiong W, Yuan J (2011) Benzo(a)pyrene-induced mitochondrial dysfunction and cell death in p53-null Hep3B cells. Mutat Res 726:75–83

Jiang Y, Chen X, Yang G, Wang Q, Wang J, Xiong W, Yuan J (2013) BaP-induced DNA damage initiated p53-independent necroptosis via the mitochondrial pathway involving Bax and Bcl-2. Hum Exp Toxicol 32:1245–1257

Jimma Y, Jimma K, Yachi M, Hakata S, Habano W, Ozawa S, Terashima J (2019) Aryl hydrocarbon receptor mediates cell proliferation enhanced by benzo[a]pyrene in human lung cancer 3D spheroids. Cancer Investig 37:367–375

Kandoth C, McLellan MD, Vandin F, Ye K, Niu B, Lu C, Xie M, Zhang Q, McMichael JF, Wyczalkowski MA, Leiserson MDM, Miller CA, Welch JS, Walter MJ, Wendl MC, Ley TJ, Wilson RK, Raphae J, Ding L (2013) Mutational landscape and significance across 12 major cancer types. Nature 502:333–339

Kang YT, Hsu WC, Wu CH, Hsin IL, Wu PR, Yeh KT, Ko JL (2017) Metformin alleviates nickel-induced autophagy and apoptosis via inhibition of hexokinase-2, activating lipocalin-2, in human bronchial epithelial cells. Oncotarget 8:105536–105552

Kanzawa T, Zhang L, Xiao L, Germano IM, Kondo Y, Kondo S (2005) Arsenic trioxide induces autophagic cell death in malignant glioma cells by upregulation of mitochondrial cell death protein BNIP3. Oncogene 24:980–991

Kim SM, Lee HM, Hwang KA, Choi KC (2017) Benzo(a)pyrene induced cell cycle arrest and apoptosis in human choriocarcinoma cancer cells through reactive oxygen species-induced endoplasmic reticulum-stress pathway. Food Chem Toxicol 107:339–348

Kim HR, Kang SY, Kim HO, Park CW, Chung BY (2020) Role of aryl hydrocarbon receptor activation and autophagy in psoriasis-related inflammation. Int J Mol Sci 21:E2195

Kimmelman AC, White E (2017) Autophagy and tumor metabolism. Cell Metab 25:1037–1043

Kroemer G, Galluzzi L, Vandenabeele P, Abrams J, Alnemri ES, Baehrecke EH, Blagosklonny MV, El-Deiry WS, Golstein P, Green DR, Hengartner M, Knight RA, Kumar S, Lipton SA, Malorni W, Nuñez G, Peter ME, Tschopp J, Yuan J, Piacentini M, Zhivotovsky B, Melino G (2009) Classification of cell death: recommendations of the Nomenclature Committee on Cell Death. Cell Death Differ 16:3–11

Kudo I, Hosaka M, Haga A, Tsuji N, Nagata Y, Okada H, Fukuda K, Kakizaki Y, Okamoto T, Grave E, Itoh H (2018) The regulation mechanisms of AhR by molecular chaperone complex. J Biochem 163:223–232

Kvitko K, Bandinelli E, Henriques JA, Heuser VD, Rohr P, da Silva FR, Schneider NB, Fernandes S, Ancines C, da Silva J (2012) Susceptibility to DNA damage in workers occupationally exposed to pesticides, to tannery chemicals and to coal dust during mining. Genet Mol Biol 35:1060–1068

Lafleur MA, Stevens JL, Lawrence JW (2013) Xenobiotic perturbation of ER stress and the unfolded protein response. Toxicol Pathol 41:235–262

Le Y, Wang Y, Zhou L, Xiong J, Tian J, Yang X, Gai X, Sun Y (2020) Cigarette smoke-induced HMGB1 translocation and release contribute to migration and NF-κB activation through inducing autophagy in lung macrophages. J Cell Mol Med 24:1319–1331

Lee J, Jeong H, Park EJ, Hwang JW, Bae EK, Ahn JK, Ahn KS, Koh EM, Cha HS (2013) A role for benzo[a]pyrene and slug in invasive properties of fibroblast-like synoviocytes in rheumatoid arthritis: a potential molecular link between smoking and radiographic progression. Joint Bone Spine 80:621–625

Levine B (2007) Cell biology: autophagy and cancer. Nature 446:745–747

Levine B, Klionsky DJ (2004) Development by self-digestion: molecular mechanisms and biological functions of autophagy. Dev Cell 6:463–477

Li Q, Gao C, Deng H, Song Q, Yuan L (2019a) Benzo[a]pyrene induces pyroptotic and autophagic death through inhibiting PI3K/Akt signaling pathway in HL-7702 human normal liver cells. J Toxicol Sci 44:121–131

Li Z, Li Q, Lv W, Jiang L, Geng C, Yao X, Shi X, Liu Y, Cao J (2019b) The interaction of Atg4B and Bcl-2 plays an important role in Cd-induced crosstalk between apoptosis and autophagy through disassociation of Bcl-2-Beclin1 in A549 cells. Free Radic Biol Med 130:76–591

Li R, Yang L, Jiang N, Wang F, Zhang P, Zhou R, Zhang J (2020) Activated macrophages are crucial during acute PM2.5 exposure-induced angiogenesis in lung cancer. Oncol Lett 19:725–734

Liao TL, Chen SC, Tzeng CR, Kao SH (2014) TCDD induces the hypoxia-inducible factor (HIF)-1α regulatory pathway in human trophoblastic JAR cells. Int J Mol Sci 15:17733–17750

Lin T, Mak NK, Yang MS (2008) MAPK regulate p53-dependent cell death induced by benzo[a]pyrene: involvement of p53 phosphorylation and acetylation. Toxicology 247:145–153

Lin YF, Chiu IJ, Cheng FY et al (2016) The role of hypoxia-inducible factor-1 alpha in zinc oxide nanoparticle-induced nephrotoxicity in vitro and in vivo. Part Fibre Toxicol 13:52

Lin H, Zhang X, Feng N, Wang R, Zhang W, Deng X, Wang Y, Yu X, Ye X, Li L, Qian Y, Yu H, Qian B (2018) LncRNA LCPAT1 mediates smoking/particulate matter 2.5-induced cell autophagy and epithelial-mesenchymal transition in lung cancer cells via RCC2. Cell Physiol Biochem 47:1244–1258

Liu F, Inageda K, Nishitai G, Matsuoka M (2006) Cadmium induces the expression of Grp78, an endoplasmic reticulum molecular chaperone, in LLC-PK1 renal epithelial cells. Environ Health Perspect 114:859–864

Liu T, Wu B, Wang Y, He H, Lin Z, Tan J, Yang L, Kamp DW, Zhou X, Tang J, Huang H, Zhang L, Bin L, Liu G (2015a) Particulate matter 2.5 induces autophagy via inhibition of the phosphatidylinositol 3-kinase/Akt/mammalian target of rapamycin kinase signaling pathway in human bronchial epithelial cells. Mol Med Rep 12:1914–1922

Liu Y, Wu YM, Yu Y, Cao CS, Zhang JH, Li K, Zhang PY (2015b) Curcumin and resveratrol in combination modulate drug-metabolizing enzymes as well as antioxidant indices during lung carcinogenesis in mice. Hum Exp Toxicol 34:620–627

Liu XJ, Wang LN, Zhang ZH, Liang C, Li Y, Luo JS, Peng CJ, Zhang XL, Ke ZY, Huang LB, Tang YL, Luo XQ (2020) Arsenic trioxide induces autophagic degradation of the FLT3-ITD mutated protein in FLT3-ITD acute myeloid leukemia cells. J Cancer 11:3476–3482

Luo YH, Wu SB, Wei YH, Chen YC, Tsai MH, Ho CC, Lin SY, Yang CS, Lin P (2013) Cadmium-based quantum dot induced autophagy formation for cell survival via oxidative stress. Chem Res Toxicol 26:662–673

Luo C, Li Y, Yang L et al (2014) Activation of Erk and p53 regulates copper oxide nanoparticle-induced cytotoxicity in keratinocytes and fibroblasts. Int J Nanomedicine 9:4763–4772

Mannino DM, Aguayo SM, Petty TL, Redd SC (2003) Low lung function and incident lung cancer in the United States: data From the First National Health and Nutrition Examination Survey follow-up. Arch Intern Med 163:1475–1480

Mao J, Ma L, Shen Y, Zhu K, Zhang R, Xi W, Ruan Z, Luo C, Chen Z, Xi X, Chen S (2018) Arsenic circumvents the gefitinib resistance by binding to P62 and mediating autophagic degradation of EGFR in non-small cell lung cancer. Cell Death Dis 9:963

Marchi S, Bittremieux M, Missiroli S, Morganti C, Patergnani S, Sbano L, Rimessi A, Kerkhofs M, Parys JB, Bultynck G, Giorgi C, Pinton P (2017) Endoplasmic reticulum-mitochondria communication through Ca(2+) signaling: the importance of mitochondria-associated membranes (MAMs). Adv Exp Med Biol 997:49–67

Martini-Stoica H, Xu Y, Ballabio A, Zheng H (2016) The autophagy-lysosomal pathway in neurodegeneration: a TFEB perspective. Trends Neurosci 39:221–234

Mathew R, White E (2011) Autophagy in tumorigenesis and energy metabolism: friend by day, foe by night. Curr Opin Genet Dev 21:113–119

Mathew R, Kongara S, Beaudoin B, Karp CM, Bray K, Degenhardt K, Chen G, Jin S, White E (2007) Autophagy suppresses tumor progression by limiting chromosomal instability. Genes Dev 21:1367–1381

McAuliffe PF, Meric-Bernstam F, Mills GB, Gonzalez-Angulo AM (2010) Deciphering the role of PI3K/Akt/mTOR pathway in breast cancer biology and pathogenesis. Clin Breast Cancer 3:S59–S65

Memmott RM, Dennis PA (2010) The role of the Akt/mTOR pathway in tobacco carcinogen-induced lung tumorigenesis. Clin Cancer Res 16:4–10

Messner B, Türkcan A, Ploner C, Laufer G, Bernhard D (2016) Cadmium overkill: autophagy, apoptosis and necrosis signalling in endothelial cells exposed to cadmium. Cell Mol Life Sci 73:1699–1713

Micale RT, La Maestra S, Di Pietro A, Visalli G, Baluce B, Balansky R, Steele VE, De Flora S (2013) Oxidative stress in the lung of mice exposed to cigarette smoke either early in life or in adulthood. Arch Toxicol 87:915–918

Mizumura K, Cloonan SM, Nakahira K, Bhashyam AR, Cervo M, Kitada T, Glass K, Owen CA, Mahmood A, Washko GR, Hashimoto S, Ryter SW, Choi AM (2014) Mitophagy-dependent necroptosis contributes to the pathogenesis of COPD. J Clin Invest 124:3987–4003

Moreau K, Luo S, Rubinsztein DC (2010) Cytoprotective roles for autophagy. Curr Opin Cell Biol 22:206–211

Naveenkumar C, Raghunandakumar S, Asokkumar S, Binuclara J, Rajan B, Premkumar T, Devaki T (2014) Mitigating role of baicalein on lysosomal enzymes and xenobiotic metabolizing enzyme status during lung carcinogenesis of Swiss albino mice induced by benzo(a)pyrene. Fundam Clin Pharmacol 28:310–322

Nebert DW, Dalton TP (2006) The role of cytochrome P450 enzymes in endogenous signalling pathways and environmental carcinogenesis. Nat Rev Cancer 6:947–960

Ng Kee Kwong F, Nicholson AG, Harrison CL, Hansbro PM, Adcock IM, Chung KF (2017) Is mitochondrial dysfunction a driving mechanism linking COPD to nonsmall cell lung carcinoma? Eur Respir Rev 26:170040

Ogata M, Hino S, Saito A, Morikawa K, Kondo S, Kanemoto S, Murakami T, Taniguchi M, Tanii I, Yoshinaga K, Shiosaka S, Hammarback JA, Urano F, Imaizumi K (2006) Autophagy is activated for cell survival after endoplasmic reticulum stress. Mol Cell Biol 26:9220–9231

Ogier-Denis E, Pattingre S, El Benna J et al (2000) Erk1/2-dependent phosphorylation of Galpha-interacting protein stimulates its GTPase accelerating activity and autophagy in human colon cancer cells. J Biol Chem 275:39090–39095

Pal D, Suman S, Kolluru V, Sears S, Das TP, Alatassi H, Ankem MK, Freedman JH, Damodaran C (2017) Inhibition of autophagy prevents cadmium-induced prostate carcinogenesis. Br J Cancer 117:56–64

Palanisamy K, Krishnaswamy R, Paramasivan P, Chih-Yang H, Vishwanadha VP (2015) Eicosapentaenoic acid prevents TCDD-induced oxidative stress and inflammatory response by modulating MAP kinases and redox-sensitive transcription factors. Br J Pharmacol 172:4726–4740

Panda PK, Mukhopadhyay S, Das DN, Sinha N, Naik PP, Bhutia SK (2015) Mechanism of autophagic regulation in carcinogenesis and cancer therapeutics. Semin Cell Dev Biol 39:43–55

Park SY, Lee SM, Ye SK, Yoon SH, Chung MH, Choi J (2006) Benzo[a]pyrene-induced DNA damage and p53 modulation in human hepatoma HepG2 cells for the identification of potential biomarkers for PAH monitoring and risk assessment. Toxicol Lett 167:27–33

Park EJ, Umh HN, Kim SW et al (2014) ERK pathway is activated in bare-FeNPs-induced autophagy. Arch Toxicol 88:323–336

Patterson AD, Gonzalez FJ, Perdew GH, Peters JM (2018) Molecular regulation of carcinogenesis: friend and foe. Toxicol Sci 165:277–283

Pelissier-Rota MA, Pelosi L, Meresse P, Jacquier-Sarlin MR (2015) Nicotine-induced cellular stresses and autophagy in human cancer colon cells: a supportive effect on cell homeostasis via up-regulation of Cox-2 and PGE(2) production. Int J Biochem Cell Biol 65:239–256

Peng H, Zhao XH, Bi TT, Yuan XY, Guo JB, Peng SQ (2017) PM (2.5) obtained from urban areas in Beijing induces apoptosis by activating nuclear factor-kappa B. Mil Med Res 4:27

Person RJ, Tokar EJ, Xu Y, Orihuela R, Ngalame NN, Waalkes MP (2013) Chronic cadmium exposure in vitro induces cancer cell characteristics in human lung cells. Toxicol Appl Pharmacol 273:281–288

Petrache Voicu SN, Dinu D, Sima C et al (2015) Silica nanoparticles induce oxidative stress and autophagy but not apoptosis in the MRC-5 cell line. Int J Mol Sci 16:29398–29416

Piao MJ, Ahn MJ, Kang KA, Ryu YS, Hyun YJ, Shilnikova K, Zhen AX, Jeong JW, Choi YH, Kang HK, Koh YS, Hyun JW (2018) Particulate matter 2.5 damages skin cells by inducing oxidative stress, subcellular organelle dysfunction, and apoptosis. Arch Toxicol 92:2077–2091

Purdue MP, Gold L, Jarvholm B, Alavanja MC, Ward MH, Vermeulen R (2007) Impaired lung function and lung cancer incidence in a cohort of Swedish construction workers. Thorax 62:51–56

Qin H, Gao F, Wang Y, Huang B, Peng L, Mo B, Wang C (2019) Nur77 promotes cigarette smoke-induced autophagic cell death by increasing the dissociation of Bcl2 from Beclin-1. Int J Mol Med 44:25–36

Rainey NE, Saric A, Leberre A, Dewailly E, Slomianny C, Vial G, Zeliger HI, Petit PX (2017) Synergistic cellular effects including mitochondrial destabilization, autophagy and apoptosis following low-level exposure to a mixture of lipophilic persistent organic pollutants. Sci Rep 7:4728

Ratovitski EA (2011) ΔNp63α/IRF6 interplay activates NOS2 transcription and induces autophagy upon tobacco exposure. Arch Biochem Biophys 506:208–215

Reyes H, Reisz-Porszasz S, Hankinson O (1992) Identification of the Ah receptor nuclear translocator protein (Arnt) as a component of the DNA binding form of the Ah receptor. Science 256:1193–1195

Rinna A, Magdolenova Z, Hudecova A et al (2015) Effect of silver nanoparticles on mitogen-activated protein kinases activation: role of reactive oxygen species and implication in DNA damage. Mutagenesis 30:59–66

Rizzuto R, Bernardi P, Pozzan T (2000) Mitochondria as all-round players of the calcium game. J Physiol 529(Pt 1):37–47

Roy R, Singh SK, Chauhan LK, Das M, Tripathi A, Dwivedi PD (2014) Zinc oxide nanoparticles induce apoptosis by enhancement of autophagy via PI3K/Akt/mTOR inhibition. Toxicol Lett 227:29–40

Ryter SW, Rosas IO, Owen CA, Martinez FJ, Choi ME, Lee CG, Elias JA, Choi AMK (2018) Mitochondrial dysfunction as a pathogenic mediator of chronic obstructive pulmonary disease and idiopathic pulmonary fibrosis. Ann Am Thorac Soc 15:S266–S272

Salem AF, Al-Zoubi MS, Whitaker-Menezes D, Martinez-Outschoorn UE, Lamb R, Hulit J, Howell A, Gandara R, Sartini M, Galbiati F, Bevilacqua G, Sotgia F, Lisanti MP (2013) Cigarette smoke metabolically promotes cancer, via autophagy and premature aging in the host stromal microenvironment. Cell Cycle 12:818–825

Shen J, Xu L, Owonikoko TK, Sun SY, Khuri FR, Curran WJ, Deng X (2012) NNK promotes migration and invasion of lung cancer cells through activation of c-Src/PKCι/FAK loop. Cancer Lett 318:106–113

Smith DM, Patel S, Raffoul F, Haller E, Mills GB, Nanjundan M (2010) Arsenic trioxide induces a beclin-1-independent autophagic pathway via modulation of SnoN/SkiL expression in ovarian carcinoma cells. Cell Death Differ 17:1867–1881

Song D, Chen Y, Wang B, Li D, Xu C, Huang H, Huang S, Liu R (2019) Bisphenol A inhibits autophagosome-lysosome fusion and lipid droplet degradation. Ecotoxicol Environ Saf 183:109492

Soto AM, Sonnenschein C (2010) Environmental causes of cancer: endocrine disruptors as carcinogens. Nat Rev Endocrinol 6:363–370

Su R, Jin X, Zhang W, Li Z, Liu X, Ren J (2017) Particulate matter exposure induces the autophagy of macrophages via oxidative stress mediated PI3K/AKT/mTOR pathway. Chemosphere 167:444–453

Sui L, Zhang RH, Zhang P, Yun KL, Zhang HC, Liu L, Hu MX (2015) Lead toxicity induces autophagy to protect against cell death through mTORC1 pathway in cardiofibroblasts. Biosci Rep 35:e00186

Tai H, Wang Z, Gong H, Han X, Zhou J, Wang X, Wei X, Ding Y, Huang N, Qin J, Zhang J, Wang S, Gao F, Chrzanowska-Lightowlers ZM, Xiang R, Xiao H (2017) Autophagy impairment with lysosomal and mitochondrial dysfunction is an important characteristic of oxidative stress-induced senescence. Autophagy 13:99–113

Tam LM, Price NE, Wang Y (2020) Molecular mechanisms of arsenic-induced disruption of DNA repair. Chem Res Toxicol 33:709–726

Tang MS, Wu XR, Lee HW, Xia Y, Deng FM, Moreira AL, Chen LC, Huang WC, Lepor H (2019) Electronic-cigarette smoke induces lung adenocarcinoma and bladder urothelial hyperplasia in mice. Proc Natl Acad Sci U S A 116:21727–21731

Tsai MJ, Wang TN, Lin YS, Kuo PL, Hsu YL, Huang MS (2015) Aryl hydrocarbon receptor agonists upregulate VEGF secretion from bronchial epithelial cells. J Mol Med (Berl) 93:1257–1269

Tsay JJ, Tchou-Wong KM, Greenberg AK, Pass H, Rom WN (2013) Aryl hydrocarbon receptor and lung cancer. Anticancer Res 33:1247–1256

Ueng TH, Chang YL, Tsai YY, Su JL, Chan PK, Shih JY, Lee YC, Ma YC, Kuo ML (2010) Potential roles of fibroblast growth factor-9 in the benzo(a)pyrene-induced invasion in vitro and the metastasis of human lung adenocarcinoma. Arch Toxicol 84:651–660

Valvezan AJ, Turner M, Belaid A, Lam HC, Miller SK, McNamara MC, Baglini C, Housden BE, Perrimon N, Kwiatkowski DJ, Asara JM, Henske EP, Manning BD (2017) mTORC1 couples nucleotide synthesis to nucleotide demand resulting in a targetable metabolic vulnerability. Cancer Cell 32:624–638.e5

Villano CM, Murphy KA, Akintobi A, White LA (2006) 2,3,7,8-Tetrachlorodibenzo-p-dioxin (TCDD) induces matrix metalloproteinase (MMP) expression and invasion in A2058 melanoma cells. Toxicol Appl Pharmacol 210:212–224

Volkov MS, Kobliakov VA (2011) Activation of transcription factor NF-kappaB by carcinogenic polycyclic aromatic hydrocarbons. Tsitologiia 53:418–422

Wang BY, Wu SY, Tang SC, Lai CH, Ou CC, Wu MF, Hsiao YM, Ko JL (2015) Benzo[a]pyrene-induced cell cycle progression occurs via ERK-induced Chk1 pathway activation in human lung cancer cells. Mutat Res 773:1–8

Wang J, Yu Y, Lu K et al (2017a) Silica nanoparticles induce autophagy dysfunction via lysosomal impairment and inhibition of autophagosome degradation in hepatocytes. Int J Nanomedicine 12:809–825

Wang G, Zhang T, Sun W, Wang H, Yin F, Wang Z, Zuo D, Sun M, Zhou Z, Lin B, Xu J, Hua Y, Li H, Cai Z (2017b) Arsenic sulfide induces apoptosis and autophagy through the activation of ROS/JNK and suppression of Akt/mTOR signaling pathways in osteosarcoma. Free Radic Biol Med 106:24–37

Wang Y, Mandal AK, Son YO, Pratheeshkumar P, Wise JTF, Wang L, Zhang Z, Shi X, Chen Z (2018) Roles of ROS, Nrf2, and autophagy in cadmium-carcinogenesis and its prevention by sulforaphane. Toxicol Appl Pharmacol 353:23–30

Wang Y, Shi L, Li J, Li L, Wang H, Yang H (2019a) Involvement of p38 MAPK pathway in benzo (a)pyrene-induced human hepatoma cell migration and invasion. Environ Sci Pollut Res Int 26:35838–35845

Wang L, Kumar M, Deng Q, Wang X, Liu M, Gong Z, Zhang S, Ma X, Xu-Monette ZY, Xiao M, Yi Q, Young KH, Ramos KS, Li Y (2019b) 2,3,7,8-Tetrachlorodibenzo-p-dioxin (TCDD) induces peripheral blood abnormalities and plasma cell neoplasms resembling multiple myeloma in mice. Cancer Lett 440–441:135–144

Wang K, Fang Zhang Q, Zhang Y, Yang Zhao Z (2019c) S-Allylcysteine as an inhibitor of benzo (a)pyrene-induced precancerous carcinogenesis in human lung cells via inhibiting activation of nuclear factor-kappa. Nat Prod Commun 14:12

Watabe Y, Nazuka N, Tezuka M, Shimba S (2010) Aryl hydrocarbon receptor functions as a potent coactivator of E2F1-dependent transcription activity. Biol Pharm Bull 33:389–397

Wei X, Qi Y, Zhang X, Qiu Q, Gu X, Tao C, Huang D, Zhang Y (2014) Cadmium induces mitophagy through ROS-mediated PINK1/Parkin pathway. Toxicol Mech Methods 24:504–511

Wu H, Lin J, Liu P et al (2015) Is the autophagy a friend or foe in the silver nanoparticles associated radiotherapy for glioma? Biomaterials 62:47–57

Wu JC, Tsai ML, Lai CS, Lo CY, Ho CT, Wang YJ, Pan MH (2018) Polymethoxyflavones prevent benzo[a]pyrene/dextran sodium sulfate-induced colorectal carcinogenesis through modulating xenobiotic metabolism and ameliorate autophagic defect in ICR mice. Int J Cancer 142:1689–1701

Wu X, Sun R, Wang H, Yang B, Wang F, Xu H, Chen S, Zhao R, Pi J, Xu Y (2019) Enhanced p62-NRF2 feedback loop due to impaired autophagic flux contributes to arsenic-induced malignant transformation of human keratinocytes. Oxidative Med Cell Longev 2019:1038932

Wu YF, Li ZY, Dong LL, Li WJ, Wu YP, Wang J, Chen HP, Liu HW, Li M, Jin CL, Huang HQ, Ying SM, Li W, Shen HH, Chen ZH (2020a) Inactivation of MTOR promotes autophagy-mediated epithelial injury in particulate matter-induced airway inflammation. Autophagy 16:435–450

Wu Y, Niu Y, Leng J, Xu J, Chen H, Li H, Wang L, Hu J, Xia D, Wu Y (2020b) Benzo(a)pyrene regulated A549 cell migration, invasion and epithelial-mesenchymal transition by up-regulating long non-coding RNA linc00673. Toxicol Lett 320:37–45

Xu C, Bailly-Maitre B, Reed JC (2005) Endoplasmic reticulum stress: cell life and death decisions. J Clin Invest 115:2656–2664

Xu XC, Wu YF, Zhou JS, Chen HP, Wang Y, Li ZY, Zhao Y, Shen HH, Chen ZH (2017) Autophagy inhibitors suppress environmental particulate matter-induced airway inflammation. Toxicology 280:206–212

Xu Z, Ding W, Deng X (2019) PM(2.5), fine particulate matter: a novel player in the epithelial-mesenchymal transition? Front Physiol 10:1404

Xueyi L, Shenyuan H, Chunxia G, Deng H, Liu Y, Li C, Yuan L, Luo Y (2019) Isoorientin attenuates benzo[a]pyrene-induced liver injury by inhibiting autophagy and pyroptosis in vitro and vivo. Food Agric Immunol 1:841–861

Yang H, Zhang H, Pan T, Wang H, Wang Y (2018) Benzo(a)pyrene promotes migration, invasion and metastasis of lung adenocarcinoma cells by upregulating TGIF. Toxicol Lett 294:11–19

Yang X, Ku T, Sun Z, Liu QS, Yin N, Zhou Q, Faiola F, Liao C, Jiang G (2019) Assessment of the carcinogenic effect of 2,3,7,8-tetrachlorodibenzo-p-dioxin using mouse embryonic stem cells to form teratoma in vivo. Toxicol Lett 312:139–147

Yecies JL, Manning BD (2011) Transcriptional control of cellular metabolism by mTOR signaling. Cancer Res 71:2815–2820

Yim WW, Mizushima N (2020) Lysosome biology in autophagy. Cell Discov 6:6

Yin L, Yu X (2018) Arsenic-induced apoptosis in the p53-proficient and p53-deficient cells through differential modulation of NFkB pathway. Food Chem Toxicol 118:849–860

Yu KN, Yoon TJ, Minai-Tehrani A, Kim JE, Park SJ, Jeong MS, Ha SW, Lee JK, Kim JS, Cho MH (2013) Zinc oxide nanoparticle induced autophagic cell death and mitochondrial damage via reactive oxygen species generation. Toxicol In Vitro 27:1187–1195

Zahedi A, Phandthong R, Chaili A, Leung S, Omaiye E, Talbot P (2019) Mitochondrial stress response in neural stem cells exposed to electronic cigarettes. iScience 16:250–269

Zhai H, Pan T, Yang H, Wang H, Wang Y (2019) Cadmium induces A549 cell migration and invasion by activating ERK. Exp Ther Med 18:1793–1799

Zhang Q, Tang X, Zhang ZF, Velikina R, Shi S, Le AD (2007) Nicotine induces hypoxia-inducible factor-1alpha expression in human lung cancer cells via nicotinic acetylcholine receptor-mediated signaling pathways. Clin Cancer Res 13:4686–4694

Zhang W, Liu N, Wang X, Jin X, Du H, Peng G, Xue J (2015) Benzo(a)pyrene-7,8-diol-9,10-epoxide induced p53-independent necrosis via the mitochondria-associated pathway involving Bax and Bak activation. Hum Exp Toxicol 34:179–190

Zhang L, Xia Q, Zhou Y, Li J (2019a) Endoplasmic reticulum stress and autophagy contribute to cadmium-induced cytotoxicity in retinal pigment epithelial cells. Toxicol Lett 311:105–113

Zhang M, Shi R, Zhang Y, Shan H, Zhang Q, Yang X, Li Y, Zhang J (2019b) Nix/BNIP3L-dependent mitophagy accounts for airway epithelial cell injury induced by cigarette smoke. J Cell Physiol 234:14210–14220

Zhao J, Tang C, Nie X, Xi H, Jiang S, Jiang J, Liu S, Liu X, Liang L, Wan C, Yang J (2016) Autophagy potentially protects against 2,3,7,8-tetrachlorodibenzo-p-dioxin induced apoptosis in SH-SY5Y cells. Environ Toxicol 31:1068–1079

Zilfou JT, Lowe SW (2009) Tumor suppressive functions of p53. Cold Spring Harb Perspect Biol 1: a001883

Zimta AA, Schitcu V, Gurzau E, Stavaru C, Manda G, Szedlacsek S, Berindan-Neagoe I (2019) Biological and molecular modifications induced by cadmium and arsenic during breast and prostate cancer development. Environ Res 178:108700

Zou H, Wang T, Yuan J, Sun J, Yuan Y, Gu J, Liu X, Bian J, Liu Z (2020) Cadmium-induced cytotoxicity in mouse liver cells is associated with the disruption of autophagic flux via inhibiting the fusion of autophagosomes and lysosomes. Toxicol Lett 321:32–43

Autophagy: An Agonist and Antagonist with an Interlink of Apoptosis in Cancer

Vanishri Chandrashekhar Haragannavar, Roopa S. Rao, Kewal Kumar Mahapatra, Srimanta Patra, Bishnu Prasad Behera, Amruta Singh, Soumya Ranjan Mishra, Chandra Sekhar Bhol, Debasna Pritimanjari Panigrahi, Prakash Priyadarshi Praharaj, Sujit Kumar Bhutia, and Shankargouda Patil

Abstract

Autophagy and apoptosis are the two evolutionarily conserved processes regulating the turnover of defective organelles and other contents inside cells and damaged whole cells inside organisms, respectively. Although apoptosis and autophagy function differently, their signaling pathways are interconnected and mediated by a toggle switch that is triggered based on the requirements of a cell and its surroundings. Suppression of apoptosis and autophagy due to uncontrolled stress is thought to be a hallmark of carcinogenesis. In general, autophagy and apoptosis mediate each other through a roller coaster of up- and downregulation of factors; that is, autophagy attenuates apoptosis induction, and caspase-dependent apoptosis turns off the autophagic machinery in cancer cells, with several exceptions. Moreover, in certain scenarios, autophagy or autophagy-associated proteins induce excessive degradation of cytoplasmic components, causing "autophagic cell death." Autophagy can also rescue cancer cells from apoptosis by modulating stress levels, determining cancer cell fate. However, the molecular signals driving the cell toward either autophagy or apoptosis remain largely unknown. Therefore, in this review, we focus on understanding the complex crossover signaling between autophagy and apoptosis pathways and their modulation in the transformation from benign proliferation to malignant carcinogenesis.

V. C. Haragannavar · R. S. Rao
Department of Oral Pathology and Microbiology, Faculty of Dental Sciences, Ramaiah University of Applied Sciences, Bangalore, Karnataka, India

K. K. Mahapatra · S. Patra · B. P. Behera · A. Singh · S. R. Mishra · C. S. Bhol · D. P. Panigrahi · P. P. Praharaj · S. K. Bhutia
Department of Life Science, National Institute of Technology Rourkela, Rourkela, Odisha, India

S. Patil (✉)
Division of Oral Pathology, Department of Maxillofacial Surgery and Diagnostic Sciences, College of Dentistry, Jazan University, Jazan, Saudi Arabia

© Springer Nature Singapore Pte Ltd. 2020
S. K. Bhutia (ed.), *Autophagy in Tumor and Tumor Microenvironment*,
https://doi.org/10.1007/978-981-15-6930-2_2

Keywords

Autophagy · Apoptosis · Autophagic cell death · Cancer

2.1 Introduction

Cell proliferation and cell death are the two wheels on which life processes roll. The proper organization and construction of a living body depend on the coordination between these two processes. Every living organism has an inherent ability to detect and remove individual damaged cells through a process of regulated cell death (RCD). RCD involves very sophisticated molecular machinery by which cells commit themselves to die after receiving a proper signal of death. The RCD process has multidimensional roles ranging from removal of damaged cells to the proper organization and formation of organs and tissues at specific developmental stages. Different forms of RCD, such as apoptosis, pyroptosis, ferroptosis, and so on, are cellular responses to stress that exceed the limit of tolerance. In the past few years, another guardian of the cellular microenvironment called autophagy was found to be extremely important for the maintenance of cellular homeostasis. Autophagy is a self-digestion process in which unwanted cellular organelles and other cellular components are entrapped within a double-layered vesicular structure known an autophagosome and degraded by the subsequent autophagosome fusion with a lysosome. Hence, the process of autophagy contributes toward the maintenance of cellular homeostasis in a varying range of stress conditions, like nutrient deprivation, an abundance of damaged and misfolded proteins, pathogens, infections, and hypoxia (Klionsky 2005; Mahapatra et al. 2019). The function of autophagy as a cytoprotective process is not limited to scavenging damaged sterile targets as it is also a very effective and regulated way of removing intracellular pathogens through xenophagy, which is the sequestration of pathogen-derived proteins and many other types of invaders (Saha et al. 2018).

The ability of a cell to prevent apoptosis is one of the important hallmarks of cancer (Hanahan and Weinberg 2011). Cancer cells disable the apoptosis mechanism either by mutation of tumor suppressors or by overexpressing anti-apoptotic signals to achieve a malignant state. Conventionally, many cancer therapeutics have been developed to kill cancer cells by promoting apoptosis. These therapeutics mainly target damaged DNA to restore mutated tumor suppressors, such as p53. The strategy of these manipulations is to reduce the uncontrolled proliferation of cancer cells and to eliminate them in a regulated and targeted manner through apoptosis. However, when cancer cells are exposed to any type of stress (such as anticancer therapeutics or nutrient deprivation in the tumor microenvironment), the cellular stress response of autophagy can be activated. Here, the tumor-promoting role of autophagy leads to eradicated cell stress that thus inhibits apoptosis, aiding in tumor growth and progression. Such cancer-cell protective autophagy is one of the main causes of therapeutic resistance shown by different cancers against the majority of anticancer drugs. However, the critical function of the role of autophagy in tumor condition is not simple, as it maintains a critical role in both tumor promotion and suppression during various stages of cancer progression. Therefore, in this review, we shed light on the interaction between the two fundamental processes, that is, autophagy and apoptosis, and discuss every possible angle in their connections to cancer progression and roles in effective cancer treatment.

2.2 Apoptosis in Cancer

Apoptosis (apo-separation, ptosis-falling off) is a highly coordinated process of controlled cell depletion described by morphological features such as a blebbing cell membrane, nuclear fragmentation, condensation of chromatin, and so on, leading to the generation of apoptotic bodies that are successively eliminated by the process of phagocytosis (Kerr et al. 1972). The key intracellular architect of apoptosis is a series of cysteine-dependent aspartate-driven proteases (caspases) (Galluzzi et al. 2008; Garrido and Kroemer 2004), but evidence of caspase-independent cell death has also been presented. Caspases are initially secreted in inactive procaspase form and, as the full name implies, the C-terminus is cleaved at an aspartate residue.

2.2.1 Mechanism of Apoptosis

Depending on the activation of caspases, the apoptotic pathways are primarily induced by either extrinsic mechanism or intrinsic/mitochondrial mechanism. The extrinsic apoptotic signaling is activated through a death receptor (DR), a special type of cell surface receptor in tumor necrosis factor (TNF) superfamily proteins (Bhardwaj and Aggarwal 2003). Signals for the extrinsic apoptotic pathways are transmitted by death ligands that interact with and are activated by DRs. Structurally, TNF family receptors possess a conserved death domain (DD) that undergoes trimerization upon binding with ligands like TRAIL (tumor necrosis factor-related apoptosis-inducing ligand) and that recruits additional DD-containing proteins like FADD (Fas-associated protein with death domain) and TRADD (TNF-R type 1-associated death domain protein) to form the death-inducing signaling complex (DISC), followed by the interaction and induction of caspase 8 and 10. Then the stimulated caspase 8 and 10 cleave caspases 3, 6, and 7, leading to the subsequent cleavage of target peptides (Fig. 2.1) (Mukhopadhyay et al. 2014; Wang and El-Deiry 2003).

The intrinsic apoptotic signaling pathway is triggered by different stresses, but the first and irreversible step in the sequence of events activating this apoptosis is the loss of the mitochondrial membrane potential (MMP) due to outer membrane permeabilization. In this process, the guardians of mitochondrial integrity belong to the Bcl-2 family of proteins containing a Bcl-2 homology (BH) domain, with the BH1–BH4 domains playing important roles. Several proteins belonging to this family have all four BH domains, including Bcl-w, Bcl-xL, Mcl-1, Bcl-B, and A1,

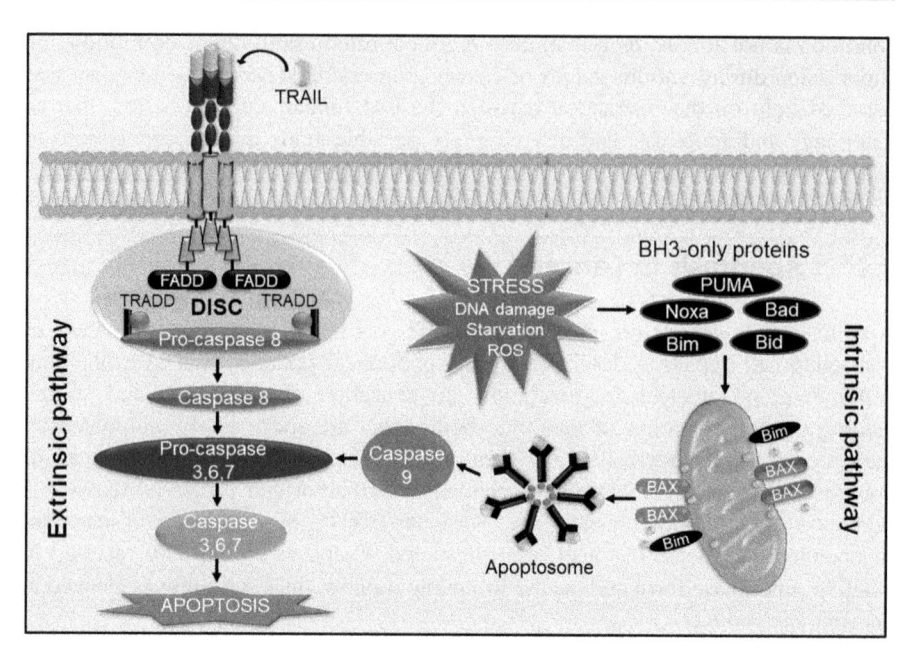

Fig. 2.1 Overview of apoptotic pathways. It is mainly regulated through extrinsic and intrinsic way to induce cell death. The extrinsic mode activates the ligands like FasL and TRAIL to bind with receptors and ligand-receptor interaction induces DISC complex consisting of FADD, TRADD, and an initiator caspase-like caspase 8 to form active caspase. The active caspase 8 then activates executioner caspases like caspase 3, 6, and 7 to regulate the apoptotic process. In contrast, the intrinsic mode in response to DNA damage, increased ROS, activate several BH3 only proteins like PUMA, NOXA, Bim, Bid, and Bad to act on mitochondria to release cytochrome c. The cytochrome c then interacts with Apaf-1 and pro-caspase 9 to form a complex called apoptosome to activate the caspase 9. The active caspase then interacts with executioner caspases to regulate the intrinsic apoptosis

which inhibit apoptosis and thus led to their identification as antiapoptotic proteins. The proapoptotic members consist of two groups: BH3-only proteins and BH123 proteins. The BH123 proteins include BCL-2-associated X protein (Bax), BCL-2 antagonist/killer (Bak), and BCL-2 ovarian killer (Bok). The BH3-only protein group includes the p53-upregulated modulator of apoptosis (PUMA), Noxa, BCL2-associated agonist of cell death (Bad), BH3-interacting domain death agonist (Bid), BCL-2-interacting mediator of cell death (Bim), and Harakiri (HRK). The BH3-only proteins perceive apoptosis signalings, such as that induced by DNA damage and ER stress, and are then translocated to the outer mitochondrial membrane, where they activate Bak and Bax or causes inhibition of anti-apoptotic proteins. Activated Bak and Bax oligomerize at the surface of mitochondria and facilitate the formation of pores in the outer membrane of mitochondria, leading to loss of membrane permeability and the release of several apoptosis-inducing proteins, such as SMAC/DIABLO and cytochrome c, from mitochondria to the cytosol. The released cytochrome-c activates with apoptotic protease-activating factor 1 (APAF1) to form the apoptosome complex and induces the activation of caspase-9, which subsequently activates caspases 3, 6, and 7 (Fig. 2.1) (Galluzzi et al. 2012; Kaya-Aksoy et al. 2019; Singh et al. 2019).

2.2.2 Role of Apoptosis in Cancer

Cancer is a complex and variable process with excessive variations in the genetic material that leads to cancer development, with several steps occurring progressively, starting with the initiation of tumorigenesis and eventually leading to metastasis. During cancer development, cells encounter byproducts of the physiological elimination of damaged cells. Therefore, cancer cells must acquire some protective machinery to prevent induced programmed cell death, that is, apoptosis. Cancer cells can regulate apoptotic pathways transcriptionally, translationally, and post-translationally to avoid stress, such as that generated by hypoxia and genomic instability. Moreover, cancer cells may suppress apoptosis by increasing anti-apoptotic genes expression or decreasing proapoptotic genes expression in a context-dependent manner, which results in higher anti-apoptotic protein expression. Generally, apoptotic pathways are restricted in cancer cells to prevent this cellular response. Several reports have indicated that attenuated interactions of the proapoptotic BIM which is a BH3-only protein with the antiapoptotic protein BCL-2 can support the survival and growth of cancer cells (Hübner et al. 2008). Furthermore, the induction of BIM induces oncogenic inactivation and apoptosis in acute lymphoblastic leukemia (Li et al. 2016b). Notably, it was demonstrated that the inhibition of a BH3-only protein or a caspase protein caused a genetic mouse model to develop resistance against certain proapoptotic signals with a commensurate increase in tumor initiation (Parsons et al. 2013). The BH3-only molecules BIM and PUMA are downregulated in breast cancer cells, and overexpression of these molecules induces HER2 inactivation, which induces apoptosis (Bean et al. 2013). Overexpressed BH3-only molecule BIM function as a cytoprotective molecule in cancer cells, and its association through phosphorylation with BCL-xL/MCL-1 block its proapoptotic functions (Gogada et al. 2013). Moreover, caspase-2-deficiency impedes apoptosis and generates genomic instability, resulting in tumorigenesis (Shalini et al. 2016). However, MCL-1 is a vital prosurvival factor in triple-negative breast cancer (TNBC), and its inhibition might be an effective strategy for treating TNBC (Li et al. 2018). Furthermore, the suppression of the tumor suppressor gene p53 mediates cell proliferation, and stabilized p53 phosphorylation activates BAK and BCL-xL, inducing apoptosis (Nieminen et al. 2013). However, suppression of the proapoptotic protein BCL-2 antagonist killer 1 (BAK1) leads to the proliferation of breast cancer cells of various lineages (Zhou et al. 2010), whereas the inhibition of BCL-xL and BCL-2 activates BAX/BAK and induces apoptosis in human myeloid leukemia cells (Rahmani et al. 2013). Thus, it is strongly said that the inhibition of apoptosis has to play a critical role in the case of cancer cell survival and tumor development. Targeting apoptosis induction is a novel strategy for cancer therapy.

2.3 Autophagy in Cancer

Autophagy (in Greek, "auto," defined as oneself, and "phagy," defined as to eat) states to a self-cannibalistic mechanism to degrades cytoplasmic components and unwanted organelles through lysosomes to maintain the homeostasis (Levine and Klionsky 2017; Mancias and Kimmelman 2016; Tan et al. 2017). Recent reports have suggested that autophagy significantly modulates various physiological processes, such as growth, development, cell division, and immunity, with any dysfunction in autophagy leading to severe pathophysiological implications, such as neurodegenerative disorders, autoimmune disorders, and cancer.

2.3.1 Mechanism of Autophagy

Autophagy is a complex multistep process regulated through the coordinated action of 42 ATG (AuTophaGy) genes that are sequentially involved in different steps, such as (a) phagophore nucleation and formation, (b) elongation of phagophores, (c) cytoplasmic cargo selection, (d) lysosomal docking and fusion, and (e) cargo degradation (Fig. 2.2) (Mancias and Kimmelman 2016; Wang and El-Deiry 2003; White 2015). Autophagy is initiated upon a shift in the extracellular milieu of a cell, mainly due to different stress signals, such as loss of growth factor signaling, nutrient deprivation, energy depletion, and hypoxia. These stress signals induce the release of reactive oxygen species (ROS) that inhibit the TOR1 (target of rapamycin complex 1)-dependent signaling pathway, which acts as a molecular sensor for autophagy initiation. First, the ATG protein complex is formed as a scaffold of regulatory proteins ATG1, ATG13, ATG17, ATG29, and ATG31, which recruits other ATG proteins to activate downstream targets via phosphorylation and a PAS (phagophore assembly site/pre-autophagosomal structure) (Bhol et al. 2019; Bhutia et al. 2010; Davies et al. 2015; Levine and Klionsky 2004). Then, the cytoplasmic contents, protein aggregates, defective organelles are sequestered through a double-membrane structure known as a phagophore. For phagophore formation, the Ulk1 kinase, in combination with ATG13 and ATG17, is activated at the same time as transmembrane protein ATG9, which extracts phospholipids. Subsequently, the class III PI-3 kinases, particularly Vps34 (vesicular protein sorting) interacts with Beclin1 leading to increase phagophore catalytic activity, in which phosphatidylinositol (PI) is utilized to produce phosphatidylinositol triphosphate (PI3P) for phagophore elongation. The formation of autophagosomes and elongation of vesicles depends on three ubiquitin-like conjugation systems: ATG12 is activated through the utilization of ATP and ubiquitin-activating enzyme E1-like ATG7, which non-covalently binds to ubiquitin-activating enzyme E2-like ATG10. Then, the E2-like action triggers the covalent binding of ATG12 and ATG5, forming the conjugated ATG5-ATG12 complex. This complex then pairs with ATG16L to form the ATG5–ATG12–ATG16L complex, which is essential for the extension of the phagophore and acquisition of the appropriate membrane curvature, and once double-membrane autophagosomes are formed, the complex starts to disintegrate. Thus, the formation

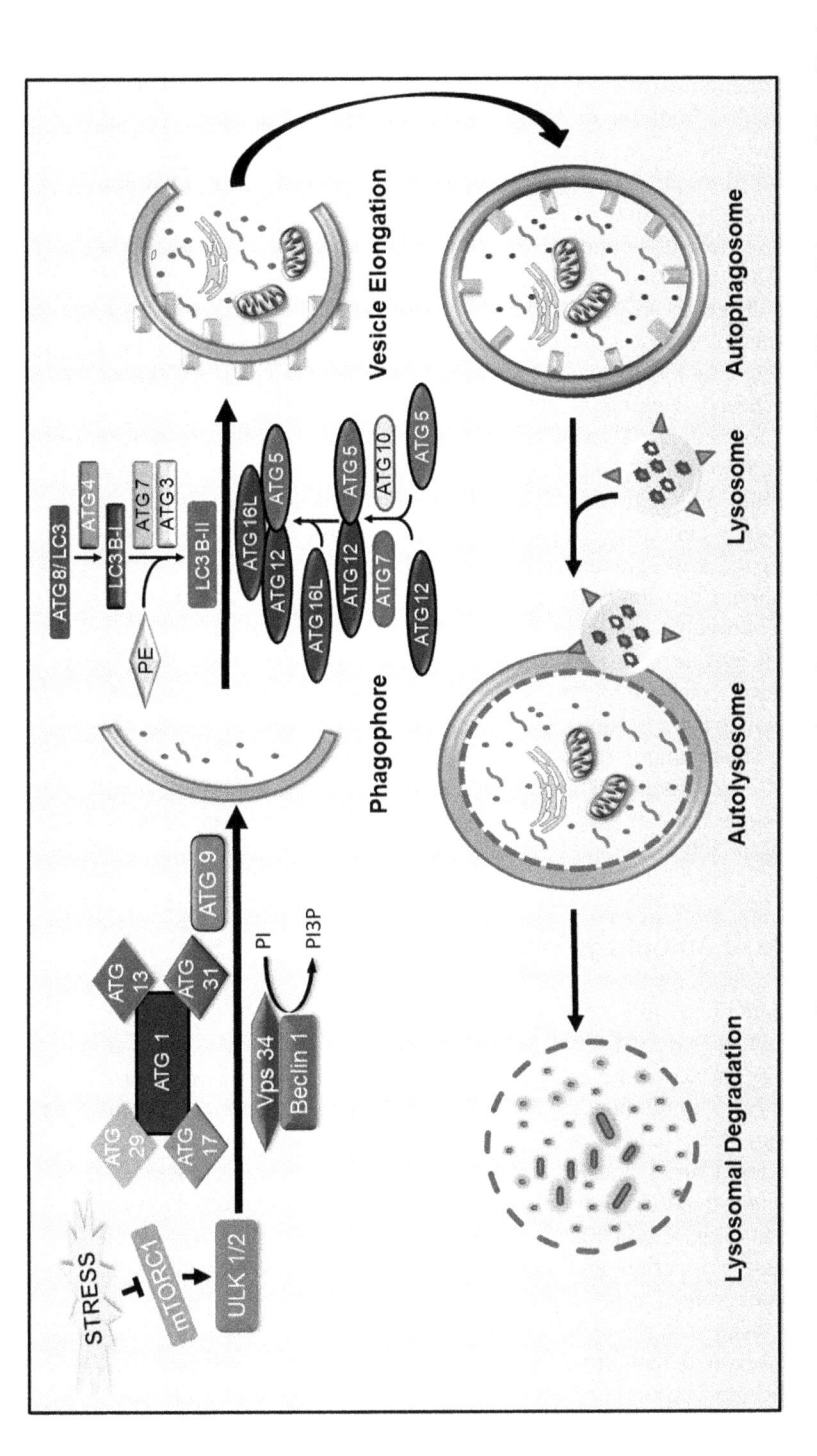

Fig. 2.2 An autophagic mechanism in the mammalian system. It involves four major orchestrated events starting with the initiation of phagophore followed by its elongation, maturation, and fusion with the lysosome. The cellular stress conditions act as inducers of phagophore initiation through activation ULK1/2 associated autophagic initiation complex. The activated ULK1/2 then triggers various ATGs like ATG1, 29, 17, 13, 31, and 9 along with Beclin1-Vps34 complex for initiating phagophore formation by importing lipid molecules from various organelles. The PI to PI3P conversion by the Beclin1-Vps34 complex precedes ATG5–ATG12 conjugation and its interaction with ATG16L. This interaction is essential for elongating phagophore and it is assisted by the processing of LC3 where the LC3 B-II is incorporated into the outer phagophore membrane. This event is followed by the cargo selection for their degradation leading to the formation of a complete phagophore called an autophagosome. The cargo loaded autophagosome then fuses with competent lysosomes to form autolysosome where the lysosomal content is used for the degradation process

of the ATG5-ATG12 conjugated complex does not depend on the activation of autophagy and is a poor marker of autophagy. A second ubiquitin-like system is required for the formation of microtubule-associated protein light chain 3 (LC3), known as the mammalian homolog of ATG8. During autophagy induction, cysteine protease ATG4 cleaves LC3 to form LC3-I. This complex is in turn stimulated in an ATP-dependent way through ATG7, E1-like enzyme. Following this activation stimulated LC3-I is transferred to ATG3 (E2-like) and ATG5-ATG12-ATG16L (E3-ligase complex) before it is conjugated with phosphatidylethanolamine (PE) to generate LC3-II. This lapidated LC3 acts as a receptor for cargo selection in coordination with different adaptor proteins (p62/SQSTM1, NBR1, TAXBP1, etc.) or organelle-specific receptor proteins (PHB2, AMBRA1, Nix, BNIP3L, etc.) contributing to selective molecule uptake and degradation (Bhol et al. 2019; Grumati and Dikic 2018). During docking of autophagosomes to preexisting lysosomes, autolysosomes are formed. In this step, the acidic constituents of the lysosomes digest the selected cargos of the autophagosomes. These organelles migrate along the side of microtubules in a bidirectional manner, with autophagosomes having an affinity for the lysosome-enriched microtubule organizing center (LEMOC). Vesicular docking and fusion are regulated through several proteins, including LAMP-2 (lysosomal membrane protein-2) and GTPases, such as class C Vps proteins, SNARE (soluble *N*-ethylmaleimide-sensitive factor activating protein receptor), ESCRT (endosomal sorting complex required for transport), and Rabs (Rab7). Any mutation in these proteins halts the progression of autophagosome maturation and fusion. In addition to these proteins, UVRAG, a Beclin1-interacting protein, also plays key roles by maintaining the fusion machinery on autophagosomes and clamping the class C Vps proteins. Thus, Rab7 is activated, which in turn enhances the fusion of lysosomes and late endosomes (Levine and Kroemer 2008; Mathew et al. 2007; Mizushima 2007; Xie and Klionsky 2007).

2.3.2 Dual Role of Autophagy in Cancer

The actual role of autophagy in cancer is still debatable because of its dual action, functioning as both a survival- and death-promoting mechanism. Survival-promoting autophagy increases the chances of cancer cell survival under adverse conditions by positively regulating the hallmarks of cancer. In contrast, death-promoting autophagy kills cancer cells by limiting mechanisms such as ROS production and the degradation of survival proteins. Therefore, the correct therapeutic approach for cancer treatment, by either promoting or inhibiting autophagy, is still a matter of discussion. Studies have also shown that autophagy is associated with tumor-suppressive function and that inhibition or defective autophagy is associated with tumor induction and malignant transformation. For instance, loss of Beclin1 in mice resulted in the development of hepatocellular carcinoma (Liang et al. 1999). Moreover, heterozygous deletion of Beclin1 was also associated with the development of breast and other human malignancies, suggesting that Beclin1 acts as a tumor suppressor (Qu et al. 2003; Yue et al. 2003). Similarly, the downregulation of ATG5 has been reported to be associated with early-stage cutaneous melanoma pathogenesis (Liu et al. 2013). Although the senescence mechanism by which autophagy acts as a tumor suppressor is controversial, a few studies have suggested that autophagy and senescence can occur simultaneously. Autophagy suppresses melanoma tumorigenesis by inducing senescence (Liu et al. 2014). A recent study in papillary thyroid carcinoma by Liu et al. suggested that BIRC7 induces the epithelial-mesenchymal transition and metastasis by limiting autophagy (Liu et al. 2020). Similarly, ATG7 induces triple-negative breast cancer progression by inhibiting the invasion, migration, and epithelial–mesenchymal transition (Li et al. 2019). In contrast, autophagy has been found to contribute to tumorigenesis by giving the advantage to tumor cells over normal cells under various stress environments and supporting cancer cell aggressiveness. For example, in a Chinese Han population, the expression of ATG12 consequently contributes to the risk of head and neck squamous cell carcinoma (Song et al. 2018). Increased expression of ATG5 induced by HIF1α has been reported to increase tumor size in prostate cancer (Yu et al. 2019). In colon cancer, RACK1-induced protective autophagy triggers cell proliferation and attenuates apoptotic cell death (Xiao et al. 2018). Similarly, in renal cell carcinoma, autophagy is critical for cell survival and the epithelial-mesenchymal transition (Singla and Bhattacharyya 2017). Guo et al. reported an oncogenic role of CCAT1 in hepatocellular carcinoma mediated through ATG7-dependent autophagy induction (Guo et al. 2019). Studies have also reported the association of autophagy with enhanced drug resistance. For example, Wnt3a is reported to promote radioresistance via autophagy in head and neck squamous cell carcinoma (Jing et al. 2019). Similarly, cisplatin resistance and osteosarcoma progression were induced by SNHG16 upregulation of ATG4B expression (Liu et al. 2019).

2.4 The Role of Autophagy and Apoptosis Crosstalk in Cancer Growth and Progression

Cells undergo apoptosis under extreme stress, whereas autophagy is well known for its involvement in cellular homeostasis. Hence, the induction of autophagy in cells undergoing stress before the point of no return can protect them from death. However, after cells are destined for death, autophagy can induce a peculiar form of cell death called autophagic cell death (Apel et al. 2009; Wong 2011; Yang et al. 2015). In this context, some reports have shown critical crosstalk between components of two highly complex processes, that is, autophagy and apoptosis (Fig. 2.3).

2.4.1 The Intersection of Autophagy and Apoptosis Molecules

One of the most important and well-known points of interaction between the autophagic and apoptotic pathway coincides with the action of antiapoptotic Bcl-2

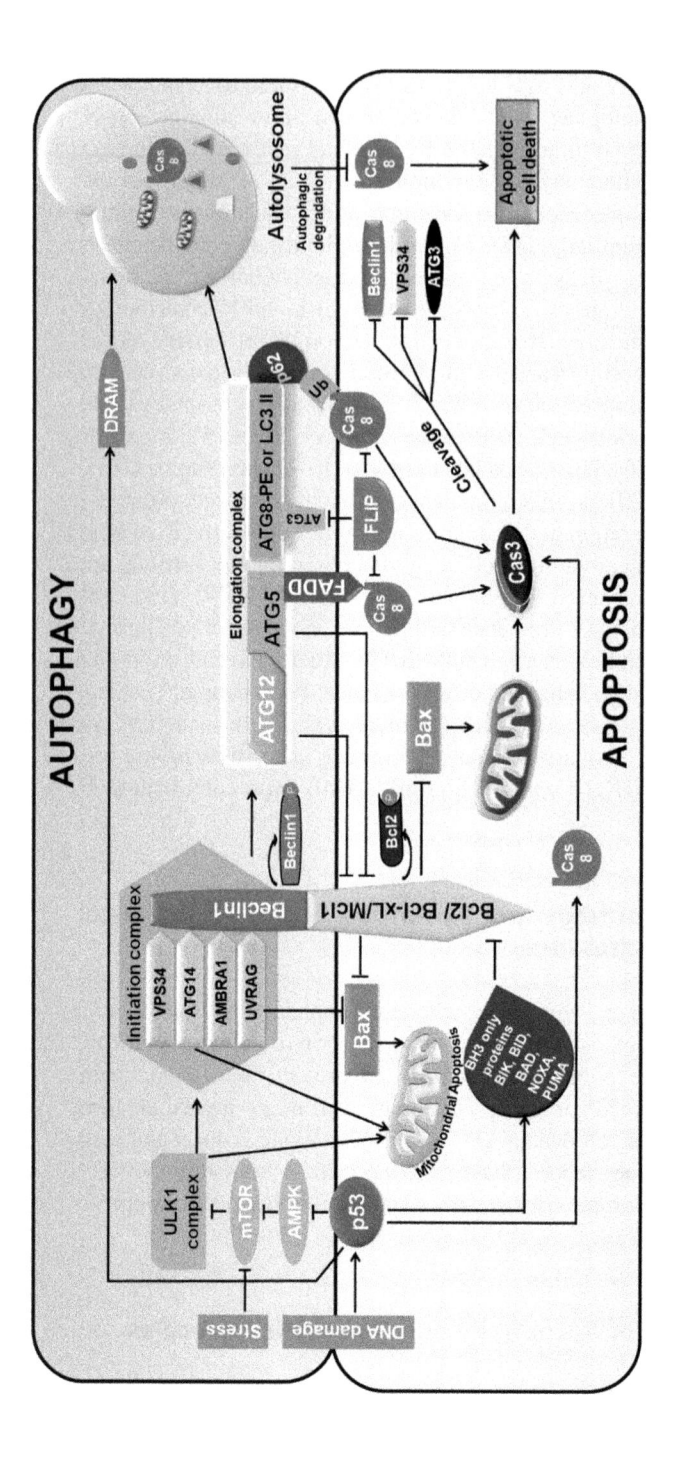

Fig. 2.3 Autophagy and apoptosis crosstalk during cellular stress. The crosstalk modulates through the interaction of Bcl2/Bcl-xL–Beclin1 and Bcl2/Bcl-xL–proapoptotic BH3 only proteins. Cellular stress like nutrient starvation activates the Ulk1 complex by inhibiting mTOR results in its interaction with the autophagic initiation complex for the onset of autophagy. This is made possible by dissociating Bcl2/Bcl-xL–Beclin1 complex through the phosphorylation of Beclin1 and Bcl2/Bcl-xL. The free Beclin1 with other autophagic proteins activates the elongation complex which directs the cell to employ either autophagy or apoptosis by inhibiting anti-apoptotic proteins through ATG5, ATG12, and by regulating pro-apoptotic proteins through ATG5-FADD interaction and caspase 8 activations through ubiquitinated p62. In contrast, the cell committed to autophagy restricts apoptosis through the lysosomal degradation of caspases. Moreover, stress like DNA damage induces p53 regulates apoptosis through inhibition of anti-apoptotic proteins through BH3 only proteins and through phosphorylating of Bcl2. It also helps in the activation of caspases and proapoptotic proteins like Bax. The p53 also regulates autophagy either by activating it through DRAM mediated mitophagy or by inhibiting through suppression of AMPK and by degrading autophagic proteins like Beclin1, Vps34, ATG3 through activated caspases. Some inhibitor of apoptotic proteins like FLIP regulates apoptosis by inhibiting caspases and autophagy by disrupting the ATG3- ATG8 of the elongation complex

protein and the conserved autophagic protein Beclin1. Antiapoptotic proteins including Bcl-2 family members (Bcl-2, Bcl-XL, and Mcl-1) with all four BH3 domains interact with the proapoptotic members through BH4 domains (Pattingre et al. 2005). The BH3 binding pocket of Bcl-2 binds with the BH3 domain of Beclin1 to inhibit Beclin1-dependent autophagy. ER-localized Bcl-2 inhibits starvation-induced Beclin1-dependent autophagy by binding with Beclin1 facilitated by NAF-1 (nutrient-deprivation autophagy factor-1) (Chang et al. 2010). The binding affinity of Beclin1 for Bcl-2 is lower than the proapoptotic proteins due to the presence of a polar threonine instead of a hydrophobic amino acid at position 119 (Feng et al. 2007; Oberstein et al. 2007). Therefore, the interaction between Beclin1 and Bcl-2 does not disturb the antiapoptotic property of Bcl-2. During stress-induced autophagy, BH3-only proteins bind to the BH3-binding pocket of Bcl-2 to disrupt their interaction. STKs (Ser/Thr kinases) including JNK (JUN N-terminal kinase), Akt, and DAPK, also make regulatory contributions to the processes of autophagy and apoptosis through their crosstalk with other Bcl-2 family members. c-Jun N-terminal protein kinase 1 (JNK1) controls both autophagy and apoptosis through the phosphorylation of Bcl-2. Under conditions of mild stress, Bcl-2 phosphorylation by JNK-1 causes the dissociation of BCl-2 and Beclin1 but not that of the Bcl-2 and Bax to initiate autophagy; however, under prolonged stress, the Bcl-2-Bax interaction is disrupted, which initiates apoptosis (Wei et al. 2008).

Many studies have shown that active caspases cleave key autophagy-related proteins and allow apoptosis to overtake autophagy. ATG5, a protein required for phagophore elongation, is cleaved by calpain, an active caspase, to form a 24 kDa protein that interacts with Bcl-XL to induce apoptosis through the release of cytochrome c (Yousefi et al. 2006). Furthermore, ATG12, a copartner in the ATG5-ATG12 conjugation system, is thought to act as a proapoptotic protein by supporting the other apoptotic proteins in activating the caspases under various stresses. Moreover, ATG12 can inhibit the anti-apoptotic proteins Bcl-2 and Mcl1 by acting specifically on the BH3-like domain of ATG12. Similarly, in acute lymphoblastic lymphoma (ALL), the cleavage of ATG3 by caspase 8 promotes apoptosis by inhibiting autophagy (Oral et al. 2012). In contrast, caspase 9, which is involved in the intrinsic pathway of apoptosis, is found to lipidated LC3 by interacting with ATG7 to promote autophagy. Hence, apoptosis induction results in the site-specific breakdown of autophagic proteins such as Beclin1, ATG5, and ATG7 to undermine the cytoprotective effects of autophagy (Marquez and Xu 2012). Interestingly, the autophagy adaptor protein p62 displays a critical role in caspase activation to induce apoptosis (Islam et al. 2018; Jung and Oh 2019). The tumor suppressor p53 also regulates crosstalk at various points. It activates autophagy through the AMPK–TSC2–mTOR axis by translocating to the nucleus upon stress, whereas in the cytoplasm, p53 interacts with FIP200 and inhibits autophagy (Vousden and Lane 2007). Cytoplasmic p53 interacts with several Bcl-2 pro-apoptotic family proteins, such as Bax, NOXA, PUMA, and others, to induce apoptosis by regulating mitochondrial outer membrane permeability (MOMP) (Vaseva et al. 2012). Moreover, p53-induced DRAM-mediated autophagy is also associated with apoptosis (Crighton et al. 2006). Furthermore, the antiapoptotic protein FLICE-like inhibitor protein (FLIP) inhibits apoptosis by inactivating death receptors and is known to regulate autophagy by interacting with autophagic proteins such as ATG3 and LC3. FLIP can also influence both apoptosis and autophagy independently by acting on the plasma membrane to inhibit apoptosis and at the site of autophagosome formation to modulate autophagy, like the action of Bcl-2 (Eisenberg-Lerner et al. 2009). Interestingly, serum starvation was shown to activate antiapoptotic protein cIAP to promote mitochondrial autophagy. It revealed that cIAP1 translocated onto mitochondria to interact with Ulk1, TOM20, and LC3 to stimulate mitophagy through the ubiquitination pathway (Mukhopadhyay et al. 2016).

An autophagosome, by itself, modulates apoptosis either through direct sequestration and removal of proapoptotic proteins or through the engulfment and subsequent elimination of damaged cellular molecules, such as those from the mitochondria-dependent apoptotic cell death. In another mechanism, the membrane of autophagosome functions as a platform for processing of apoptotic proteins and thus contributes to the process of apoptosis; for example, in the presence of bortezomib and pan-sphingosine kinase inhibitor, SKI-I, there is autophagy-dependent activation of the extrinsic apoptotic pathway through caspase-8. Caspase 8 in association with FADD is recruited to the membrane of autophagosomes in a p62-dependent manner by interacting with ATG5 (Mukhopadhyay et al. 2014; Young et al. 2012).

2.4.2 The Role of Autophagy and Apoptosis Crosstalk in Cancer

Although the interconnection between apoptosis and autophagy has unique complexities, many studies that explain how apoptosis and autophagy are interlinked and induce cell death or sustain tumor growth and proliferation with common regulators (Table 2.1). For example, p53 was found to modulate autophagy and apoptosis in context-dependent way in different types of cancer. It showed that autophagy degrades p53 to maintain the hepatic cancer stem cells (Liu et al. 2017). Recently, the induction of autophagy by activating the AMPK-ULK1 axis and inhibiting mTOR was found to induce apoptosis through caspase activation, which reduced tumor proliferation in triple-negative breast cancer (Cao et al. 2018). Another study also showed that the activation of autophagy via PI3K/AKT/mTOR signaling reduced the viability of prostate cancer cells by inducing cytotoxicity in conjunction with apoptosis-mediated cell death (Tian et al. 2017). Besides, the autophagy adaptor protein p62 plays central signaling for tumor initiation as well as suppression of tumor progression in the stromal cells (Moscat et al. 2016; Zhang et al. 2013). During cisplatin-mediated ER stress, the induction of cell death occurs through caspase-mediated apoptosis, but Beclin1 mediates autophagy to eliminate excessive stress in human lung cancer cells (Shi et al. 2016). Bax-negative colon cancer cells can undergo TRAIL-induced cell death under compromised autophagy conditions (Li et al. 2016a; Mariño et al. 2014). Moreover, it has been reported that the inhibition of autophagy causes an elevation in NOXA expression, which is a

Table 2.1 Autophagy and apoptosis crosstalk proteins in cancer

Protein	Role in autophagy	Role in apoptosis	Relevance in cancer
Bcl-2	Anti-autophagic through Beclin1	Antiapoptotic	Overexpression in cancer and acts as an oncogene
Bad, Bak, BNIP3, Nix	Pro-autophagic inhibits Beclin1/Bcl-2 interaction	Proapoptotic	Deletion in cancer and involves in tumor suppression
Bax/PUMA	Pro-autophagic	Proapoptotic	Deletion in cancer and involves in tumor suppression
NOXA	Pro-autophagic inhibits Mcl-1/Beclin1 interaction	Proapoptotic	Degradation through autophagy limits tumor suppression
cIAP	Mitophagy through ubiquitination	Antiapoptotic	Overexpression in cancer and promotes proliferation
Caspase 9	Lipidation of LC3 by interacting with ATG7	Proapoptotic	Deletion in cancer and involves in tumor suppression
p53	Context-dependent, cytoplasmic p53 inhibits and nuclear p53 promotes autophagy	Proapoptotic	Deletion in cancer and degradation in cancer stem cells through autophagy
Ulk1	Nucleation	Proapoptotic	Context-dependent
Beclin1	Phagophore nucleation	Cleaved C-fragment induces mitochondrial apoptosis	Context-dependent
ATG5	Phagophore elongation	Antiapoptotic through FADD, cleaved N-terminal involves in mitochondrial apoptosis	Context-dependent
ATG12	Phagophore elongation	Inhibit Bcl-2 and Mcl-1 interaction	Context-dependent
ATG14	Phagophore elongation	Proapoptotic	Context-dependent
UVRAG	Activates Vps34–Beclin1 interaction	Prevent translocation of Bax to mitochondria	Tumor suppressor in cancer
p62	Autophagic adaptor protein	Caspase activation	Overexpression in cancer
mTOR	Inhibit autophagy, dephosphorylation involves in initiation	mTOR regulates apoptosis	mTOR inhibitors in cancer therapy
FOXO3	Autophagy transcription factor	Binds with pro-apoptotic PUMA	Tumor suppressor in cancer

Bcl2 family protein with a BH3-only domain. Generally, NOXA is degraded in the autophagic pathway by sequestration onto autophagosomes through the action of the adaptor protein p62. However, the blockage of autophagy leads to the accumulation of NOXA, which subsequently activates apoptosis and acts as a bridge between autophagy and apoptosis in cancer cells (Wang et al. 2018). In addition, proapoptotic protein PUMA and Bax promote autophagy to contribute to apoptosis (Yee et al. 2009). Interestingly, our study showed that in response to anticancer therapy, mitophagy was induced through PUMA leading apoptosis in glioma cells (Panda et al. 2018; Yee et al. 2009). Moreover, the autophagy protein Ulk1 and ATG14 have found to have an important role in the induction of apoptosis. The upregulation of Ulk1 translocated to mitochondria to inhibit the activity of manganese superoxide dismutase resulting in the production of ROS causing to cell death the cancer cells (Mukhopadhyay et al. 2015, 2017). Furthermore ATG14 along with Ulk1 induced lipophagy, selective autophagy resulting in free fatty acid accumulation leading to ER stress-mediated apoptosis (Mukhopadhyay et al. 2017. On other hand, UVRAG also can act as an antiapoptotic protein by preventing the translocation of Bax to mitochondria, where it initiates mitochondrial apoptosis in response to therapy (Eisenberg-Lerner et al. 2009; Maiuri et al. 2009). Interestingly, UVRAG with truncating mutation displayed higher inflammatory response through NLRP3-inflammasome hyperactivation and exhibited significant spontaneous tumorigenesis through β-catenin stabilization and centrosome amplification (Quach et al. 2019) establishing complex crosstalk between autophagy and apoptosis during tumor growth and progression.

During chemotherapeutic stress, cancer cells trigger autophagy to eradicate stress and support tumor growth and progression. Such protective autophagy is the major cause of therapy resistance that develops against a majority of anticancer drugs used for different cancers. For example, cisplatin-induced autophagy protects cancer cells against drug-induced apoptosis (Harhaji-Trajkovic et al. 2009). Therefore, autophagy inhibitors in combination with apoptosis-inducing drugs might increase the efficacy of anticancer therapy. Chloroquine, a potent autophagy inhibitor, is reported to enhance existing chemotherapeutics without inducing toxicity in cells (Amaravadi et al. 2007). Hence, the use of chloroquine and its analog hydroxychloroquine for inhibiting autophagy following anticancer therapy is widely accepted and is currently under clinical trial (Cudjoe et al. 2019). Docetaxel-induced autophagy (Zhang et al. 2019) and paclitaxel-induced autophagy (Kim et al. 2013) play tumor protective roles in cancer cells, leading to treatment failure. Pretreatment with 3-MA along with paclitaxel significantly enhances cytochrome C release and the subsequent induction of the mitochondrial apoptotic pathway (Xi et al. 2011). Furthermore, the expression of FOXO3, an autophagic transcription factor, is increased after autophagy is blocked. Then, FOXO3 at increased levels can directly bind to the promoter of the proapoptotic protein PUMA and hence trigger the apoptotic pathway in osteosarcoma and other cancer cells (Fitzwalter and Thorburn 2018; Fitzwalter et al. 2018; Jiang et al. 2017). Collectively, these studies implicate the antiapoptotic function of protective autophagy against therapy-induced apoptosis in cancer cell lines that leads to therapy resistance. However, contradictory evidence is presented for cases of cancer cells induced toward death upon radiation treatment, which attenuates apoptosis by knocking out proapoptotic proteins (such as Bax and

Bak), and the treatment-associated cell death is not attributed to apoptosis but type-II programmed cell death, that is, autophagy (Kim et al. 2006).

2.5 Modulation of Apoptosis and Autophagy in Potential Cancer Therapeutics

The present era of targeted cancer therapy is rationally based on specific signaling pathways that have high expression in specific tumor types. The paradigm suggests that target-specific therapeutic agents sensitize cancer cells toward chemotherapy through a closed circuit of signals shuffled between apoptosis and autophagy components (Table 2.2).

2.5.1 Distinguishing Autophagic Cell Death from Apoptosis for Cancer Therapeutics

Since cancer cells can block apoptosis and increase their resistance to chemotherapeutic agents, targeting autophagy as an alternative cell death pathway is an attractive strategy for anticancer therapy (Jain et al. 2013). Arsenic trioxide and EB1089 (a vitamin D analog) have been reported to induce dynamic changes in lysosomal activity that provoke Beclin1-mediated autophagic cell death (Lo-Coco et al. 2013; Qian et al. 2007). Several chemotherapeutic agents, such as dexamethasone with fenretinide or etoposide, along with key dietary phytochemicals, such as resveratrol and fisetin, have also been reported to induce autophagic cell death in a type of non-apoptotic programmed cell death mediated by Bcl2-dependent autophagy genes (Fazi et al. 2008; Jain et al. 2013; Laane et al. 2009). Sodium selenite, a mitophagy inducer, has also emerged as a therapeutic agent against malignant glioma cells that subsequently promotes cell death through superoxide-mediated mitochondrial damage (Kim et al. 2007). Imatinib and cannabinoid have also exhibited autophagic cell death in glioma cells through the inhibition of autophagy, leading to the stimulation of ER stress (Salazar et al. 2009). Spautin-1, a potent autophagy inhibitor, triggers the inhibition of autophagy in chronic myeloid leukemia cells to induce autophagic cell death through the inhibition of class III PI3K and targeting the deubiquitination of USP10- and USP13-mediated degradation of Beclin1 (Liu et al. 2011; Shao et al. 2014; Wilde et al. 2018). Moreover, β-lapachone and elisidepsin also induce marked levels of autophagic cell death in lung cancer. Tamoxifen, bortezomib, trastuzumab, and sulforaphane have been used against breast cancer and provoke cell death by inhibiting autophagy. Bortezomib, 5-FU, and sulforaphane have also been used in colorectal cancer treatment. Temozolomide, 4-HPR, imatinib, rapamycin, and PI-103 are effective in the treatment of colorectal cancer. The inhibition of autophagy at the early stage attenuates the cytotoxicity induced by chemotherapeutic drugs, and this effect is augmented when autophagy is inhibited at later stages. Chronic myeloid leukemia responds to treatment with SAHA and OSI-027 in an autophagy-dependent cell death manner through the epigenetic modulation of

Table 2.2 Anticancer drugs in the modulation of autophagy and apoptosis for potential cancer therapy

Cell death mode	Compounds	Expression in cancer	Regulatory signaling pathways
Autophagy as an alternative cell death mechanism	β-Lapachone and Elisidepsin	Lung cancer	Induction of autophagic cell death through inhibition of the Akt/mTOR signaling pathway
	Arsenic trioxide and EB1089	Glioma cells and leukemia cells	Dynamic alterations in lysosomal activity for subsequent activation of beclin1 mediated autophagic cell death
	Dexamethasone, Fenretinide, and Etoposide	Lymphoma and leukemia cells	Along with resveratrol and fisetin regulates Bcl2 mediated apoptosis independent autophagic cell death
	Imatinib	Glioma cells	Autophagy inhibition leads to ER stress-associated autophagic cell death
	Cannabinoid	U87MG	Inhibition of autophagy flux, activation of ER stress, autophagic cell death modulation induction via TRB3 dependent inhibition of Akt/mTOR
	Spautin-1	Chronic myelogenous leukemia (CML)	Inhibition of class III PI3K and targeting deubiquitination of USP10 and USP13 degrade Beclin1 to induce autophagic cell death
	Sodium selenite	Glioma cells	Superoxide-mediated mitochondrial damage leading to subsequent mitophagic cell death
	Bortezomib, 5-FU, and Sulforaphane	Colorectal cancer	Autophagy inhibition in combination with CQ to mediated change in lysosomal activity for the onset of autophagic cell death
	AZD2014	Breast cancer and ALL	Induction of autophagic cell death through modulation of PI3K/AKT/mTORC2 signaling
	Quercetin	Gastric cancer	Induction of autophagic cell death through inhibition of PI3K/AKT signaling
	Tamoxifen, Bortezomib, and Sulforaphane	Breast cancer	Autophagy inhibition in combination with CQ to mediated change in lysosomal activity for the onset of autophagic cell death

(continued)

Table 2.2 (continued)

Cell death mode	Compounds	Expression in cancer	Regulatory signaling pathways
	Voacamine	Osteosarcoma	Chemo sensitization of doxorubicin multidrug resistance cells through inhibition of P-glycoprotein activity and autophagy induction in
Inhibition of autophagy promotes apoptotic cell death	Gallic acid	Oral cancer	Autophagic flux inhibition, ROS generation provokes caspase-dependent apoptosis induction
	Verteporfin	Acute promyelocytic leukemia, prostate, and colon cancer	Inhibition of autophagy leads to ROS dependent induction of apoptosis dependent cancer cell death
	Elaiophylin	Ovarian cancer	Inhibit autophagic flux that associates ER stress to promote apoptosis
	Deguelin	PNAC1, breast, gastric, and prostate cancer	Inhibits autophagosome maturation, in conjunction with doxorubicin induce caspase-dependent apoptosis, autophagy inhibition through via modulation of PI3K/Akt signaling pathway lead to subsequent activation of caspase-dependent apoptosis
	Withaferin A	Breast cancer	Inhibition of autophagic flux, disturbance in lysosomal proteolytic activity, accumulation of autophagosome, ROS induction leads to apoptotic cell death
	Ginsenoside	Esophageal cancer	Caspase dependent apoptosis induction after inhibition of autophagy
	Liensinine	Breast cancer	Inhibition of autolysosome formation via inhibition of RAB7A recruitment. Chemo sensitizes of cancer cells to anticancer drugs for apoptosis induction by activating mitochondrial fission

(continued)

Table 2.2 (continued)

Cell death mode	Compounds	Expression in cancer	Regulatory signaling pathways
Apoptosis-autophagy coexist to mediate cell death	Arsenic trioxide	Acute promyelocytic leukemia	In conjunction with all-trans retinoic acid modulate apoptotic and autophagic cell death mechanism
	Plumbagin	SMMC-7721	Excessive ROS accumulation leads to caspase-dependent apoptosis and enhanced autophagosome to autolysosome formation trigger autophagic cell death
	SB202190 and SB203580	–	MAPK inhibition leads to modulate apoptotic and autophagic cell death mechanism
	Curcumin	CML, Colon cancer and glioblastoma	Induction of autophagy via inhibition of AKT/mTOR/p70S6K, Bcl2 downregulation, LC3 lipidation and ROS induction leading to the intrinsic onset of apoptosis
	Conconavalin A	Lung cancer and melanoma cells	Downregulate PI3K/Akt/mTOR signaling for autophagy induction and ROS accumulation for caspase-dependent apoptosis
	Abrus agglutinin	Oral, prostate and Colon cancer	PUMA dependent mitophagy contributes toward apoptotic cell death via ceramide generation, NRF2 downregulation leads to apoptosis cell death

histone proteins (Choi 2012). ADI-PEG20 and saracatinib have been reported to be effective against prostate cancer. AZD2014, another small-molecule autophagy inhibitor, has shown promise in the clinical treatment of breast cancer and acute lymphoblastic leukemia through the modulation of PI3K/AKT/mTORC2 signaling (Tabe et al. 2013). A dietary phytochemical, quercetin also displays potent anticancer efficacy in gastric cancer. Voacamine, a bisindole alkaloid, has been shown to induce autophagic cell death in an apoptosis-independent manner (Panda et al. 2015). In preclinical trials with patients diagnosed with melanoma, pancreatic adenocarcinoma or bladder cancer, CQ and HCQ, as single agents, have been shown to have potent anti-cancer properties, as exhibited through the inhibition of autophagy. Furthermore, treatments based on CQ or HCQ combined with metabolic stressors have been found to potentiate autophagy-mediated cell death (Amaravadi et al. 2011; Jain et al. 2013; Panda et al. 2015; Wilde et al. 2018).

2.5.2 Autophagic Facilitation of Apoptosis in Cancer Therapeutics

Autophagy-mediated facilitation of apoptosis in several cancer cells displays a remarkable therapeutic avenue against cancer. Both induction and inhibition of autophagy drive cellular mechanisms toward apoptosis induction. Recent studies have shown that the gallic acid in *Terminalia bellirica* extracts inhibits autophagy and can be used to fuel the induction of apoptosis in a ROS-dependent manner in oral squamous cell carcinoma (Patra et al. 2020). Verteporfin, a benzoporphyrin derivative, in conjunction with gemcitabine, inhibits autophagy and promotes apoptosis in acute promyelocytic leukemia and exhibits clinical potency against glioma, prostate, and colon cancer (Donohue et al. 2011, 2013). Furthermore, elaiophylin has also been reported to inhibit the autophagic flux that is associated with the endoplasmic reticulum stress to promote apoptosis in ovarian cancer (Zhao et al. 2015). Deguelin, another autophagy inhibitor, triggers apoptosis and enhances the chemosensitization of several types of cancer cells to doxorubicin (Xu et al. 2017). Also, withaferin A and ginsenoside have been reported to induce apoptosis after autophagy inhibition in breast and esophageal cancer cells (Muniraj et al. 2019; Zheng et al. 2016). Similarly, liensinine, an isoquinoline alkaloid, has also been reported to inhibit autophagosome-lysosome fusion to provoke the induction of mitochondrial fission and apoptosis in triple-negative breast cancer (Zhou et al. 2015).

2.5.3 Apoptosis-Autophagy Links in Cancer Therapeutics

Apoptosis and autophagy may induce cell death during chemotherapy in a parallel or sequential manner. The cooperation of apoptotic and autophagic machinery is required for the induction of cell death in the mature tumor environment (Jain et al. 2013). Arsenic trioxide in combination with all-trans retinoic acid has evolved as a potent drug against acute promyelocytic leukemia by modulating both apoptotic and autophagic pathways (Lo-Coco et al. 2013; Qian et al. 2007). Furthermore, plumbagin (a naphthoquinone derivative) has also been reported to induce apoptosis and autophagy with two long synthetic MAPK inhibitors, SB202190 and SB203580 (Li et al. 2014). Several dietary phytochemicals, including resveratrol, curcumin, quercetin, lutein, lycopene, catechin, and β-carotene, have demonstrated proapoptotic and autophagic potential owing to their antioxidant properties in several cancer cells (Choi 2012). Recently, different plant lectins, such as *Abrus* agglutinin, a lectin from *Abrus precatorius*, was implicated in the onset of autophagy and is being considered as a means to induce apoptotic cell death in prostate cancer and oral squamous cell carcinoma (Panda et al. 2020; Panigrahi et al. 2020). Another lectin, concanavalin A, has also been reported to downregulate PI3K/Akt/mTOR signaling, thus contributing to autophagic cell death (Roy et al. 2014). Finally, concanavalin A has also induced caspase-dependent apoptosis in human melanoma cells (Liu et al. 2009).

2.6 Conclusion and Future Perspective

Being a double-edged sword, autophagy regulates other modes of cell death in cancer cells. Several of its key regulators act as tumor suppressors or promoters, depending on the threshold level of the cellular or tumor microenvironmental stress in the cancer cell milieu. Moreover, the inhibition of autophagy disrupts cancer cell metabolism, interferes with differentiation, and destabilizes anticancer immunosurveillance. At later stages of tumorigenesis, restoration of autophagy leads to the development of chemo- and/or radioresistance in cancer cells. As explained earlier, the majority of research has focused on understanding the coordinated regulation of autophagy and apoptosis, which sensitizes cancer cells toward death. However, several issues remain unresolved, such as the mechanism by which apoptotic activation of effector caspases turns off the autophagic machinery and the identities of key autophagy proteins that drive cells from being in a pro-autophagic state to acquiring a proapoptotic phenotype during this period, causing various pathophysiological consequences. Several clinical trials using autophagy inhibitors, such as chloroquine and hydroxychloroquine, have also been used for targeted cancer therapy. Besides, several new anticancer drugs have been formulated to modulate both autophagy and apoptosis. However, altering only autophagy in cancer cells might not be an ideal approach; although it is beneficial at low levels, overactive autophagy becomes detrimental, leading to tumor development. Hence, anticancer or antidegenerative drugs modulating autophagy and targeting apoptosis may work more effectively in the clinic when combined. Altogether, it will be interesting to reveal the mechanism and thereby understand autophagy and apoptosis in cancer, which will help to leverage their functional interrelation for developing new targets for the possible effective therapeutic intervention of cancer therapy.

References

Amaravadi RK, Yu D, Lum JJ, Bui T, Christophorou MA, Evan GI, Thomas-Tikhonenko A, Thompson CB (2007) Autophagy inhibition enhances therapy-induced apoptosis in a Myc-induced model of lymphoma. J Clin Invest 117:326–336

Amaravadi RK, Lippincott-Schwartz J, Yin XM, Weiss WA, Takebe N, Timmer W, DiPaola RS, Lotze MT, White E (2011) Principles and current strategies for targeting autophagy for cancer treatment. Clin Cancer Res 17:654–666

Apel A, Zentgraf H, Buchler MW, Herr I (2009) Autophagy—a double-edged sword in oncology. Int J Cancer 125:991–995

Bean GR, Ganesan YT, Dong Y, Takeda S, Liu H, Chan PM, Huang Y, Chodosh LA, Zambetti GP, Hsieh JJD, Cheng EHY (2013) PUMA and BIM are required for oncogene inactivation-induced apoptosis. Sci Signal 6:ra20

Bhardwaj A, Aggarwal BB (2003) Receptor-mediated choreography of life and death. J Clin Immunol 23:317–332

Bhol CS, Panigrahi DP, Praharaj PP, Mahapatra KK, Patra S, Mishra SR, Behera BP, Bhutia SK (2019) Epigenetic modifications of autophagy in cancer and cancer therapeutics. Semin Cancer Biol. https://doi.org/10.1016/j.semcancer.2019.05.020

Bhutia SK, Dash R, Das SK, Azab B, Su ZZ, Lee SG, Grant S, Yacoub A, Dent P, Curiel DT, Sarkar D, Fisher PB (2010) Mechanism of autophagy to apoptosis switch triggered in prostate cancer cells by antitumor cytokine melanoma differentiation-associated gene 7/interleukin-24. Cancer Res 70:3667–3676

Cao C, Huang W, Zhang N, Wu F, Xu T, Pan X, Peng C, Han B (2018) Narciclasine induces autophagy-dependent apoptosis in triple-negative breast cancer cells by regulating the AMPK-ULK1 axis. Cell Prolif 51:e12518

Chang NC, Nguyen M, Germain M, Shore GC (2010) Antagonism of Beclin 1-dependent autophagy by BCL-2 at the endoplasmic reticulum requires NAF-1. EMBO J 29:606–618

Choi KS (2012) Autophagy and cancer. Exp Mol Med 44:109–120

Crighton D, Wilkinson S, O'Prey J, Syed N, Smith P, Harrison PR, Gasco M, Garrone O, Crook T, Ryan KM (2006) DRAM, a p53-induced modulator of autophagy, is critical for apoptosis. Cell 126:121–134

Cudjoe EK, Lauren Kyte S, Saleh T, Landry JW, Gewirtz DA (2019) Chapter 12: Autophagy inhibition and chemosensitization in cancer therapy. In: Johnson DE (ed) Targeting cell survival pathways to enhance response to chemotherapy. Academic Press, London, pp 259–273

Davies CW, Stjepanovic G, Hurley JH (2015) How the ATG1 complex assembles to initiate autophagy. Autophagy 11:185–186

Donohue E, Tovey A, Vogl AW, Arns S, Sternberg E, Young RN, Roberge M (2011) Inhibition of autophagosome formation by the benzoporphyrin derivative verteporfin. J Biol Chem 286:7290–7300

Donohue E, Thomas A, Maurer N, Manisali I, Zeisser-Labouebe M, Zisman N, Anderson HJ, Ng SS, Webb M, Bally M, Roberge M (2013) The autophagy inhibitor verteporfin moderately enhances the antitumor activity of gemcitabine in a pancreatic ductal adenocarcinoma model. J Cancer 4:585–596

Eisenberg-Lerner A, Bialik S, Simon HU, Kimchi A (2009) Life and death partners: apoptosis, autophagy and the cross-talk between them. Cell Death Differ 16:966–975

Fazi B, Bursch W, Fimia GM, Nardacci R, Piacentini M, Di Sano F, Piredda L (2008) Fenretinide induces autophagic cell death in caspase-defective breast cancer cells. Autophagy 4:435–441

Feng W, Huang S, Wu H, Zhang M (2007) Molecular basis of Bcl-xL's target recognition versatility revealed by the structure of Bcl-xL in complex with the BH3 domain of Beclin1. J Mol Biol 372:223–235

Fitzwalter BE, Thorburn A (2018) FOXO3 links autophagy to apoptosis. Autophagy 14:1467–1468

Fitzwalter BE, Towers CG, Sullivan KD, Andrysik Z, Hoh M, Ludwig M, O'Prey J, Ryan KM, Espinosa JM, Morgan MJ, Thorburn A (2018) Autophagy inhibition mediates apoptosis sensitization in cancer therapy by relieving FOXO3a turnover. Dev Cell 44:555–565.e553

Galluzzi L, Joza N, Tasdemir E, Maiuri MC, Hengartner M, Abrams JM, Tavernarakis N, Penninger J, Madeo F, Kroemer G (2008) No death without life: vital functions of apoptotic effectors. Cell Death Differ 15:1113–1123

Galluzzi L, Kepp O, Kroemer G (2012) Mitochondria: master regulators of danger signalling. Nat Rev Mol Cell Biol 13:780–788

Garrido C, Kroemer G (2004) Life's smile, death's grin: vital functions of apoptosis-executing proteins. Curr Opin Cell Biol 16:639–646

Gogada R, Yadav N, Liu J, Tang S, Zhang D, Schneider A, Seshadri A, Sun L, Aldaz CM, Tang DG, Chandra D (2013) Bim, a proapoptotic protein, up-regulated via transcription factor E2F1-dependent mechanism, functions as a prosurvival molecule in cancer. J Biol Chem 288:368–381

Grumati P, Dikic I (2018) Ubiquitin signaling and autophagy. J Biol Chem 293:5404–5413

Guo J, Ma Y, Peng X, Jin H, Liu J (2019) LncRNA CCAT1 promotes autophagy via regulating ATG7 by sponging miR-181 in hepatocellular carcinoma. J Cell Biochem 120:17975–17983

Hanahan D, Weinberg RA (2011) Hallmarks of cancer: the next generation. Cell 144:646–674

Harhaji-Trajkovic L, Vilimanovich U, Kravic-Stevovic T, Bumbasirevic V, Trajkovic V (2009) AMPK-mediated autophagy inhibits apoptosis in cisplatin-treated tumour cells. J Cell Mol Med 13:3644–3654

Hübner A, Barrett T, Flavell RA, Davis RJ (2008) Multisite phosphorylation regulates Bim stability and apoptotic activity. Mol Cell 30:415–425

Islam MA, Sooro MA, Zhang P (2018) Autophagic regulation of p62 is critical for cancer therapy. Int J Mol Sci 19:1405

Jain MV, Paczulla AM, Klonisch T, Dimgba FN, Rao SB, Roberg K, Schweizer F, Lengerke C, Davoodpour P, Palicharla VR, Maddika S, Los M (2013) Interconnections between apoptotic, autophagic and necrotic pathways: implications for cancer therapy development. J Cell Mol Med 17:12–29

Jiang K, Zhang C, Yu B, Chen B, Liu Z, Hou C, Wang F, Shen H, Chen Z (2017) Autophagic degradation of FOXO3a represses the expression of PUMA to block cell apoptosis in cisplatin-resistant osteosarcoma cells. Am J Cancer Res 7:1407–1422

Jing Q, Li G, Chen X, Liu C, Lu S, Zheng H, Ma H, Qin Y, Zhang D, Zhang S, Ren S, Huang D, Tan P, Chen J, Qiu Y, Liu Y (2019) Wnt3a promotes radioresistance via autophagy in squamous cell carcinoma of the head and neck. J Cell Mol Med 23:4711–4722

Jung K-T, Oh S-H (2019) Polyubiquitination of p62/SQSTM1 is a prerequisite for Fas/CD95 aggregation to promote caspase-dependent apoptosis in cadmium-exposed mouse monocyte RAW264.7 cells. Sci Rep 9:12240

Kaya-Aksoy E, Cingoz A, Senbabaoglu F, Seker F, Sur-Erdem I, Kayabolen A, Lokumcu T, Sahin GN, Karahuseyinoglu S, Bagci-Onder T (2019) The pro-apoptotic Bcl-2 family member Harakiri (HRK) induces cell death in glioblastoma multiforme. Cell Death Discov 5:64

Kerr JF, Wyllie AH, Currie AR (1972) Apoptosis: a basic biological phenomenon with wide-ranging implications in tissue kinetics. Br J Cancer 26:239–257

Kim KW, Mutter RW, Cao C, Albert JM, Freeman M, Hallahan DE, Lu B (2006) Autophagy for cancer therapy through inhibition of pro-apoptotic proteins and mammalian target of rapamycin signaling. J Biol Chem 281:36883–36890

Kim EH, Sohn S, Kwon HJ, Kim SU, Kim MJ, Lee SJ, Choi KS (2007) Sodium selenite induces superoxide-mediated mitochondrial damage and subsequent autophagic cell death in malignant glioma cells. Cancer Res 67:6314–6324

Kim HJ, Lee SG, Kim YJ, Park JE, Lee KY, Yoo YH, Kim JM (2013) Cytoprotective role of autophagy during paclitaxel-induced apoptosis in Saos-2 osteosarcoma cells. Int J Oncol 42:1985–1992

Klionsky DJ (2005) The molecular machinery of autophagy: unanswered questions. J Cell Sci 118:7–18

Laane E, Tamm KP, Buentke E, Ito K, Kharaziha P, Oscarsson J, Corcoran M, Bjorklund AC, Hultenby K, Lundin J, Heyman M, Soderhall S, Mazur J, Porwit A, Pandolfi PP, Zhivotovsky B, Panaretakis T, Grander D (2009) Cell death induced by dexamethasone in lymphoid leukemia is mediated through initiation of autophagy. Cell Death Differ 16:1018–1029

Levine B, Klionsky DJ (2004) Development by self-digestion: molecular mechanisms and biological functions of autophagy. Dev Cell 6:463–477

Levine B, Klionsky DJ (2017) Autophagy wins the 2016 Nobel Prize in Physiology or Medicine: breakthroughs in baker's yeast fuel advances in biomedical research. Proc Natl Acad Sci U S A 114:201–205

Levine B, Kroemer G (2008) Autophagy in the pathogenesis of disease. Cell 132:27–42

Li YC, He SM, He ZX, Li M, Yang Y, Pang JX, Zhang X, Chow K, Zhou Q, Duan W, Zhou ZW, Yang T, Huang GH, Liu A, Qiu JX, Liu JP, Zhou SF (2014) Plumbagin induces apoptotic and autophagic cell death through inhibition of the PI3K/Akt/mTOR pathway in human non-small cell lung cancer cells. Cancer Lett 344:239–259

Li F, Zheng X, Liu Y, Li P, Liu X, Ye F, Zhao T, Wu Q, Jin X, Li Q (2016a) Different roles of CHOP and JNK in mediating radiation-induced autophagy and apoptosis in breast cancer cells. Radiat Res 185:539–548, 510

Li Y, Deutzmann A, Choi PS, Fan AC, Felsher DW (2016b) BIM mediates oncogene inactivation-induced apoptosis in multiple transgenic mouse models of acute lymphoblastic leukemia. Oncotarget 7:26926–26934

Li H, Liu L, Chang H, Zou Z, Xing D (2018) Downregulation of MCL-1 and upregulation of PUMA using mTOR inhibitors enhance antitumor efficacy of BH3 mimetics in triple-negative breast cancer. Cell Death Dis 9:137–137

Li M, Liu J, Li S, Feng Y, Yi F, Wang L, Wei S, Cao L (2019) Autophagy-related 7 modulates tumor progression in triple-negative breast cancer. Lab Investig 99:1266–1274

Liang XH, Jackson S, Seaman M, Brown K, Kempkes B, Hibshoosh H, Levine B (1999) Induction of autophagy and inhibition of tumorigenesis by beclin 1. Nature 402:672–676

Liu B, Min MW, Bao JK (2009) Induction of apoptosis by Concanavalin A and its molecular mechanisms in cancer cells. Autophagy 5:432–433

Liu J, Xia H, Kim M, Xu L, Li Y, Zhang L, Cai Y, Norberg HV, Zhang T, Furuya T, Jin M, Zhu Z, Wang H, Yu J, Li Y, Hao Y, Choi A, Ke H, Ma D, Yuan J (2011) Beclin1 controls the levels of p53 by regulating the deubiquitination activity of USP10 and USP13. Cell 147:223–234

Liu H, He Z, von Rutte T, Yousefi S, Hunger RE, Simon HU (2013) Down-regulation of autophagy-related protein 5 (ATG5) contributes to the pathogenesis of early-stage cutaneous melanoma. Sci Transl Med 5:202ra123

Liu H, He Z, Simon HU (2014) Autophagy suppresses melanoma tumorigenesis by inducing senescence. Autophagy 10:372–373

Liu K, Lee J, Kim JY, Wang L, Tian Y, Chan ST, Cho C, Machida K, Chen D, Ou JJ (2017) Mitophagy controls the activities of tumor suppressor p53 to regulate hepatic cancer stem cells. Mol Cell 68:281–292.e285

Liu Y, Gu S, Li H, Wang J, Wei C, Liu Q (2019) SNHG16 promotes osteosarcoma progression and enhances cisplatin resistance by sponging miR-16 to upregulate ATG4B expression. Biochem Biophys Res Commun 518:127–133

Liu K, Yu Q, Li H, Xie C, Wu Y, Ma D, Sheng P, Dai W, Jiang H (2020) BIRC7 promotes epithelial-mesenchymal transition and metastasis in papillary thyroid carcinoma through restraining autophagy. Am J Cancer Res 10:78–94

Lo-Coco F, Avvisati G, Vignetti M, Thiede C, Orlando SM, Iacobelli S, Ferrara F, Fazi P, Cicconi L, Di Bona E, Specchia G, Sica S, Divona M, Levis A, Fiedler W, Cerqui E, Breccia M, Fioritoni G, Salih HR, Cazzola M, Melillo L, Carella AM, Brandts CH, Morra E, von Lilienfeld-Toal M, Hertenstein B, Wattad M, Lubbert M, Hanel M, Schmitz N, Link H, Kropp MG, Rambaldi A, La Nasa G, Luppi M, Ciceri F, Finizio O, Venditti A, Fabbiano F, Dohner K, Sauer M, Ganser A, Amadori S, Mandelli F, Dohner H, Ehninger G, Schlenk RF, Platzbecker U (2013) Retinoic acid and arsenic trioxide for acute promyelocytic leukemia. N Engl J Med 369:111–121

Mahapatra KK, Panigrahi DP, Praharaj PP, Bhol CS, Patra S, Mishra SR, Behera BP, Bhutia SK (2019) Molecular interplay of autophagy and endocytosis in human health and diseases. Biol Rev Camb Philos Soc 94:1576–1590

Maiuri MC, Tasdemir E, Criollo A, Morselli E, Vicencio JM, Carnuccio R, Kroemer G (2009) Control of autophagy by oncogenes and tumor suppressor genes. Cell Death Differ 16:87–93

Mancias JD, Kimmelman AC (2016) Mechanisms of selective autophagy in normal physiology and cancer. J Mol Biol 428:1659–1680

Mariño G, Niso-Santano M, Baehrecke EH, Kroemer G (2014) Self-consumption: the interplay of autophagy and apoptosis. Nat Rev Mol Cell Biol 15:81–94

Marquez RT, Xu L (2012) Bcl-2:Beclin 1 complex: multiple, mechanisms regulating autophagy/apoptosis toggle switch. Am J Cancer Res 2:214–221

Mathew R, Karantza-Wadsworth V, White E (2007) Role of autophagy in cancer. Nat Rev Cancer 7:961–967

Mizushima N (2007) Autophagy: process and function. Genes Dev 21:2861–2873

Moscat J, Karin M, Diaz-Meco MT (2016) p62 in cancer: signaling adaptor beyond autophagy. Cell 167:606–609

Mukhopadhyay S, Panda PK, Sinha N, Das DN, Bhutia SK (2014) Autophagy and apoptosis: where do they meet? Apoptosis 19:555–566

Mukhopadhyay S, Das DN, Panda PK, Sinha N, Naik PP, Bissoyi A, Pramanik K, Bhutia SK (2015) Autophagy protein Ulk1 promotes mitochondrial apoptosis through reactive oxygen species. Free Radic Biol Med 89:311–321

Mukhopadhyay S, Naik PP, Panda PK, Sinha N, Das DN, Bhutia SK (2016) Serum starvation induces anti-apoptotic cIAP1 to promote mitophagy through ubiquitination. Biochem Biophys Res Commun 479:940–946

Mukhopadhyay S, Schlaepfer IR, Bergman BC, Panda PK, Praharaj PP, Naik PP, Agarwal R, Bhutia SK (2017) ATG14 facilitated lipophagy in cancer cells induce ER stress mediated mitoptosis through a ROS dependent pathway. Free Radic Biol Med 104:199–213

Muniraj N, Siddharth S, Nagalingam A, Walker A, Woo J, Gyorffy B, Gabrielson E, Saxena NK, Sharma D (2019) Withaferin A inhibits lysosomal activity to block autophagic flux and induces apoptosis via energetic impairment in breast cancer cells. Carcinogenesis. https://doi.org/10.1093/carcin/bgz015

Nieminen AI, Eskelinen VM, Haikala HM, Tervonen TA, Yan Y, Partanen JI, Klefström J (2013) Myc-induced AMPK-phospho p53 pathway activates Bak to sensitize mitochondrial apoptosis. Proc Natl Acad Sci 110:E1839

Oberstein A, Jeffrey PD, Shi Y (2007) Crystal structure of the Bcl-XL-Beclin 1 peptide complex: Beclin 1 is a novel BH3-only protein. J Biol Chem 282:13123–13132

Oral O, Oz-Arslan D, Itah Z, Naghavi A, Deveci R, Karacali S, Gozuacik D (2012) Cleavage of ATG3 protein by caspase-8 regulates autophagy during receptor-activated cell death. Apoptosis 17:810–820

Panda PK, Mukhopadhyay S, Das DN, Sinha N, Naik PP, Bhutia SK (2015) Mechanism of autophagic regulation in carcinogenesis and cancer therapeutics. Semin Cell Dev Biol 39:43–55

Panda PK, Naik PP, Meher BR, Das DN, Mukhopadhyay S, Praharaj PP, Maiti TK, Bhutia SK (2018) PUMA dependent mitophagy by Abrus agglutinin contributes to apoptosis through ceramide generation. Biochim Biophys Acta Mol Cell Res 1865:480–495

Panda PK, Patra S, Naik PP, Praharaj PP, Mukhopadhyay S, Meher BR, Gupta PK, Verma RS, Maiti TK, Bhutia SK (2020) Deacetylation of LAMP1 drives lipophagy-dependent generation of free fatty acids by Abrus agglutinin to promote senescence in prostate cancer. J Cell Physiol 235:2776–2791

Panigrahi DP, Bhol CS, Nivetha R, Nagini S, Patil S, Maiti TK, Bhutia SK (2020) Abrus agglutinin inhibits oral carcinogenesis through inactivation of NRF2 signaling pathway. Int J Biol Macromol 155:1123–1132

Parsons MJ, McCormick L, Janke L, Howard A, Bouchier-Hayes L, Green DR (2013) Genetic deletion of caspase-2 accelerates MMTV/c-neu-driven mammary carcinogenesis in mice. Cell Death Differ 20:1174–1182

Patra S, Panda PK, Naik PP, Panigrahi DP, Praharaj PP, Bhol CS, Mahapatra KK, Padhi P, Jena M, Patil S, Patra SK, Bhutia SK (2020) Terminalia bellirica extract induces anticancer activity through modulation of apoptosis and autophagy in oral squamous cell carcinoma. Food Chem Toxicol 136:111073

Pattingre S, Tassa A, Qu X, Garuti R, Liang XH, Mizushima N, Packer M, Schneider MD, Levine B (2005) Bcl-2 antiapoptotic proteins inhibit Beclin 1-dependent autophagy. Cell 122:927–939

Qian W, Liu J, Jin J, Ni W, Xu W (2007) Arsenic trioxide induces not only apoptosis but also autophagic cell death in leukemia cell lines via up-regulation of Beclin1. Leuk Res 31:329–339

Qu X, Yu J, Bhagat G, Furuya N, Hibshoosh H, Troxel A, Rosen J, Eskelinen EL, Mizushima N, Ohsumi Y, Cattoretti G, Levine B (2003) Promotion of tumorigenesis by heterozygous disruption of the beclin 1 autophagy gene. J Clin Invest 112:1809–1820

Quach C, Song Y, Guo H, Li S, Maazi H, Fung M, Sands N, O'Connell D, Restrepo-Vassalli S, Chai B, Nemecio D, Punj V, Akbari O, Idos GE, Mumenthaler SM, Wu N, Martin SE, Hagiya A, Hicks J, Cui H, Liang C (2019) A truncating mutation in the autophagy gene UVRAG drives inflammation and tumorigenesis in mice. Nat Commun 10:5681

Rahmani M, Aust MM, Attkisson E, Williams DC Jr, Ferreira-Gonzalez A, Grant S (2013) Dual inhibition of Bcl-2 and Bcl-xL strikingly enhances PI3K inhibition-induced apoptosis in human myeloid leukemia cells through a GSK3- and Bim-dependent mechanism. Cancer Res 73:1340–1351

Roy B, Pattanaik AK, Das J, Bhutia SK, Behera B, Singh P, Maiti TK (2014) Role of PI3K/Akt/mTOR and MEK/ERK pathway in Concanavalin A induced autophagy in HeLa cells. Chem Biol Interact 210:96–102

Saha S, Panigrahi DP, Patil S, Bhutia SK (2018) Autophagy in health and disease: a comprehensive review. Biomed Pharmacother 104:485–495

Salazar M, Carracedo A, Salanueva IJ, Hernandez-Tiedra S, Lorente M, Egia A, Vazquez P, Blazquez C, Torres S, Garcia S, Nowak J, Fimia GM, Piacentini M, Cecconi F, Pandolfi PP, Gonzalez-Feria L, Iovanna JL, Guzman M, Boya P, Velasco G (2009) Cannabinoid action induces autophagy-mediated cell death through stimulation of ER stress in human glioma cells. J Clin Invest 119:1359–1372

Shalini S, Nikolic A, Wilson CH, Puccini J, Sladojevic N, Finnie J, Dorstyn L, Kumar S (2016) Caspase-2 deficiency accelerates chemically induced liver cancer in mice. Cell Death Differ 23:1727–1736

Shao S, Li S, Qin Y, Wang X, Yang Y, Bai H, Zhou L, Zhao C, Wang C (2014) Spautin-1, a novel autophagy inhibitor, enhances imatinib-induced apoptosis in chronic myeloid leukemia. Int J Oncol 44:1661–1668

Shi S, Tan P, Yan B, Gao R, Zhao J, Wang J, Guo J, Li N, Ma Z (2016) ER stress and autophagy are involved in the apoptosis induced by cisplatin in human lung cancer cells. Oncol Rep 35:2606–2614

Singh R, Letai A, Sarosiek K (2019) Regulation of apoptosis in health and disease: the balancing act of BCL-2 family proteins. Nat Rev Mol Cell Biol 20:175–193

Singla M, Bhattacharyya S (2017) Autophagy as a potential therapeutic target during epithelial to mesenchymal transition in renal cell carcinoma: an in vitro study. Biomed Pharmacother 94:332–340

Song X, Yuan Z, Yuan H, Wang L, Ji P, Jin G, Dai J, Ma H (2018) ATG12 expression quantitative trait loci associated with head and neck squamous cell carcinoma risk in a Chinese Han population. Mol Carcinog 57:1030–1037

Tabe Y, Jin L, Iwanami H, Matsushita H, Kazuno S, Fujimura T, Ueno T, Miida T, Weinstock DM, Thomas DA, Andreeff M, Konopleva M (2013) Efficacy and mechanisms of the mTOR inhibitor AZD2014 combined with L-Asparaginase or JAK2 inhibitor TG101348 in ALL. Blood 122:1282–1282

Tan YQ, Zhang J, Zhou G (2017) Autophagy and its implication in human oral diseases. Autophagy 13:225–236

Tian X, Song HS, Cho YM, Park B, Song Y-J, Jang S, Kang SC (2017) Anticancer effect of Saussurea lappa extract via dual control of apoptosis and autophagy in prostate cancer cells. Medicine 96:e7606

Vaseva AV, Marchenko ND, Ji K, Tsirka SE, Holzmann S, Moll UM (2012) p53 opens the mitochondrial permeability transition pore to trigger necrosis. Cell 149:1536–1548

Vousden KH, Lane DP (2007) p53 in health and disease. Nat Rev Mol Cell Biol 8:275–283

Wang S, El-Deiry WS (2003) TRAIL and apoptosis induction by TNF-family death receptors. Oncogene 22:8628–8633

Wang J, Cui D, Gu S, Chen X, Bi Y, Xiong X, Zhao Y (2018) Autophagy regulates apoptosis by targeting NOXA for degradation. Biochim Biophys Acta Mol Cell Res 1865:1105–1113

Wei Y, Sinha S, Levine B (2008) Dual role of JNK1-mediated phosphorylation of Bcl-2 in autophagy and apoptosis regulation. Autophagy 4:949–951

White E (2015) The role for autophagy in cancer. J Clin Invest 125:42–46

Wilde L, Tanson K, Curry J, Martinez-Outschoorn U (2018) Autophagy in cancer: a complex relationship. Biochem J 475:1939–1954

Wong RS (2011) Apoptosis in cancer: from pathogenesis to treatment. J Exp Clin Cancer Res 30:87

Xi G, Hu X, Wu B, Jiang H, Young CYF, Pang Y, Yuan H (2011) Autophagy inhibition promotes paclitaxel-induced apoptosis in cancer cells. Cancer Lett 307:141–148

Xiao T, Zhu W, Huang W, Lu SS, Li XH, Xiao ZQ, Yi H (2018) RACK1 promotes tumorigenicity of colon cancer by inducing cell autophagy. Cell Death Dis 9:1148

Xie Z, Klionsky DJ (2007) Autophagosome formation: core machinery and adaptations. Nat Cell Biol 9:1102–1109

Xu XD, Zhao Y, Zhang M, He RZ, Shi XH, Guo XJ, Shi CJ, Peng F, Wang M, Shen M, Wang X, Li X, Qin RY (2017) Inhibition of autophagy by Deguelin sensitizes pancreatic cancer cells to doxorubicin. Int J Mol Sci 18:370

Yang X, Yu DD, Yan F, Jing YY, Han ZP, Sun K, Liang L, Hou J, Wei LX (2015) The role of autophagy induced by tumor microenvironment in different cells and stages of cancer. Cell Biosci 5:14

Yee KS, Wilkinson S, James J, Ryan KM, Vousden KH (2009) PUMA- and Bax-induced autophagy contributes to apoptosis. Cell Death Differ 16:1135–1145

Young MM, Takahashi Y, Khan O, Park S, Hori T, Yun J, Sharma AK, Amin S, Hu C-D, Zhang J, Kester M, Wang H-G (2012) Autophagosomal membrane serves as platform for intracellular death-inducing signaling complex (iDISC)-mediated caspase-8 activation and apoptosis. J Biol Chem 287:12455–12468

Yousefi S, Perozzo R, Schmid I, Ziemiecki A, Schaffner T, Scapozza L, Brunner T, Simon H-U (2006) Calpain-mediated cleavage of ATG5 switches autophagy to apoptosis. Nat Cell Biol 8:1124–1132

Yu K, Xiang L, Li S, Wang S, Chen C, Mu H (2019) HIF1alpha promotes prostate cancer progression by increasing ATG5 expression. Anim Cells Syst (Seoul) 23:326–334

Yue Z, Jin S, Yang C, Levine AJ, Heintz N (2003) Beclin 1, an autophagy gene essential for early embryonic development, is a haploinsufficient tumor suppressor. Proc Natl Acad Sci U S A 100:15077–15082

Zhang YB, Gong JL, Xing TY, Zheng SP, Ding W (2013) Autophagy protein p62/SQSTM1 is involved in HAMLET-induced cell death by modulating apoptosis in U87MG cells. Cell Death Dis 4:e550–e550

Zhang M, Zhang W, Tang G, Wang H, Wu M, Yu W, Zhou Z, Mou Y, Liu X (2019) Targeted codelivery of docetaxel and ATG7 siRNA for autophagy inhibition and pancreatic cancer treatment. ACS Appl Bio Mater 2:1168–1176

Zhao X, Fang Y, Yang Y, Qin Y, Wu P, Wang T, Lai H, Meng L, Wang D, Zheng Z, Lu X, Zhang H, Gao Q, Zhou J, Ma D (2015) Elaiophylin, a novel autophagy inhibitor, exerts antitumor activity as a single agent in ovarian cancer cells. Autophagy 11:1849–1863

Zheng K, Li Y, Wang S, Wang X, Liao C, Hu X, Fan L, Kang Q, Zeng Y, Wu X, Wu H, Zhang J, Wang Y, He Z (2016) Inhibition of autophagosome-lysosome fusion by ginsenoside Ro via the ESR2-NCF1-ROS pathway sensitizes esophageal cancer cells to 5-fluorouracil-induced cell death via the CHEK1-mediated DNA damage checkpoint. Autophagy 12:1593–1613

Zhou M, Liu Z, Zhao Y, Ding Y, Liu H, Xi Y, Xiong W, Li G, Lu J, Fodstad O, Riker AI, Tan M (2010) MicroRNA-125b confers the resistance of breast cancer cells to paclitaxel through suppression of pro-apoptotic Bcl-2 antagonist killer 1 (Bak1) expression. J Biol Chem 285:21496–21507

Zhou J, Li G, Zheng Y, Shen HM, Hu X, Ming QL, Huang C, Li P, Gao N (2015) A novel autophagy/mitophagy inhibitor liensinine sensitizes breast cancer cells to chemotherapy through DNM1L-mediated mitochondrial fission. Autophagy 11:1259–1279

Cross-Talk Between DNA Damage and Autophagy and Its Implication in Cancer Therapy

3

Ganesh Pai Bellare, Pooja Gupta, Saikat Chakraborty, Mrityunjay Tyagi, and Birija Sankar Patro

Abstract

Chemotherapy and radiotherapy regimens are designed primarily to induce DNA damage to kill cancer cells. DNA damage response (DDR) proteins recognize and repair a variety of DNA damages. In response to DNA damage, a well-orchestrated autophagy program, comprising of than 30 autophagy-related genes (ATG), are triggered to degrade and recycle damaged proteins and cellular components for aiding DNA repair process. Recently, several interesting reports have showed the pivotal role of DDR proteins in regulating dozens of autophagy proteins and vice versa. Cross-talk between these two functionally different cellular processes may immensely contribute towards the understanding of resistance or sensitization of cancer cells in response to chemotherapy and radiotherapy. Nevertheless, the precise molecular link between DDR and autophagy still remains obscure and elusive. In the current review, we provide comprehensive insights into the underlying mechanisms involved in the molecular crosstalk between DDR and autophagy, which differentially regulate cancer cell fate in response to DNA damaging chemotherapeutics and radiotherapeutics or chemotherapy and radiotherapy.

Keywords

Autophagy · DNA repair · DNA damage response (DDR) · Chemotherapy · Radiotherapy · Cisplatin · Radiosensitization · DNA damage response · Cancer

---

Ganesh Pai Bellare, Pooja Gupta, Saikat Chakraborty and Mrityunjay Tyagi contributed equally to this work.

G. P. Bellare · P. Gupta · S. Chakraborty · M. Tyagi · B. S. Patro (✉)
Bio-Organic Division, Bhabha Atomic Research Centre, Mumbai, India

Homi Bhabha National Institute, Mumbai, India
e-mail: bisank@barc.gov.in

© Springer Nature Singapore Pte Ltd. 2020
S. K. Bhutia (ed.), *Autophagy in Tumor and Tumor Microenvironment*,
https://doi.org/10.1007/978-981-15-6930-2_3

Abbreviations

AMPK	Adenosine monophosphate Kinase
ATG	Autophagy related gene
ATM	Ataxia telangiectasia mutated
ATR	Ataxia telangiectasia and rad3-related Protein
BAK	Bcl-2 homologous antagonist killer
BAX	Bcl-2-associated X protein
CMA	Chaperone mediated autophagy
DDR	DNA damage response
HR	Homologous recombination
HSP	Heat shock protein
LAMP2A	Lysosome associated membrane protein 2A
LC3/MAP1LC3	Microtubule-associated protein 1 light chain 3
LKB1	Liver kinase B1, also known as serine/threonine kinase 11 (STK11)
MMR	Mismatch repair
mTOR	Mechanistic target of rapamycin kinase
NER	Nucleotide excision repair
NHEJ	Nonhomologous end joining
PARP1	Poly(ADP–ribose) polymerase 1
PCD	Programmed cell death
PI3K	Phosphatidylinositol-4,5-bisphosphate 3-kinase
ROS	Reactive oxygen species

3.1 Introduction

Cancer is one of the leading causes of death in many developing countries, including India. In recent years, advancements in the chemotherapeutic regimes, especially the development of novel drugs or a combination of drugs, and radiotherapy provide better therapeutic outcomes and enhance disease-free survival (Jemal et al. 2011). However, the development of inherent and adaptive resistance to therapeutics is the key feature of therapeutic failure in oncology (Luqmani 2005). Resistance to chemo and radio-therapeutics have been attributed to multiple factors like evading growth suppressors, avoiding immune destruction, enabling replicative immortality, tumor promoting inflammation, activating invasion and metastasis, inducing angiogenesis, genome instability, mutation, resisting cell death and deregulating cellular energetics (Hanahan and Weinberg 2011). In the recent past, several evidences have shown that cellular autophagy is yet another mode of resistance, linking to therapeutic failure (Abedin et al. 2007).

Cellular autophagy is an evolutionarily conserved process of packaging damaged or aged organelles or misfolded proteins into autophagosome and their fusion with lysosome for degradation. Subsequently, degraded materials can be recycled for renewal (Mizushima and Komatsu 2011). Autophagy is categorized into (1) macro-autophagy, (2) micro-autophagy, and (3) chaperone-mediated autophagy (CMA). While macro-autophagy is an autophagosome mediated process, micro-autophagy is direct engulfment of cytosolic materials by lysosomes. CMA is involved in the lysosomal delivery of unfolded proteins through multimerization of lysosomal membrane-associated protein (LAMP2A) and heat shock protein 70 (HSP70) complex. Autophagy can behave dichotomously by inducing the pro-survival or death process in a context-dependent manner (Buszczak and Kramer 2019). Controlled induction of autophagy plays a vital role in cell survival, while the hyperactivation of autophagy is linked to autophagic cell death (Nyfeler and Eng 2016). Many chemotherapeutics and radiotherapy treatment kill cancer cells by primarily inducing DNA damage and additional genomic instability. Cancer resistance to DNA damaging therapeutics might also stem from processing additional sources of genomic instability, including micronuclei (Bartsch et al. 2017), chromatin fragments (Ivanov et al. 2013), and endogenous retrotransposons (Guo et al. 2014). Although the mechanism of DNA repair in cancer resistance is well established, autophagy inhibition was also shown to abolish resistance in cancer cells in response to chemotherapeutics and radiation therapy. Therefore, a better understanding of the crosstalk between DNA damage/repair and autophagy in the context of chemotherapeutics and radiotherapy is required. This article is focused on reviewing several such findings to shed light on how key players of the DNA repair process are involved in autophagy regulation and vice versa in response to DNA damaging therapeutics.

3.1.1 Role of Autophagy in Response to Cisplatin Treatment

Cisplatin is mainly used for lung-cancer treatment. Cisplatin primarily causes DNA damages through intra-strand crosslinking. Nucleotide excision repair (NER), mismatch repair (MMR), homologous recombination (HR) and non-homologous end-joining (NHEJ) are involved in repairing cisplatin-induced DNA damage (Rocha et al. 2018). Interestingly, the formation of autophagosomes in response to cisplatin treatment was observed in the 1980s (Nilsson 1988). Later, autophagy was detected as early as 2–4 h after cisplatin exposure and co-treatment with an autophagy inhibitor (3-methyladenine) led to an increase in caspase activation and cell death in renal proximal tubular cells (Yang et al. 2008). In the mouse renal proximal tubular cells, cisplatin was found to induce cytoprotective autophagy in p53 (tumor suppressor protein) dependent manner as the use of p53-inhibitor (pifithrin-α) partially suppressed the autophagosome formation (Periyasamy-Thandavan et al. 2008). Induction of p53 in response to cisplatin has also been shown to activate microRNA dependent survival of mouse proximal tubular cells. In this study, pifithrin-α or specific antisense oligonucleotides for miR-32 increased cell death by reducing miR-34a induction (Bhatt et al. 2010). The DNA damage-dependent activation of p53 can have a dual effect on autophagy. It may upregulate autophagy through its transcriptional activity or downregulate through its cytoplasmic functions (Budanov and Karin 2008; Green and Kroemer 2009). Upregulation of Beclin1 after cisplatin treatment is reported to be responsible for cisplatin-induced autophagy in human bladder cancer cells (Lin et al. 2017). Low-dose cisplatin also induced autophagy and the inhibition of autophagy using 3-methyladenine resulted in apoptosis (Yang et al. 2012). This study suggests that even a low amount of DNA damage may also induce pro-survival autophagy. ISG20L1 another regulator protein of the p53 protein family has also been identified as a regulator of autophagy after DNA damage induction by cisplatin and etoposide (Eby et al. 2010). The knockdown of ISG20L1 suppresses autophagy in response to cisplatin. In glioma and fibrosarcoma cells, inhibition of autophagic response after cisplatin treatment was found to increase the ROS production. Autophagy induction is also reported to precede adenosine monophosphate-activated protein kinase (AMPK) activation, which switches signaling AMP/ATP ratio to ATP-generating catabolic pathways and concomitant down-regulation of mammalian target of rapamycin (mTOR)-mediated phosphorylation of p70 S6 kinase (Harhaji-Trajkovic et al. 2009) (Fig. 3.1). Activated AMPK (phosphorylated at Thr-172) is known to activate TSC2 (Tuberous sclerosis complex 2) and subsequent inhibition of mTOR function (Fig. 3.1). The use of both early-stage autophagy inhibitors (wortmannin) and late-stage blockers (bafilomycin and chloroquine, CQ) augmented cell death by cisplatin, indicating a role for autophagy in suppressing cisplatin-triggered apoptotic death (Harhaji-Trajkovic et al. 2009). Recently, AMPK activation in nutrient-deficient cells has been linked to Poly(ADP-ribosyl)ation (PARylation) dependent spatial and temporal regulation leading to nuclear export followed by autophagy induction (Rodríguez-Vargas et al. 2016). Since cisplatin treatment leads to PARylation of various proteins (Prasad et al. 2017; Schaaf et al. 2016), it may be plausible that PARylated AMPK plays a role in the induction of autophagy (Fig. 3.1).

3.1.2 Role of Autophagy in Response to Topoisomerase Inhibitor Treatment

Inhibitors of topoisomerase I (topotecan, irinotecan) and topoisomerase II (VP-16 or etoposide) are extensively used for the treatment of the different types of cancers. These drugs cause stalled replication fork mediated DNA double-strand breaks. In contrast to the survival role of autophagy, the embryonic fibroblasts from BAX/BAK double knockout mice, resistant to apoptosis were found to display an autophagy-dependent non-apoptotic cell death in response to DNA-damaging agent like etoposide (Shimizu et al. 2004). Alexander et al. reported the activation of ATM/ATR in response to etoposide (Alexander et al. 2010). In this study, it has been shown a cytoplasmic function of ATM in activating a tumor suppressor, TSC2 via the LKB1/AMPK metabolic pathway to repress mTORC1 and activate autophagy (Fig. 3.1). Further, the dysregulation of mTORC1 in ATM-deficient cells was inhibited by rapamycin (Alexander et al. 2010).

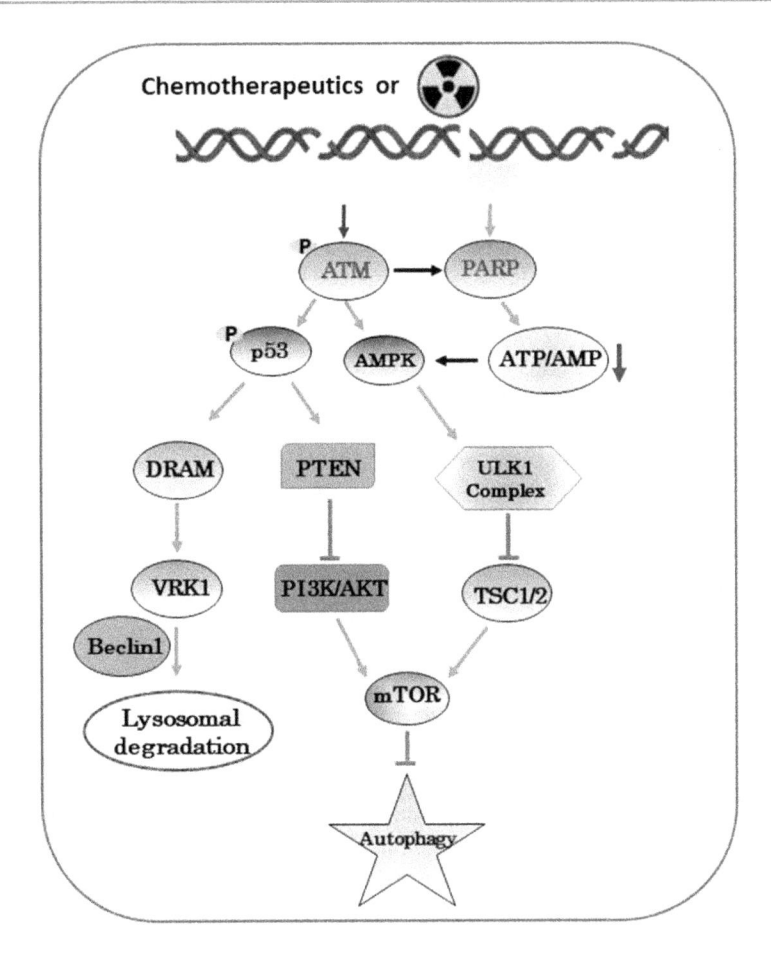

Fig. 3.1 Crosstalk between DNA damage and autophagy. Autophagy in response to DNA damaging agents (chemotherapeutics and radiation) mostly protects cancer cells from death. Key role of various DDR proteins, in the activation of autophagy, is shown in the above illustration

Another mode of autophagy was observed in ATG5 or ATG7 knockout cells. Although LC3 puncta formation, which requires lipid modification, was not observed, the autophagosome associated membranes were seen in ATG5/ATG7 deficient cells under few conditions (Nishida et al. 2009). Interestingly, etoposide induced the formation of autophagic vacuoles in ATG5 knockout mouse embryonic fibroblasts cells while the same was abrogated in ATG5-p53 double knock out cells in response to etoposide. This suggested a role of p53 in alternate autophagy. Later DRAM1, a downstream protein of p53 was found to be both necessary and sufficient to induce alternative autophagy (Nagata et al. 2018) (Fig. 3.1). DRAM1 was also found to co-localizes at the LC3-positive puncta indicating its role in conventional autophagy too. In hepatoma cell (HepG2), inhibition of AMPK also triggered apoptosis through suppression of autophagy. In contrast, augmentation of autophagy was observed after p53 inactivation leading to cell survival (Xie et al. 2011). Recently, a Ser/Thr protein phosphatase Mg^{2+}/Mn^{2+} dependent 1D (PPM1D), which is transactivated by p53, was identified as a factor that dephosphorylates serine-637 of ULK1 (unc-51 like autophagy activating kinase) (Torii et al. 2016). ULK1 is a subunit of the ATG1-complex that functions at the most upstream position in ATG signaling and the dephosphorylation of this complex is well-known to be essential for the induction of autophagy during starvation. This study links the possibility of ULK1 dephosphorylation in response to p53 activation as a trigger for the initiation of autophagosome formation in response to genotoxic stress (Fig. 3.1). Topoisomerase I inhibitor, topotecan, the treatment also leads to autophagy induction in terms of LC3 puncta formation, LC3 I/II conversions, and p62 degradation in colon carcinoma cells (Li et al. 2012). Topotecan induces DNA damage-dependent cytoprotective autophagy in p53 positive colon cancer cells while autophagic death was observed in p53 knock out cells. This suggests a role of p53 in switching the fate of autophagy from death to cell survival. DNA damage-dependent activation of p53 upregulates expression of Sestrin 2, enhances phosphorylation of AMPKα, and inhibits mTORC1, leading to activation of autophagy (Li et al. 2012).

3.1.3 Role of Autophagy in Response to Doxorubicin

Doxorubicin is a DNA intercalating drug and used for the treatment of breast cancer, bladder cancer, Kaposi's sarcoma, lymphoma, and acute lymphocytic leukemia. Doxorubicin is known to activate genotoxin stress-induced autophagy (GTA), which involves ATM-p53-mTOR signaling axis. The role of p53, a protein known to get induced during DNA damage in autophagy was determined through high-throughput sequencing via analyzing global p53 transcriptional networks in primary mouse embryo fibroblasts (Kenzelmann Broz et al. 2013). This study demonstrated that p53 is activated in an ATM/ATR-dependent manner and can bind the promoters of various autophagy genes leading to their transcriptional upregulation (Fig. 3.1). This p53-mediated transcriptional upregulation was found to be important for GTA as p53−/− cells were unable to induce autophagy after doxorubicin-induced DNA damage. Chromatin immunoprecipitation and RNA sequencing led to the identification of p53-bound and regulated genes, involved in multiple steps of autophagy, including upstream (TSC2, FOXO3a, mTOR, LKB1, and AMPK), core machinery-encoding genes (ULK1, ATG4a, ATG7, ULK2, and UVRAG) and lysosomal protein-encoding genes (*Ctsd*, *Laptm4a*, and *Vmp4*).

3.2 Linkage of Starvation-Induced Autophagy with DNA Damage

Rodríguez-Vargas et al. demonstrated DNA damage is an early event of starvation-induced autophagy. Here accumulation of both γH2AX and comet tails were found to be due to ROS generated in response to starvation. Further, ROS-induced DNA damage activates PARP-1, leading to ATP depletion and thus activation of AMPK-autophagy network (Rodríguez-Vargas et al. 2012). PARP-1 knockout cells blunted AMPK activation, leading to a delay in autophagy (pro-survival role) in starved cells. Recently, Poly-ADP-ribosylation (PARylation) of proteins was found to regulate autophagy in both spatial and temporal manner by modulating AMPK subcellular localization and activation (Rodríguez-Vargas et al. 2016). Here, the nutrient deprivation induces PARP-1 catalyzed PARylation, leading to the dissociation of the PARP-1/AMPK complex followed by the export of free PARylated nuclear AMPK to the cytoplasm to activate autophagy. DRAM (damage-regulated autophagy modulator) is a lysosomal protein essential for p53-mediated apoptosis and also reported to mediate a specific DNA damage responsive branch of the autophagy pathway (Crighton et al. 2006) p53 can activate autophagy via activation of the protein death-associated protein kinase (DAPK). The activated form of DAPK triggers autophagy in a Beclin-1-dependent manner. DAPK phosphorylates Beclin 1 on Thr 119 located at a crucial position within its BH3 domain, and thus promotes the dissociation of Beclin 1 from BCL-XL and the induction of autophagy (Zalckvar et al. 2009). Another DNA damage response protein p73 belongs to the p53 family of transcription factors, is known to regulate DRAM and autophagy during starvation. However, further studies revealed that p73-mediated autophagy is DRAM-independent (Crighton et al. 2007). Interestingly, p73 also modulates many mTOR regulated autophagy-associated genes. Besides, endogenous p73 binds to the regulatory regions of several autophagy genes such as ATG5, ATG7, and UVRAG and is an important regulator of autophagy (Rosenbluth and Pietenpol 2009).

3.3 Role of Autophagy in Response to Radiation Treatment

Ionizing radiation can damage DNA directly and indirectly by ROS generation, resulting into single-strand breaks (SSBs), base oxidation, apurinic, or apyrimidinic (AP) sites, and particularly, double-strand breaks (DSBs). Radiotherapy is one of the major treatment modality for cancer therapy but often fails to control tumor growth due to the development of resistance and dose-limiting side effects. It is reported that apoptotic death comprises less than 20% of radiation-induced cell death. So, it is imperative to explore other pathways of cell death to gain the therapeutic index by radiation. Radiation-induced activation of autophagy is well known in both cancer and normal cells (Zois and Koukourakis 2009). In response to radiation treatment, autophagy plays a dual role in promoting resistance or sensitization, depending upon severity and duration of stress, also the type and stage of tumor.

3.3.1 Radioresistance Due to Autophagy

Ionizing radiation-induced DNA damage sites are recognized by PARP1 leading to PARylation of various DDR proteins and recruitment to DNA sites. However, PARP mediated PARylation of proteins occurs at the expense of its substrate NAD^+ leading to ATP depletion (Aguilar-Quesada et al. 2007). At DSB sites, ATM is activated and PARylated by PARP1 which further leads to activation of the energy sensor AMP-activated protein kinase (AMPK), leading to autophagy progression by inhibiting the mTORC1 complex (Fig. 3.1). Thus, the activation of autophagy provides sustained energy required for DNA repair processes that lead to radioresistance and delayed apoptotic cell death. This may be the reason for the accumulation of DNA damage and genomic instability in autophagy-deficient cells. For instance, radioresistant breast tumor cells show a strong post-irradiation induction of autophagy, which thus serves as a protective and pro-survival mechanism (Chaachouay et al. 2011). In addition to this, ATM binds to FOXO3a (transcription factor), which regulates the expression of autophagy-related genes like LC3 and BNIP3 and upregulates autophagy (Nazio et al. 2017). Normal tissues, which are late responding, are benefited more from prolonged fraction delivery time (FDTs) than acute-responding tissues because of ATM-AMPK mediated autophagy process (Yao et al. 2015). In response to IR, ATM is also known to activate autophagy through three pathways: the MAPK14 pathway, mTOR pathway, and Beclin1/PI3KIII complexes and modulate radiosensitivity (Liang et al. 2019).

Similarly, autophagy also induces cell survival in esophageal squamous cell carcinoma and bladder cancer, which was abrogated by autophagy inhibitor (CQ) in response to radiation treatment (Chen et al. 2015; Wang et al. 2018). A recent study demonstrated the protective mechanism of radiation-induced autophagy in hematopoietic cells by activation of STAT3 signaling, which upregulated the expression of BRCA1 via ATG–KAP1–STAT3–BRCA1 pathways and increases DNA repair ability (Xu et al. 2017b). In thyroid cancer, radiotherapy induces autophagy by increasing expression of autophagy-associated proteins Beclin-1 and LC3, which is blocked by either 3-methyladenine or Beclin-1 siRNA, leading to upregulation p53 and then apoptosis (Gao et al. 2019). This shows that p53 acts as a switch between protective autophagy and apoptosis in thyroid cancer in response to radiotherapy.

Further, it is known that due to poor vascularization, a certain population of tumor cells (known as hypoxic cells) is deprived of oxygen, nutrient supply, and waste removal caused to stimulate autophagy and inhibit apoptosis (He et al. 2012). A previous study has demonstrated that the induction of BNIP3, a downstream target of HIF-1α, in hypoxic cells disrupts the Beclin1-BCL2 complex and releases Beclin1. This in turn induces autophagy as an adaptive survival mechanism during prolonged hypoxia in different cell lines like MEF, MCF, PC3, and LS174 (Bellot et al. 2009). In a similar context, radioresistance was observed in osteosarcoma cancer cells overexpressing HIF-1α which induces protective autophagy (Feng et al. 2016). Hypoxia leads to an increase in ROS production due to its effect on ETC of mitochondria. This ROS production by hypoxia causes DNA damage which can also stimulate autophagy by mitochondrial production and providing energy for cell survival (Zhang et al. 2008). Nevertheless, hypoxia-induced autophagy leads to a marked accumulation of autophagosomes along with RNA induction of autophagy-related genes such as Beclin-1, ATG5, and ATG12, leading to radioresistance (He et al. 2012). Thus, tumor cells create a more protective intracellular environment by glycolytic reprogramming, and the presence of mitochondrial defects, accompanied by the adaptation to hypoxic conditions, provide radioresistant properties, as well as survival and growth benefits. Apart from this, radiation also causes an increased formation of the acidic vesicle, which will induce autophagy to protect the damage, although the detailed mechanism of autophagy induced radioresistance is yet to be characterized for different tumor type and stage.

There are cases, where the induced autophagy exhibits neither cytoprotective nor cytotoxic functions, which we have termed as dormant autophagy. A study has shown that ATG7 and LC3 silencing lead to the sensitization of tumor cells but that is independent of autophagy (Schaaf et al. 2015). They showed that both chloroquine and knockdown of the essential autophagy genes, ATG7 and LC3b, effectively inhibit autophagy; however, only knockdown of LC3b or ATG7 but not CQ reduced survival. This indicates a radioprotective role of these autophagy-associated genes. However, the radioresistant effect is independent of autophagic degradation through lysosomes, and thus unrelated to canonical autophagy.

Ultraviolet (UVB) radiation is efficiently absorbed by DNA within the epidermis and damages DNA directly to form photoproducts. UV-induced DNA photoproducts induce the stabilization of p53. The anti-apoptotic Beclin1-binding protein BCL-2 is downregulated following UVB exposure, which may free Beclin1 to bind UVRAG (UV-irradiation-resistance-associated gene) and induce autophagy. UVRAG plays a dual role in autophagy (autophagosome formation and maturation) and DNA repair (chromosome stability); later process is autophagy-independent. In autophagy, UVRAG is responsible for the activation of PI(3) class III (PI(3)KC3) kinase through Beclin 1 interaction (Su et al. 2013). During NHEJ, UVRAG interacts and helps the assembly of the upstream protein kinase of the NHEJ pathway, DNA-PK. Moreover, UVRAG is found to be associated with centrosomes by its interaction with CEP63 (Zhao et al. 2012). Affecting the UVRAG-centrosome interaction destabilizes centrosomes, resulting in extensive aneuploidy. UVRAG is a key factor in suppressing proliferation after UVB, independent of its function in autophagy activation. For instance, a mutation of exon 8 of UVRAG reduced autophagy and promoted in colorectal and gastric cancer types (Tam et al. 2017). In response to UV or DNA alkylating agent (methyl methanesulphonate) induced DNA damage, ATR is also known to activate autophagy through ATR/Chk1/RhoB mediated lysosomal recruitment of tuberous sclerosis complex (TSC complex) and subsequent mTORC1 inhibition (Liu et al. 2018).

3.3.2 Radio-Sensitization Due to Autophagy

Recent evidence showed that autophagy regulating ATG proteins has a tumor-suppressive role because down-regulation of certain ATG proteins can promote tumorigenesis. Previous studies have confirmed that radiation-induced autophagy leads to increased radiosensitivity in BAX/BAK double knockout cells in comparison to parent cells (Kim et al. 2006). Increased radiosensitivity is due to ER stress, which is activated by unfolded protein response (UPR). Moreover, they found that PERK is essential for radiation-induced autophagy leading to increased cell death in apoptotic deficient breast cancer cells (Kim et al. 2006). Radiation-induced autophagic cell death is also mediated through the p53/DRAM signaling pathway in breast cancer cells (Cui et al. 2016) (Fig. 3.1). Various in vivo and in vitro studies have demonstrated that irradiation and rapamycin-induced autophagy lead to promote premature senescence and restrict cell proliferation in radiation-resistant glioblastoma and parotid carcinoma cells (Tam et al. 2017). In addition to this, it has been demonstrated the role of autophagy in sensitizing glioblastoma cells (SU2) by using dual PI3K/mTOR inhibitor NVP-BEZ235 (Wang et al. 2013). In similar lines, increased radiosensitivity was also observed in cisplatin-resistant NSCLC tumor cells using NVP-BEZ235 (Kim et al. 2014). Although the detailed mechanism of induced cell death is not clear yet, one recent report showed autophagy induced by ionizing radiation promotes cell death in human colorectal cancer cells in hypoxia and nutrient-depleted condition and silencing of ATG7 or Beclin1 increases the survival under oxygen and glutamine starvation (Classen et al. 2019).

3.3.3 Unfolded Protein Response (UPR) Activates Autophagy

Radiation also causes damage to protein, leading to the activation of UPR mediated ER stress. The ER membrane-associated proteins, PKR-like eIF2α kinase (PERK) and activating transcription factor-6 (ATF6) act as autophagy inducers. The PERK contributes to hypoxia tolerance by phosphorylating eIF2α and stops general protein synthesis to lessen the protein load in the ER (Liang et al. 2015). However, UPR upregulates certain transcriptions factors like NF-E2-related factor 2 (NRF2), nuclear factor κB (NF-κB), and activating transcription factor 4 (ATF4) (Tam et al. 2017). NRF2 and NF-kB contribute in cytoprotective and antiapoptotic pathways and provides radioresistance, while ATF4 allows the restoration of normal ER function through the induction of CEBP homologous protein (CHOP), DNA damage-inducible protein 34 (GADD34) and lysosome-associated membrane protein 3 (LAMP3). CHOP is the pro-apoptotic component of the UPR and mediates cell death when the cell adaptation fails to withstand the ER stress (Moretti et al. 2007).

3.4 Role of Autophagy Proteins in DNA Damage Repair

Autophagy plays an important role in DNA damage repair upon genotoxic stresses and insults (Eapen et al. 2017). Although the functional significance of autophagy in DNA damage repair and response is well known, the molecular mechanisms involved are obscure. Several reports have shown that the deficiency of autophagy results in the impairment of DNA damage response and also causes replication related complexities (Liu et al. 2015; Vanzo et al. 2020; Gillespie and Ryan 2015). Cells deficient in key autophagic proteins, for example, BECN1, ATG5, ATG7 have been shown to have impaired DNA damage response (Xu et al. 2017a). The absence of these gene products and the consequential autophagic defect has also been implicated in tumorigenesis, tumor progression, and survival (Karantza-Wadsworth et al. 2007). Recently, Liu et al. have shown that loss of autophagy causes a synthetic lethal deficiency in DNA repair. It was observed that the mouse embryonic fibroblasts deficient in ATG7 showed diminished levels of phosphorylated CHK1 upon irradiation, indicating lower levels of DNA repair by HR, leading to greater dependency on error-prone NHEJ pathway (Liu et al. 2015). SQSTM1/p62, an autophagic adapter protein, plays a pivotal role in the DNA repair process (Hewitt et al. 2016; Hewitt and Korolchuk 2017). P62 protein shuttles continuously between the nucleus and the cytoplasm (Fig. 3.2). Upon exposure to ionizing radiation, it was observed that p62 accumulates in the cell, localizes to the nucleus, and binds RNF168, a ubiquitin ligase, preventing the histone ubiquitination that signals the DNA damage, hampering overall DDR (Hewitt et al. 2016). In a similar context, upon X-ray irradiation, it was observed that the p62 protein transiently associated with the DNA damage-induced foci (DDF), accumulated in the nucleus and aids in the degradation of RAD51 and Filamin A (FLNA) (Hewitt et al. 2016; Wang et al. 2016) (Fig. 3.2). This work also showed that the HR efficiency increased with p62 depletion, thus showing an inverse correlation between p62 accumulation in the nucleus and DNA damage repair. This evidence suggests that autophagic clearance of p62 is essential for the optimal and error-free repair of DNA (Fig. 3.2).

Several autophagy-independent roles of core autophagic proteins have also been reported. Beclin1, a core component of the class III phosphatidylinositol 3-kinase (PI3K-III) that aids in the formation of the autophagosomal membrane, has been found to localize in the nucleus consequent to DNA damage and promote DNA repair directly. It was found to interact with DNA topoisomerase IIβ and get recruited at the sites of double-strand breaks due to this interaction (Xu et al. 2017a) (Fig. 3.2). In the absence of BECN1, the ability of the cells treated with ionizing radiation to repair the DNA was found to be hindered (Xu et al. 2017b).

ATG5, an important protein component of the ubiquitin-like conjugation system that leads to the formation of lipidated LC3 form—LC3-II, has been shown to be induced upon DNA damage, promoting mitotic catastrophe, independent of its role in autophagy (Maskey et al. 2013) (Fig. 3.2). In response to the treatment with DNA damaging agents like cisplatin and etoposide, it was observed that ATG5 translocated to the nucleus and induced a G2/M phase arrest (Maskey et al. 2013) (Fig. 3.2). Displacement of the chromosomal passenger protein (CPC) consequent to

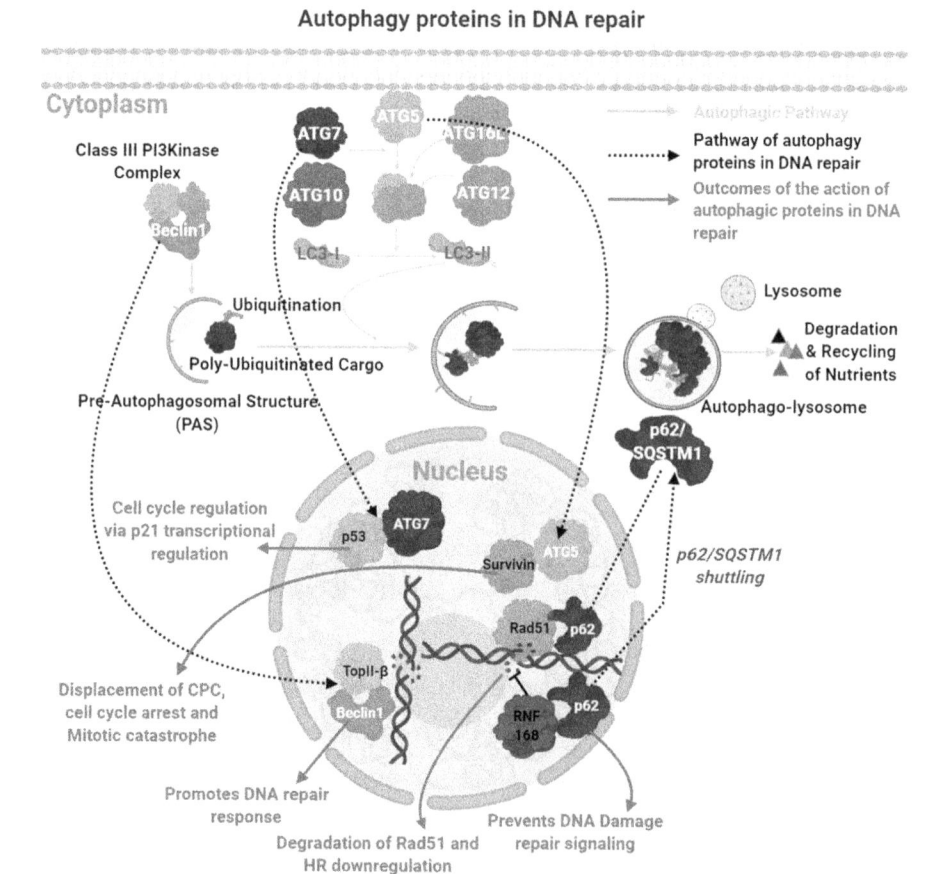

Fig. 3.2 Role of autophagy proteins in DNA repair. In response to oxidative stress, chemotherapeutic, and radiation therapy-induced DNA damage, autophagy proteins are activated and play a critical role in the DNA repair process. Above illustration was made in biorender.com

the physical interaction of ATG5 with Survivin was found to be responsible for the arrest and the ensuing mitotic catastrophe (Fig. 3.2). This activity was found to be independent of its role in autophagy and assigns two distinct functions for ATG5 based on its localization in the cell nucleus or the cytoplasm. ATG7 is another important E1 (ubiquitin-activating enzyme) like protein involved in the induction of autophagy. In one of the seminal papers, Lee et al. showed that p53 and ATG7 proteins interact with each other and aid in the arrest of cells by regulating the transcription of cell cycle inhibitor $p21^{CDKN1A}$ under starvation, in mouse embryonic fibroblasts (Lee et al. 2012) (Fig. 3.2). Withdrawal from cycling is an important response to starvation. It was also observed that the mouse embryonic fibroblasts deficient in ATG7 showed diminished levels of phosphorylated CHK1 upon irradiation, indicating lower levels of DNA repair by HR, leading to greater dependency on error-prone NHEJ pathway (Liu et al. 2015). These works highlight the importance of autophagy and its constituent proteins in the process in the DNA damage repair and response either in the composite form of autophagy or independently as proteins (Fig. 3.2).

3.5 Conclusion and Future Perspectives

DDR and autophagy are two distinct cellular functions but they are complementing each other for protecting cells by relieving DNA damage related stress or inducing cell death under higher stress conditions. Several reports advocate that all three types of autophagy (macroautophagy, microautophagy, and CMA) are being unanticipatedly linked to DDR pathways or genes. Intriguingly, autophagy-associated proteins also seem to play an unorthodox role in DDR and DNA repair. The role of autophagy in response to chemotherapy and radiotherapy is intriguingly dichotomous; leading to cell survival or death.

However, extensive work is still required to unravel the induction of autophagy at a precise molecular level in response to different DNA damages. Considering the fact that DDR and autophagy play a crucial role in cancer resistance or sensitization in response to various DNA damaging therapy, this review article raises several concerns that ought to be addressed in the future. Whether autophagy induction is differentially regulated in cancer and normal cells in response to DNA damage? What decides the induction of pro-survival or pro-death functions of autophagy? Does different DNA damages (base damage, SSBs, DSBs, etc.) induce a common or different signaling pathways to induce autophagy? What could be the precise role of autophagy induction in DNA repair or vice versa? Whether "autophagy-dependent" and/or "autophagy-independent" role of autophagy associated proteins play a crucial role in DNA repair in response to chemotherapy or radiotherapy of cancer? The focused research in this area may further foster the development of novel cancer therapeutics.

References

Abedin MJ, Wang D, McDonnell MA et al (2007) Autophagy delays apoptotic death in breast cancer cells following DNA damage. Cell Death Differ 14:500–510

Aguilar-Quesada R, Muñoz-Gámez JA, Martín-Oliva D et al (2007) Interaction between ATM and PARP-1 in response to DNA damage and sensitization of ATM deficient cells through PARP inhibition. BMC Mol Biol 8:29. https://doi.org/10.1186/1471-2199-8-29

Alexander A, Cai S-L, Kim J et al (2010) ATM signals to TSC2 in the cytoplasm to regulate mTORC1 in response to ROS. Proc Natl Acad Sci 107:4153–4158

Bartsch K, Knittler K, Borowski C et al (2017) Absence of RNase H2 triggers generation of immunogenic micronuclei removed by autophagy. Hum Mol Genet 26:3960–3972

Bellot G, Garcia-Medina R, Gounon P et al (2009) Hypoxia-induced autophagy is mediated through hypoxia-inducible factor induction of BNIP3 and BNIP3L via their BH3 domains. Mol Cell Biol 29:2570–2581. https://doi.org/10.1128/MCB.00166-09

Bhatt K, Zhou L, Mi Q-S et al (2010) MicroRNA-34a is induced via p53 during cisplatin nephrotoxicity and contributes to cell survival. Mol Med 16:409–416

Budanov AV, Karin M (2008) p53 target genes Sestrin1 and Sestrin2 connect genotoxic stress and mTOR signaling. Cell 134:451–460

Buszczak M, Kramer H (2019) Autophagy keeps the balance in tissue homeostasis. Dev Cell 49:499–500

Chaachouay H, Ohneseit P, Toulany M et al (2011) Autophagy contributes to resistance of tumor cells to ionizing radiation. Radiother Oncol 99:287–292. https://doi.org/10.1016/j.radonc.2011.06.002

Chen Y, Li X, Guo L et al (2015) Combining radiation with autophagy inhibition enhances suppression of tumor growth and angiogenesis in esophageal cancer. Mol Med Rep 12:1645–1652. https://doi.org/10.3892/mmr.2015.3623

Classen F, Kranz P, Riffkin H et al (2019) Autophagy induced by ionizing radiation promotes cell death over survival in human colorectal cancer cells. Exp Cell Res 374:29–37. https://doi.org/10.1016/j.yexcr.2018.11.004

Crighton D, Wilkinson S, O'Prey J et al (2006) DRAM, a p53-induced modulator of autophagy, is critical for apoptosis. Cell 126:121–134

Crighton D, O'Prey J, Bell HS et al (2007) p73 regulates DRAM-independent autophagy that does not contribute to programmed cell death. Cell Death Differ 14:1071–1079

Cui L, Song Z, Liang B et al (2016) Radiation induces autophagic cell death via the p53/DRAM signaling pathway in breast cancer cells. Oncol Rep 35:3639–3647. https://doi.org/10.3892/or.2016.4752

Eapen VV, Waterman DP, Bernard A et al (2017) A pathway of targeted autophagy is induced by DNA damage in budding yeast. Proc Natl Acad Sci 114:E1158–E1167

Eby KG, Rosenbluth JM, Mays DJ et al (2010) ISG20L1 is a p53 family target gene that modulates genotoxic stress-induced autophagy. Mol Cancer 9:95

Feng H, Wang J, Chen W et al (2016) Hypoxia-induced autophagy as an additional mechanism in human osteosarcoma radioresistance. J Bone Oncol 5:67–73. https://doi.org/10.1016/j.jbo.2016.03.001

Gao P, Hao F, Dong X, Qiu Y (2019) The role of autophagy and Beclin-1 in radiotherapy-induced apoptosis in thyroid carcinoma cells. Int J Clin Exp Pathol 12:885–892

Gillespie DA, Ryan KM (2015) Autophagy is critically required for DNA repair by homologous recombination. Mol Cell Oncol 3:e1030538–e1030538

Green DR, Kroemer G (2009) Cytoplasmic functions of the tumour suppressor p53. Nature 458:1127–1130

Guo H, Chitiprolu M, Gagnon D et al (2014) Autophagy supports genomic stability by degrading retrotransposon RNA. Nat Commun 5:5276

Hanahan D, Weinberg RA (2011) Hallmarks of cancer: the next generation. Cell 144:646–674

Harhaji-Trajkovic L, Vilimanovich U, Kravic-Stevovic T et al (2009) AMPK-mediated autophagy inhibits apoptosis in cisplatin-treated tumour cells. J Cell Mol Med 13:3644–3654

He W-S, Dai X-F, Jin M et al (2012) Hypoxia-induced autophagy confers resistance of breast cancer cells to ionizing radiation. Oncol Res 20:251–258

Hewitt G, Korolchuk VI (2017) Repair, reuse, recycle: the expanding role of autophagy in genome maintenance. Trends Cell Biol 27:340–351

Hewitt G, Carroll B, Sarallah R et al (2016) SQSTM1/p62 mediates crosstalk between autophagy and the UPS in DNA repair. Autophagy 12:1917–1930

Ivanov A, Pawlikowski J, Manoharan I et al (2013) Lysosome-mediated processing of chromatin in senescence. J Cell Biol 202:129–143

Jemal A, Bray F, Center MM et al (2011) Global cancer statistics. CA Cancer J Clin 61:69–90

Karantza-Wadsworth V, Patel S, Kravchuk O et al (2007) Autophagy mitigates metabolic stress and genome damage in mammary tumorigenesis. Genes Dev 21:1621–1635

Kenzelmann Broz D, Spano Mello S, Bieging K et al (2013) Global genomic profiling reveals an extensive p53-regulated autophagy program contributing to key p53 responses. Genes Dev 27:1016–1031

Kim KW, Mutter RW, Cao C et al (2006) Autophagy for cancer therapy through inhibition of pro-apoptotic proteins and mammalian target of rapamycin signaling. J Biol Chem 281:36883–36890. https://doi.org/10.1074/jbc.M607094200

Kim KW, Myers CJ, Jung DK, Lu B (2014) NVP-BEZ-235 enhances radiosensitization via blockade of the PI3K/mTOR pathway in cisplatin-resistant non-small cell lung carcinoma. Genes Cancer 5:293–302

Lee IH, Kawai Y, Fergusson MM et al (2012) Atg7 modulates p53 activity to regulate cell cycle and survival during metabolic stress. Science 336:225–228. https://doi.org/10.1126/science.121839

Li D-D, Sun T, Wu X-Q et al (2012) The inhibition of autophagy sensitises colon cancer cells with wild-type p53 but not mutant p53 to topotecan treatment. PLoS One 7:e45058

Liang DH, El-Zein R, Dave B (2015) Autophagy inhibition to increase radiosensitization in breast cancer. J Nucl Med Radiat Ther 6:254. https://doi.org/10.4172/2155-9619.1000254

Liang N, He Q, Liu X, Sun H (2019) Multifaceted roles of ATM in autophagy: from nonselective autophagy to selective autophagy. Cell Biochem Funct 37:177–184

Lin J-F, Lin Y-C, Tsai T-F et al (2017) Cisplatin induces protective autophagy through activation of BECN1 in human bladder cancer cells. Drug Des Devel Ther 11:1517–1533

Liu EY, Xu N, O'Prey J et al (2015) Loss of autophagy causes a synthetic lethal deficiency in DNA repair. Proc Natl Acad Sci 112:773–778

Liu M, Zeng T, Zhang X et al (2018) ATR/Chk1 signaling induces autophagy through sumoylated RhoB-mediated lysosomal translocation of TSC2 after DNA damage. Nat Commun 9:4139. https://doi.org/10.1038/s41467-018-06556-9

Luqmani YA (2005) Mechanisms of drug resistance in cancer chemotherapy. Med Princ Pract 14:35–48

Maskey D, Yousefi S, Schmid I et al (2013) ATG5 is induced by DNA-damaging agents and promotes mitotic catastrophe independent of autophagy. Nat Commun 4:2130

Mizushima N, Komatsu M (2011) Autophagy: renovation of cells and tissues. Cell 147:728–741

Moretti L, Cha YI, Niermann KJ, Lu B (2007) Switch between apoptosis and autophagy: radiation-induced endoplasmic reticulum stress? Cell Cycle 6:793–798. https://doi.org/10.4161/cc.6.7.4036

Nagata M, Arakawa S, Yamaguchi H et al (2018) Dram1 regulates DNA damage-induced alternative autophagy. Cell Stress 2:55–65

Nazio F, Maiani E, Cecconi F (2017) The cross talk among autophagy, ubiquitination, and DNA repair: an overview. In: Ubiquitination governing DNA repair—implications in health and disease. IntechOpen, London. https://doi.org/10.5772/intechopen.71404

Nilsson JR (1988) Cytotoxic effects of cisplatin, cis-dichlorodiammineplatinum(II), on Tetrahymena. J Cell Sci 90:707–716

Nishida Y, Arakawa S, Fujitani K et al (2009) Discovery of Atg5/Atg7-independent alternative macroautophagy. Nature 461:654–658

Nyfeler C, Eng H (2016) Revisiting autophagy addiction of tumor cells. Autophagy 12:1206–1207

Periyasamy-Thandavan S, Jiang M, Wei Q et al (2008) Autophagy is cytoprotective during cisplatin injury of renal proximal tubular cells. Kidney Int 74:631–640

Prasad CB, Prasad SB, Yadav SS et al (2017) Olaparib modulates DNA repair efficiency, sensitizes cervical cancer cells to cisplatin and exhibits anti-metastatic property. Sci Rep 7:12876

Rocha CRR, Silva MM, Quinet A et al (2018) DNA repair pathways and cisplatin resistance: an intimate relationship. Clinics (Sao Paulo) 73:e478s

Rodríguez-Vargas JM, Ruiz-Magaña MJ, Ruiz-Ruiz C et al (2012) ROS-induced DNA damage and PARP-1 are required for optimal induction of starvation-induced autophagy. Cell Res 22:1181–1198

Rodríguez-Vargas JM, Rodríguez MI, Majuelos-Melguizo J et al (2016) Autophagy requires poly (adp-ribosyl)ation-dependent AMPK nuclear export. Cell Death Differ 23:2007–2018

Rosenbluth JM, Pietenpol JA (2009) mTOR regulates autophagy-associated genes downstream of p73. Autophagy 5:114–116

Schaaf MBE, Jutten B, Keulers TG et al (2015) Canonical autophagy does not contribute to cellular radioresistance. Radiother Oncol 114:406–412. https://doi.org/10.1016/j.radonc.2015.02.019

Schaaf L, Schwab M, Ulmer C et al (2016) Hyperthermia synergizes with chemotherapy by inhibiting PARP1-dependent DNA replication arrest. Cancer Res 76:2868–2875

Shimizu S, Kanaseki T, Mizushima N et al (2004) Role of Bcl-2 family proteins in a non-apoptotic programmed cell death dependent on autophagy genes. Nat Cell Biol 6:1221–1228

Su M, Mei Y, Sinha S (2013) Role of the crosstalk between autophagy and apoptosis in cancer. J Oncol 2013:102735. https://www.hindawi.com/journals/jo/2013/102735/. Accessed 22 Nov 2019

Tam SY, Wu VWC, Law HKW (2017) Influence of autophagy on the efficacy of radiotherapy. Radiat Oncol 12:57. https://doi.org/10.1186/s13014-017-0795-y

Torii S, Yoshida T, Arakawa S et al (2016) Identification of PPM1D as an essential Ulk1 phosphatase for genotoxic stress-induced autophagy. EMBO Rep 17:1552–1564

Vanzo R, Bartkova J, Merchut-Maya JM et al (2020) Autophagy role(s) in response to oncogenes and DNA replication stress. Cell Death Differ 27:1134. https://doi.org/10.1038/s41418-019-0403-9

Wang W, Long L, Yang N et al (2013) NVP-BEZ235, a novel dual PI3K/mTOR inhibitor, enhances the radiosensitivity of human glioma stem cells in vitro. Acta Pharmacol Sin 34:681–690. https://doi.org/10.1038/aps.2013.22

Wang Y, Zhang N, Zhang L et al (2016) Autophagy regulates chromatin ubiquitination in DNA damage response through elimination of SQSTM1/p62. Mol Cell 63:34–48

Wang F, Tang J, Li P et al (2018) Chloroquine enhances the radiosensitivity of bladder cancer cells by inhibiting autophagy and activating apoptosis. CPB 45:54–66. https://doi.org/10.1159/000486222

Xie B-S, Zhao H-C, Yao S-K et al (2011) Autophagy inhibition enhances etoposide-induced cell death in human hepatoma G2 cells. Int J Mol Med 27:599–606

Xu F, Fang Y, Yan L et al (2017a) Nuclear localization of Beclin 1 promotes radiation-induced DNA damage repair independent of autophagy. Sci Rep 7:45385

Xu F, Li X, Yan L et al (2017b) Autophagy promotes the repair of radiation-induced DNA damage in bone marrow hematopoietic cells via enhanced STAT3 signaling. Radiat Res 187:382–396. https://doi.org/10.1667/RR14640.1

Yang C, Kaushal V, Shah SV, Kaushal GP (2008) Autophagy is associated with apoptosis in cisplatin injury to renal tubular epithelial cells. Am J Physiol Renal Physiol 294:F777–F787

Yang SH, Lee KK, Moon SR (2012) Autophagy induction by low dose cisplatin; the role of p53 in autophagy. Eur Respir J 40:1246

Yao Q, Zheng R, Xie G et al (2015) Late-responding normal tissue cells benefit from high-precision radiotherapy with prolonged fraction delivery times via enhanced autophagy. Sci Rep 5:9119. https://doi.org/10.1038/srep09119

Zalckvar E, Berissi H, Mizrachy L et al (2009) DAP-kinase-mediated phosphorylation on the BH3 domain of beclin 1 promotes dissociation of beclin 1 from Bcl-XL and induction of autophagy. EMBO Rep 10:285–292

Zhang H, Bosch-Marce M, Shimoda LA et al (2008) Mitochondrial autophagy is an HIF-1-dependent adaptive metabolic response to hypoxia. J Biol Chem 283:10892–10903

Zhao Z, Oh S, Li D et al (2012) A dual role for UVRAG in maintaining chromosomal stability independent of autophagy. Dev Cell 22:1001–1016. https://doi.org/10.1016/j.devcel.2011.12.027

Zois CE, Koukourakis MI (2009) Radiation-induced autophagy in normal and cancer cells: towards novel cytoprotection and radio-sensitization policies? Autophagy 5:442–450. https://doi.org/10.4161/auto.5.4.7667

miRNAs and Its Regulatory Role on Autophagy in Tumor Microenvironment

Assirbad Behura, Abtar Mishra, Ashish Kumar, Lincoln Naik, Debraj Manna, and Rohan Dhiman

Abstract

Autophagy is an intracellular catabolic process that helps in maintaining cellular homeostasis. Generally, it is involved in the recycling of unwanted proteins and damaged organelles but upon cellular stress, it helps in the survival of the cells. It is a tightly regulated process and any discrepancy in its regulation leads to the generation of many pathological abnormalities. During the early phase of cancer, it functions as a tumor suppressor whereas, at later stages, it facilitates tumor growth and helps in generating resistance to cancerous cells. Due to this functional switch of the pathway, many studies have been undertaken to find the mechanism behind its regulation in different cancer types and microRNAs (miRNAs) have been recently explored to be one of the regulatory factors. miRNAs are short non-coding RNAs that regulate the gene expression of most protein-coding genes post-transcriptionally. They control many important biological pathways including autophagic response in cancer. Their expression also gets dysregulated during different stages of cancer and thus gives a promising window of their utility as an attractive target during tumor therapy. Therefore, considering the potential of autophagy regulating miRNAs as future drug targets, this review is focused on recent advances in linking miRNAs to the regulation of autophagy pathway and their role in cancer and their implications in cancer treatment.

A. Behura · A. Mishra · A. Kumar · L. Naik · R. Dhiman (✉)
Laboratory of Mycobacterial Immunology, Department of Life Science, National Institute of Technology, Rourkela, Odisha, India
e-mail: dhimanr@nitrkl.ac.in

D. Manna
Laboratory of Mycobacterial Immunology, Department of Life Science, National Institute of Technology, Rourkela, Odisha, India

Department of Biochemistry, Indian Institute of Science, Bengaluru, Karnataka, India

© Springer Nature Singapore Pte Ltd. 2020
S. K. Bhutia (ed.), *Autophagy in Tumor and Tumor Microenvironment*,
https://doi.org/10.1007/978-981-15-6930-2_4

Keywords

miRNA · Autophagy · Cancer · Epigenetics · Tumor microenvironment

Abbreviations

ATG	Autophagy-related
CLL	Chronic lymphocyte leukemia
CoL10A1	Collagen α-1(X) chain
ELAVL1	Embryonic lethal abnormal vision-like protein-1
EOC	Epithelial ovarian cancer
FIP 2000	Focal adhesion kinase
hnRNP A1	Heterogenous unclear ribonucleoprotein A1
LAMP1	Lysosomal-associated membrane protein 1
LAMP2	Lysosomal-associated membrane protein 2
miRNA	microRNA
mTOR	Mammalian target for rapamycin
NSCLC	Non-small cell lung cancer
PE	Phosphatidyl ethanolamine
PIK3C3	Phosphatidyl inositol 3 kinase catalytic subunit type 3
PKM1	Pyruvate kinase muscle isoform 1
PKM2	Pyruvate kinase muscle isoform 2
PTB1	Polypyrimidine tsat binding protein 1
RISC	RNA-induced silencing complex
RLC	RISC-loading complex
ROS	Reactive oxygen species
TBCC	Tumor-binding cofactor C
TIGAR	TP53 inducible glycolysis and apoptosis regulator
TRPM3	Transient receptor potential melastatin 3
ULK	Unc 51 like kinase
UTR	Untranslated region
UVRAG	UV radiation resistance-associated gene protein
VHL	Van hippel lindeu

4.1 Introduction

MicroRNAs (miRNAs) are small noncoding RNA molecules of 18–25 nucleotides that have a crucial role in gene regulation at the post-transcriptional level by controlling the stability and translation of mRNAs. They are produced as primary miRNAs (pri-miRNAs) and are subsequently processed to generate mature miRNAs through precursor miRNAs (pre-miRNAs). The mature miRNAs mainly interact with the 3′ untranslated region (UTR) of the target gene to regulate their expression (Ha and Kim 2014). But their interaction is not limited to 3′ UTR only, as many reports have suggested that they can bind to either 5′ UTR or different locations like promotor or gene coding regions as well (Broughton et al. 2016; O'Brien et al. 2018). miRNAs regulate many key biological processes like cell differentiation, growth, autophagy, migration, apoptosis, and so on, and are tightly regulated because their abnormal expression has been shown to be responsible in the development of many diseases (Fu et al. 2013; Paul et al. 2018; Tüfekci et al. 2014). They are secreted out of the cells to aid in signaling between cells and act as biomarkers for various diseases including cancer (Hayes et al. 2014; Huang 2017; Wang et al. 2016b). Upregulation or downregulation of specific miRNAs is reported in all cancer cell types like colon cancer, leukemia, breast cancer, lung cancer, and so on. (O'Brien et al. 2018). Dysregulation of miRNA biogenesis or expression is reported in various stages of cancer progression and regulates resistance to anti-cancer drugs (O'Brien et al. 2018).

Autophagy is a highly conserved cellular process involved in the recycling and digestion of damaged organelles, misfolded proteins, and intracellular pathogens by lysosomal degradation to maintain cell survival (He and Klionsky 2009; Mizushima et al. 2008). It is a continuous process undergoing in the cell at the basal level under normal conditions to maintain cellular homeostasis, but under stress conditions like starvation, infection, hypoxia, and so on, it gets upregulated (Mizushima et al. 2008). During stress conditions, autophagy plays a protective role by degrading damaged and unwanted cellular contents and recycling proteins to generate energy and free amino acids but hyperactivation of autophagy leads to death of the cell under stress, known as autophagic cell death (Mizushima et al. 2008). Being a key process, it is tightly regulated but abnormalities in the pathway arise and it leads to the development of many health issues including cancer (Frankel and Lund 2012; Jing et al. 2015). Many reports have been published showing the role of miRNA in autophagy regulation and cancer development (Gozuacik et al. 2017). In this review, we will briefly summarize the emerging connection between different miRNAs and autophagy pathways and how this regulation decides the fate of cancer cells.

4.2 miRNA Biogenesis

miRNAs play an important role in various physiological processes including cellular proliferation, differentiation, maturation, host–pathogen interaction, and many more (Demirci et al. 2016; Saçar et al. 2014). miRNAs are pervasive in the genome and originate from both coding genes as well as noncoding regions as primary transcripts by the cellular machinery (Grund and Diederichs 2010; Kim et al. 2009). The biogenesis of miRNA initiates in the nucleus followed by its transport into the cytoplasm where miRNA processing takes place to generate mature miRNA. The majority of miRNA genes are transcribed by RNA polymerase II or III to form long primary transcripts called pri-miRNA that contain hairpin (stem-loop) structure with some bulges formed due to base–pair mismatch (Krishnan and Damaraju 2018). The pri-miRNA possesses a 5′ 7-methylguanosine cap and a poly-A tail at the 3′ end and are in turn cleaved by a cellular RNAase Class II endonuclease III enzymes called

Drosha (Gregory and Shiekhattar 2005). Drosha along with its cofactor DGC28/Pasha forms a microprocessor complex that specifically recognizes and cleaves the stem of pri-miRNA to liberate nearly 70–120 nucleotides shorter hairpin structure called pre-miRNA (Seitz and Zamore 2006; Shomron and Levy 2009). Following the formation of pre-miRNAs, they are transported to cytoplasm using exportin-5 (XPO5) in the presence of guanosine 5′ triphosphate bound Ras-related nuclear protein (RanGTP). XPO5 is a member from the karyopherin family and is mostly engaged in nuclear transport of structured RNAs including tRNAs, human Y1 RNA, and adenovirus VA1 RNA that possess 3′ overhang structure. Attachment of XPO5 to its cargo needs a minimum of 16 bp and a short 3′ overhang. Therefore, pre-miRNAs are properly processed by Drosha in order to be recognized and exported from the nucleus (Okada et al. 2009). Apart from its role in nucleocytoplasmic transport, Exportin-5 also stabilizes the pre-miRNA as well as prevents its degradation (Yi et al. 2003; Zeng and Cullen 2004). Once inside the cytoplasm, further processing of pre-miRNA takes place with the help of a double-stranded RNAase III enzyme Dicer. Dicer along with trans activation response RNA binding protein (TRBP) binds to 5′ phosphate and 3′ overhang at the base of stem loop of pre-miRNA and cuts both strands of the duplex at about two helical turns away from the base of stem loop. This cleavage by Dicer releases a double-stranded miRNA of ~21–24 nucleotides length called mature miRNA (Bartel 2004). One strand called the passenger strand in the newly generated double-stranded RNA undergoes degradation whereas the other strand known as guide RNA or mature RNA is loaded onto an Argonaute containing RNA induced silencing complex (RISC) by the help of RISC loading complex (RLC) that directs gene silencing. Selection of guide strand mostly depends on the thermodynamic stability and strand with less stability is selected by RISC (Khvorova et al. 2003). RISC is composed of mature miRNA, Dicer, TRBP, Argonaute protein 2 (AGO2), and protein kinase R activator (PACT) that possesses a regulatory role in both nucleus as well as cytoplasm. In the cytoplasm, the RISC complex targets the 3′UTR region of mRNA thereby results in translational repression (MacRae et al. 2008; Park and Shin 2014). Further, some of the miRNAs having nuclear localization signal are imported back to the nucleus. Mature miRNA along with Ago2 returns to the nucleus using a member of the karyopherin beta family protein called Importin-8 (Hwang et al. 2007; Wei et al. 2014). miRNA inside nucleus exhibits regulatory functions by targeting gene promoter region with the help of Argonaute proteins as well as through recruitment of epigenetic modifier proteins such as chromobox protein homolog 3 (CBX3), transcriptional intermediary factor 1beta (TIF1β), suppressor of variegation 3–9 homolog 1 (SUV39H1), euchromatic histone lysine methyltransferase 2 (EHMT2) that results in transcriptional gene silencing or gene activation (Kim et al. 2008; Liang et al. 2013; Salmanidis et al. 2014; Winter et al. 2009, Fig. 4.1).

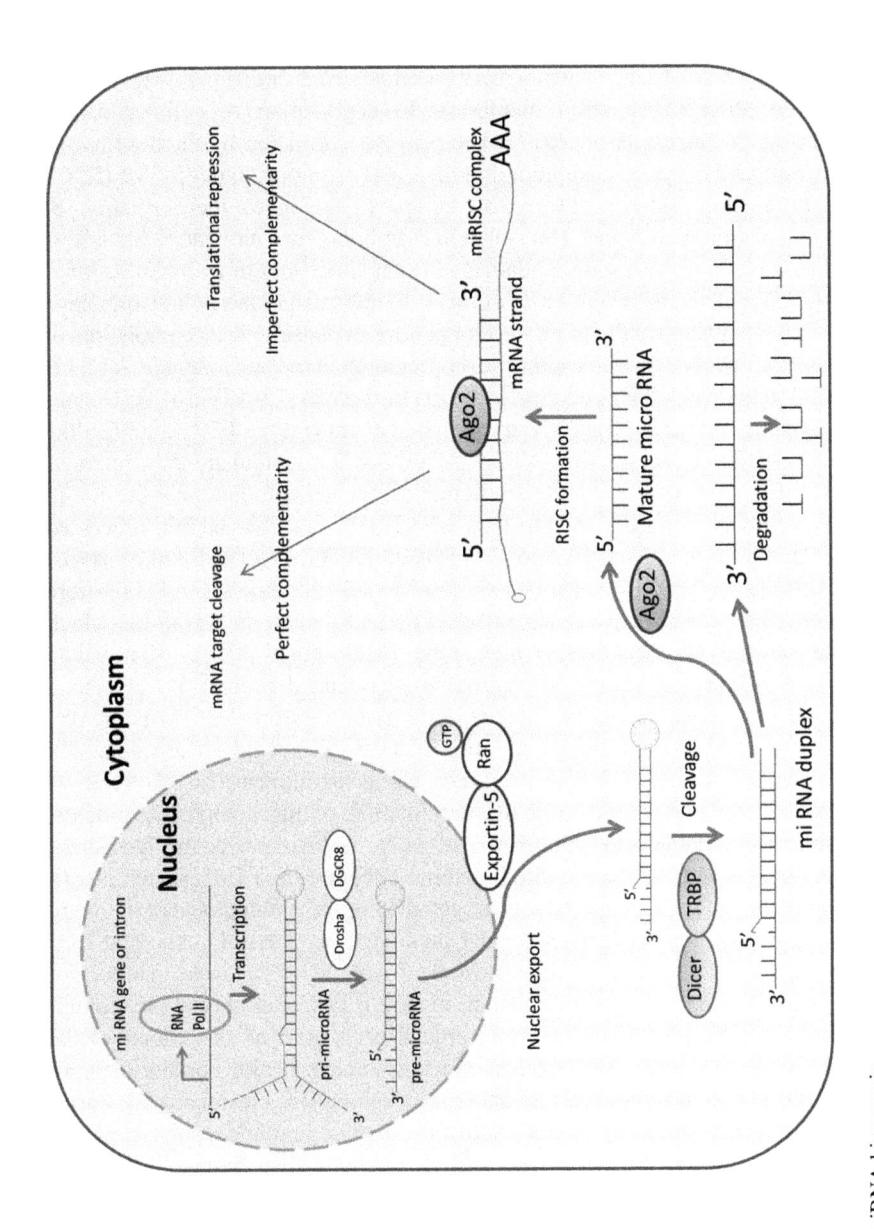

Fig. 4.1 miRNA biogenesis

4.3 Autophagy

It is a lysosome mediated catabolic process that mainly occurs in response to nutrient starvation and stress conditions to ensure cell survival (Schneider and Cuervo 2014). This process is highly complex and fundamental in eukaryotes involving 20 dedicated autophagy-related (*ATG*) genes that coordinate the entire pathway starting from the formation of isolation membrane to degradation of cellular cargos (Mizushima 2019; Suzuki et al. 2017). The process is initiated by uncoordinated-51 like kinase (ULK, mammalian homolog of ATG1) complex and Vps34/PIK3C3 phosphatidylinositol 3-kinase (PtdIns3K) complex leading to the formation of cup-shaped isolation membrane. The isolation membrane then elongates, sequesters cytoplasmic targets, cellular cargos, damaged organelles, protein aggregates, long-lived proteins, and finally encloses to form a double membrane-bound organelle called an autophagosome. Phagophore elongation, expansion, and completion of autophagosome involve two ubiquitin-like conjugation complexes namely ATG12-ATG5-ATG16 and ATG8 conjugation systems. In the ATG12-ATG5-ATG16 conjugation system, ATG12 is catalyzed by E1-like enzyme, ATG7, and E2-like enzyme ATG10 to conjugate with ATG5. ATG12-ATG5 conjugate then interacts with ATG16L1 and forms a conglomerate that associates with autophagosome. In the ATG8 conjugation system, the pro-form of ATG8 is first cleaved into processed form by ATG4 leading to its activation by ATG7 and ATG12 (act as E3 like enzyme with ATG5). Active ATG8 is then transferred to E2-like enzyme ATG3 before conjugation with phosphatidylethanolamine (PE) and named as ATG8-PE which is present on the autophagosome membrane (Mizushima 2019). Mammalian homologs of ATG8 are known as microtubule-associated protein 1 light chain 3 (LC3) and gamma-aminobutyric acid receptor-associated protein (GABARAP). Its unlipidated (ATG8-I) or lipidated forms (ATG8-PE, ATG8-II) are generally referred to as LC3-I and LC3-II respectively. Upon recruitment, LC3-II stays on the autophagosome membrane until the culmination of the autophagic process. Completion of autophagosome biogenesis leads to its fusion with the lysosomes resulting in degradation of the engulfed material (Reggiori and Ungermann 2017). Cytoplasmic cargos are recognized and targeted to nascent autophagosome membrane by an interaction between molecular tags (such as polyubiquitin) and LC3 through adaptor proteins such as sequestosome 1 (p62/SQSTM1) and neighbor of BRCA1 gene 1 (NBR1) formation (Songane et al. 2012). Under normal physiological conditions, autophagy plays different crucial roles such as restoration of the amino acid pool and cellular ATP levels during nutrient deprivation condition, tumor growth inhibition, anti-aging, pre-implantation development, clearance of intracellular microbes and modulation of the innate and adaptive immune response, and so on. (Cecconi and Levine 2008; Deretic and Levine 2009; Mizushima and Komatsu 2011). Moreover, defects in the autophagic machinery have also been reported to be associated with numerous disease conditions including neurodegeneration, cancer, cardiovascular disorders, and infectious or inflammatory conditions (Choi et al. 2004). Because of the earlier multi-dimensional role, autophagy has been exploited in the past few years as a front-runner of host-directed therapy to get rid of different pathophysiological conditions including cancer.

4.4 Interplay Between miRNA and Autophagy (Fig. 4.2)

4.4.1 Autophagy Induction

Autophagy induction commences with activation of the ULK complex, consisting of the focal adhesion kinase family interacting protein of 200 kDa (FIP200), ULK1/ULK2, ATG101, and ATG13. ULK1 protein kinase is crucial to initiate autophagy whereas mTOR complex presents upstream acts as a suppressor of autophagy. Upon nutrient abundance, mTOR associates and dephosphorylates ATG13 and ULK1 leading to inhibition of ULK1 kinase activity and autophagy. But under starvation, mTOR dissociates from ULK1 that leads to its phosphorylation and subsequent activation of autophagy. A number of miRNAs directly or indirectly target the mTOR protein complex or many other proteins in the pathway. In hepatocellular carcinoma cells, miR-7 precisely targets mTOR and P70S6K (Fang et al. 2012). miR-199a and miR-101 are reported to target mTOR in different cancer cell types (Chen et al. 2012a; Fornari et al. 2010; Wang et al. 2013; Wu et al. 2013). ULK2 is a direct target of miR-885-3p to inhibit autophagy. It is reported that in squamous cell carcinoma, miR-885-3p gets upregulated upon cisplatin treatment. Aberrant expression of this miRNA leads to cell death and its suppression reverses the cisplatin-mediated reduction in cell viability (Huang et al. 2011). In prostate cancer cells, miR-26b also targets ULK2 to inhibit autophagy (Clotaire et al. 2016).

In melanoma cells, the miR-290-295 cluster targets ULK1 and ATG7, leading to suppression of glucose starvation mediated autophagic death. In C2C12 myoblast cells, miR-106b and miR-20a have shown to target and suppress ULK1 expression that upregulates the transcription factor c-Myc to inhibit leucine deprivation mediated autophagy (Wu et al. 2012). Transfecting cells with miR-106b and miR-20a inhibitors were found to restore the leucine deprivation mediated autophagy (Wu et al. 2012). Another study has found that miR-595 and miR-4487 target ULK1 to curb autophagy in neuroblastoma cells. In MCF7 cells, ULK1 is the direct target of miR-25 to inhibit autophagy (Wang et al. 2014). In multi drug-resistant MCF7 cells, isoliquiritigenin induces cell death, and autophagy by suppressing the expression of miR-25 and activating ULK1 (Wang et al. 2014).

4.4.2 Vesicle Nucleation

Vesicle nucleation involves recruiting proteins and lipids to form the autophagosome membrane. It starts with the activation of class III phosphatidylinositol 3-kinase-Beclin1 (class III PI3K-BECN1) complex. Human vacuolar protein sorting 34 (hVPS34), bax-interacting factor 1 (BIF-1), UV radiation resistance-associated gene (UVRAG), ATG14L, and RUN domain and cysteine-rich

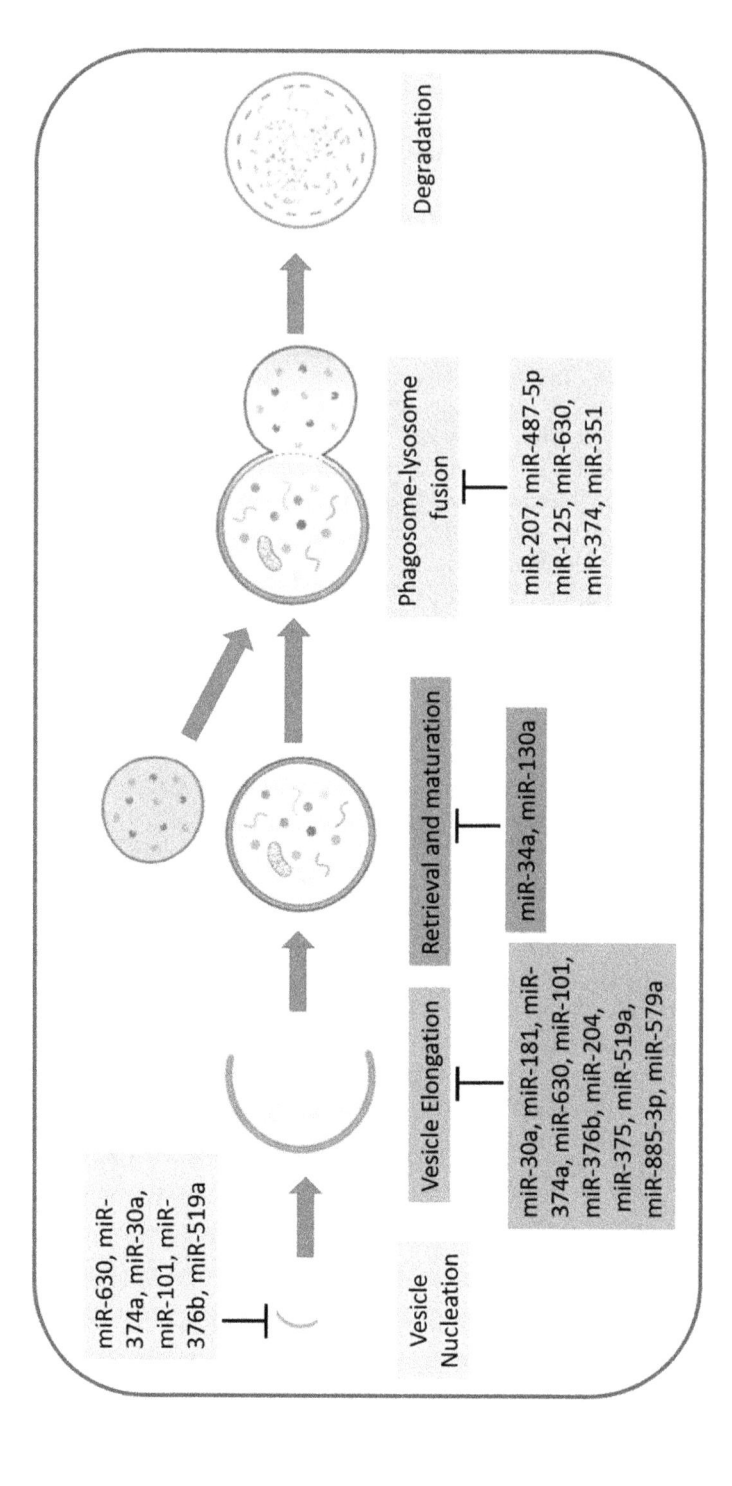

Fig. 4.2 Overview of autophagy regulation by different miRNAs (created with BioRender.com)

domain-containing, Beclin1-interacting *protein* (Rubicon) are considered as the main binding partners of the complex. Zhu et al. report for the first time, the role of miRNAs in regulating autophagy at this step (Zhu et al. 2009). They had shown that miR-30a directly targets BECN1 to inhibit autophagy. In this study, chronic myeloid leukemia cells were treated with imatinib or taxol and drug sensitivity amongst the cells was observed by miR-30a through BECN1 regulation (Zhu et al. 2009). In MCF-7 cells, the upregulation of miR-30a led to a reduction in Rapamycin mediated autophagy (Zhu et al. 2009; Zou et al. 2012). It is also reported to attenuate cisplatin-induced autophagy in Hela cells and sensitized them to chemotherapy (Zou et al. 2012). Recently it has been shown that chemoresistant osteosarcoma cells show decreased levels of miR-30a indicating their potential to inhibit autophagy (Xu et al. 2016). The role of miR-30a as an autophagy inducer or inhibitor is debatable and our group checked the effect of miR-30a-3p and -5p on autophagy and found that miR-30a-5p promotes autophagy whereas miR-30a-3p inhibits autophagy in human monocytic leukemic cell-line (THP-1 cells) upon infection with *Mycobacterium tuberculosis* (Behura et al. 2019).

In colon cancer cells, miR-409-3p is shown to block oxaliplatin mediated autophagy by targeting BECN1 and increases the sensitization of the cancerous cells to chemotherapy (Tan et al. 2016). miR-376a and miR-376b are also reported to target 3'UTR of BECN1 and ATG4C to inhibit autophagy activated by Rapamycin in lung and breast cancer cells (Korkmaz et al. 2012, 2013). This led to the proposal of "gas and brake model" stating that the autophagy activating stress signals can subsequently upregulate expression of various miRNAs that inhibit autophagy and these inhibitory miRNAs limit the hyperactivation of autophagy and ensure survival during prolonged stress conditions (Korkmaz et al. 2013; Tekirdag et al. 2016). Huang et al. have shown that miR-519a gets downregulated upon cisplatin treatment and overexpression of miR-519a blocks Cisplatin mediated autophagy by targeting BECN1 at its 3'UTR in squamous cell carcinoma cells (Huang et al. 2011). In breast cancer cells, irradiation mediated autophagy is reported to be blocked by miR-199-5p by targeting BECN1 and DRAM1 (Yi et al. 2013). In pancreatic cancer cells, miR-216a is shown to block irradiation mediated autophagy by targeting BECN1 (Zhang et al. 2015b). miR-384-5p is also reported to inhibit BECN1 expression to reduce autophagy in macrophages during atherosclerosis (Wang et al. 2016a). Huang et al. found that miR-374a and miR-630 can regulate levels of UVRAG, a BECN1 binding factor to inhibit autophagy (Huang et al. 2012). Another binding factor of class III PI3K-Beclin1 complex is identified as the direct target of miR-195 (Shi et al. 2013). miR-101 is reported to inhibit basal level autophagy and rapamycin-induced autophagy by directly targeting ras-related in brain 5A (RAB5A) protein (Frankel et al. 2011). RAB5A is a small GTPase molecule that induces the formation of autophagosome upon interaction with hVPS34 and BECN1 (Ravikumar et al. 2008). In human dermal fibroblasts, miR-23a inhibits activating molecule in BECN1-regulated autophagy *protein* 1 (AMBRA1) expression, a regulator of BECN1 and VPS34 complex to inhibit autophagy upon UV-B irradiation whereas transfection with miR-23a inhibitors restored the autophagy levels (Zhang et al. 2016).

4.4.3 Regulation of Elongation

Vesicle expansion involves two unique ubiquitin-like conjugation systems. In the initial reaction, ATG12 conjugates with ATG5 aided by ATG10 and ATG7 followed by the formation of a large multimeric conglomerate upon the interaction of ATG16L with ATG12-ATG5 complex. In the second reaction, LC3 lipidation begins with the conjugation of PE. This involves ATG4 mediated cleavage of LC3 at its C-terminal end to get cytosolic LC3-I. LC3-I further gets conjugated to PE to form LC3-II aided by ATG7 and E2 like enzyme ATG3 and E3 like enzyme ATG5-ATG12 and ATG16L complex.

In breast and liver cancer cells, miR-101, miR-376a, and miR-376b are shown to suppress autophagy by targeting the homologs ATG4D and ATG4C (Frankel et al. 2011; Korkmaz et al. 2012). ATG4 family of proteases are not only involved in the cleavage of LC3 but are also involved in the autophagosome closure, the fusion of lysosome with autophagosome, and LC3 recycling, thus they act as a crucial regulatory component of autophagy (Fujita et al. 2008; Kaminskyy and Zhivotovsky 2012). RAB5A is also involved in ATG5-ATG12 conjugation which has been shown to be inhibited by miR-101 (Frankel et al. 2011; Ravikumar et al. 2008). ATG5-ATG12 conjugation is also regulated by miR-630, miR-30a, miR-374a, and miR-181a (Huang et al. 2012; Yu et al. 2012).

Another study has shown that miR-204 can also inhibit autophagy in cardiomyocytes and renal clear cell carcinoma by regulating the conversion of LC3B by targeting its 3'UTR. miR-204 acts as a tumor suppressor gene and is generally downregulated in renal clear cell carcinoma (Mikhaylova et al. 2012; Xiao et al. 2011). In cervical cancer cells, miR-211 was found to downregulate LC3-I to LC3-II conversion (Liu et al. 2020). ATG5 is also reported to be a direct target of miR-224-3p, miR-374a, miR-181a, and miR-30a (Guo et al. 2015; Huang et al. 2012; Yu et al. 2012). Furthermore, ATG12 is shown to be suppressed by miR-30d, miR-630, and miR-200b (Huang et al. 2012; Yang et al. 2013). Huang et al. have also shown that miR-885-3p could affect the levels of ATG16L1 (Huang et al. 2011). miR 519a is suggested to inhibit the expression of both ATG16 and ATG10 (Huang et al. 2012). Many studies in the literature have shown ATG7 as the most targeted gene to inhibit autophagy by miRNAs. In hepatocellular carcinoma cells, miR-375 is shown to inhibit LC3-I to LC3-II conversion and regulate the expression of ATG7 inhibiting hypoxia-mediated autophagy thus protecting the cells against hypoxic stress and exerting tumor-suppressive activity (Chang et al. 2012). In a separate study, miR-199a-5p is reported to modulate autophagy by regulating ATG7 in hepatocellular carcinoma cells during hypoxia thus reducing their viability (Xu et al. 2012). Regulatory property of miR-20a on autophagy has been explained through ATG7 suppression leading to a further reduction in the levels of ATG16L1 (Sun et al. 2015a). In glioblastoma cells, the negative effect of miR-17 and miR-137 on starvation-induced autophagy was observed to be due to a decrease in ATG7 expression (Comincini et al. 2013; Zeng et al. 2015). The role of miR-96 in regulating autophagy during hypoxia stress in prostate cancer cells was found because of aberrant ATG7 levels (Ma et al. 2014).

4.4.4 Regulation of Retrieval

The process of retrieval is governed by ATG9 complex, a multi-spanning transmembrane protein. During retrieval, various membrane proteins and lipids are recruited to the growing autophagosome membrane (Frankel and Lund 2012). ATG9 is involved in the successful trafficking of endosomes containing lipids and proteins from the trans-Golgi network to autophagosome. ATG2A and ATG2B are involved in the closure of autophagosome vesicle (Velikkakath et al. 2012). miR-130a is reported to directly target ATG2B, inhibiting autophagy, and cell viability in human chronic lymphocytic leukemia (CLL) cells. miR-130a interferes with the retrieval of proteins and lipids to the growing phagophore membrane and inhibits the formation of ATG9-ATG2-ATG18 complex, leading to inefficient closure (Kovaleva et al. 2012). A tumor suppressor gene, miR-34a acts as an autophagic flux inhibitor by regulating ATG9A levels in mammalian cells (Kovaleva et al. 2012).

4.4.5 Regulation of Fusion

The fusion of the outer membrane of autophagosome and lysosome, leading to the formation of autophagolysosome is the final step of autophagy. This process is governed by a variety of RAB proteins. RAB7 along with LAMP1 and LAMP2 are the main players of the fusion process. In the prostate cancer cell line, RAB27A and LAMP3 are reported to be the targets of miR-205 (Pennati et al. 2014). miR-207 and miR-487-5p are shown to directly target LAMP2 (Bao et al. 2016; Tao et al. 2015). Bioinformatics analysis has shortlisted a series of miRNAs having potential involvement in the fusion step. This involves miR-142, miR-204, miR-98, miR-124, and miR-130 (Jegga et al. 2011). The predicted targets of these miRNAs are v-SNARE protein, LAMP1, and LAMP2. miR-351, miR-125, miR-630, and miR-374 are reported to regulate the levels of UVRAG (Fader et al. 2009). UVRAG is involved in the regulation of membrane curvature and endosomal trafficking leading to the maturation of autophagosomes through its interaction with BECN1 (Fader et al. 2009).

4.4.6 Regulation of miRNA Biogenesis Pathway by Autophagy

The levels of miRNAs in cancer cells are generally low due to the downregulation of major miRNA processing enzymes like Drosha and Dicer. These enzymes along with AGO2 are reported to be directly targeted for autophagosomal degradation. Inhibition of Dicer through the siRNA approach decreases LC3I and LC3II levels regardless of the presence or absence of Bafilomycin A1 (Kovaleva et al. 2012). Gibbings et al. have found that Dicer and AGO2 associate with the autophagy receptor NDP52 leading to their degradation (Gibbings et al. 2012). The autophagy-deficient cells show an increase in inactive Dicer-AGO2 complex and decreased ability of AGO2 to bind to miRNAs leading to a decrease in miRNA levels (Gibbings et al. 2012; Wampfler et al. 2015). They hypothesized that reduction in autophagy leads to an accumulation of inactive Dicer-AGO2 complexes and these inactive complexes can suppress the activity of active Dicer–AGO2 complexes. So, the degradation of the inactive Dicer-AGO2 complex is necessary for the proper functioning of Dicer–AGO2 complexes (Gibbings et al. 2012).

4.5 Role of Autophagy Regulating miRNAs and Cancer

Many autophagy regulating miRNAs are shown to regulate tumor growth, metabolism, migration, hypoxia, and response to drugs and radiotherapy. Some of the miRNAs modulating autophagy are being used as cancer biomarkers or anti-cancer agents due to their efficacy in regulating gene expression. Many groups have suggested the central role of miRNA in deciphering the outcome of cancer.

4.5.1 Cancer Cell Survival and Their Growth

Autophagy plays a major role in the growth of tumor cells and thus an important function of miRNAs is to regulate autophagy in cancer cells and control their growth and proliferation. In H1299, non-small lung cancer cells, overexpression of miR-143 by using mimics reduced their proliferation significantly by directly targeting ATG2b leading to autophagy inhibition (Wei et al. 2015). In medullary thyroid cancer cell lines, upregulation of miR-9-3p arrested cells at the G2 phase of cell cycle leading to inhibition of autophagy by decreasing the expression of major autophagy-related proteins like mTOR, ATG5, LAMP-1, and PIK3C3 which subsequently causes cell death (Gundara et al. 2015). Zhai et al. have shown that miR-502 inhibits autophagy by targeting RAB1B and p53 and overexpression of miR-502 reduced cell cycle progression and cell growth in vitro in colon cancer cells (Zhai et al. 2013). In another study, miR-204 is reported to suppress the growth of renal clear cell carcinoma cells in vitro and in mice by regulating LC3 expression. miR-204 is also known to target a tumor suppressor gene, von Hippel-Lindeu (VHL, Hall et al. 2014; Mikhaylova et al. 2012). VHL in turn suppresses the expression of transient receptor potential melastatin 3 (TRPM3) involved in renal clear cell carcinoma progression (Hall et al. 2014).

Feng et al. induced nutrient starvation in breast cancer cells and found that miR-372 expression is suppressed by Yin Yang 1 (YY1) leading to an increase in autophagy. Upregulating miR-372 leads to inhibition of autophagy and subsequent growth in cancer cells (Feng et al. 2014). Overexpression of miR-100 has been shown to reduce the cell viability of hepatocellular carcinoma cells. miR-100 directly targets the 3'UTR of mTOR to regulate its expression and activate ATG7, leading to autophagy induction that kills the cancer cells (Ge et al. 2014).

4.5.2 Cancer Cell Metabolism

The autophagy regulating miRNAs are also reported to regulate the metabolism and metabolic stress responses toward the tumor cells.

In colorectal adenoma, miR-124 was found to be downregulated (Taniguchi et al. 2015). Further studies on target validation showed that miR-124 targets polypyrimidine tsat binding protein 1 (PTB1). It regulates the splicing of pyruvate kinase muscle to isoform 1 (PKM1) or isoform 2 (PKM2, Taniguchi et al. 2015). PKM1 stimulates oxidative phosphorylation and is only found in normal tissues and cells, whereas PKM2 is exclusively present on constantly proliferating cells like cancer cells. In cancer cells, PKM2 boosts glycolysis even in the presence of abundant oxygen, thus helping the cancer cell metabolism and growth. So, overexpressing miR-124 suppresses PTB1 leading to a switch between the PKM isoforms. miR-124 preferentially favors PKM1 over PKM2 thus increasing ROS accumulation and oxidative phosphorylation in tumor cells (Taniguchi et al. 2015).

In lung cancer cells, A549 and H460, expression of miR-144 is suppressed. Upregulating its expression by using mimics blocked the proliferation of cancer cells and induced autophagy and apoptosis (Chen et al. 2015). Chen et al. showed that miR-144 targets TIGAR, a glycolysis and apoptosis regulator which is responsible for a reduction in oxidative burden and regulates cell energy for metabolism (Chen et al. 2015).

In melanoma cells, another set of miRNAs, that is, miR-290-295 cluster of miRNAs target the 3'UTR of ATG7 and ULK1 (Chen et al. 2012b). These sets of miRNAs confer resistance to metabolic stress-induced cell death in B16F1 melanoma cells by inhibiting autophagy. Glucose starvation-induced cell death in these cells by upregulating autophagy but overexpressing miR-290-295 cluster of miRNAs reversed this effect and helped in the survival of tumor cells (Chen et al. 2012b).

In malignant mesothelioma tissues, expression of miR-126 is found to be downregulated and it is shown to suppress cancer cell growth (Tomasetti et al. 2016). miR-126 suppressed IRS1 and decreased glucose uptake by the cells causing energy deprivation amongst the cancer cells. AMPK gets activated upon energy deprivation and further activates ULK1. miR-126 is also reported to alter the expression of other key proteins involved in the metabolism like acetyl co-A citrate and pyruvate dehydrogenase kinase (Tomasetti et al. 2016).

4.5.3 Hypoxia Responses

Due to irregular blood supply and abnormal vascularization, tumor cells develop a hypoxic environment. The hypoxic tumor cells rely on autophagy for their survival. Thus, a number of miRNAs regulate autophagy to control hypoxia-induced responses in cancer cells.

Upon hypoxia in prostate cancer cell lines DU145 and PC3, miR-124, and miR-144 were found to be downregulated (Gu et al. 2016). Overexpression of these miRNAs in prostate cancer cells led to a reduction in hypoxia-mediated autophagy and increased radiation-induced cell death (Gu et al. 2016). Another study has shown that miR-96 under moderate levels enhances hypoxia-mediated autophagy by suppressing mTOR (Ma et al. 2014). Whereas, overexpression of the same miRNA inhibited hypoxia-induced autophagy by targeting ATG7 (Ma et al. 2014).

miR-375 is known to be downregulated following hypoxia in hepatocellular carcinoma cell lines compared to normal liver tissues. It was found to target the 3'UTR of ATG7 to suppress hypoxia-mediated autophagy leading to cell death (Chang et al. 2012). In glioblastoma cells, hypoxia leads to the downregulation of miR-224-3p but on the contrary, overexpression of miR-224-3p reduced hypoxia-mediated autophagy leading to cell death. miR-224-3p is shown to directly target ATG5 and FIP200 to inhibit autophagy, and inhibition of miR-224-3p increased hypoxia-mediated autophagy (Guo et al. 2015). Hypoxia also leads to the upregulation of IL-6 production in glioblastoma cells. Previous study has shown that overexpression of IL-6 leads to the activation of autophagy (Xue et al. 2016). Increased IL-6 secretion leads to an increase in the level of miR-155-3p through STAT3 dependent signaling pathway. miR-155-3p is reported to directly target the CREB3 regulatory factor (CREBRF) leading to an increase in the expression of ATG5 and autophagy that enhanced the survival of the cells. Blocking IL-6 reduced autophagy, increased cancer cell death, and decreased the tumor burden. On the contrary, complementary strand miR-155-5p is reported to block autophagy by downregulating mTOR and causing cell cycle arrest. Therefore, miR-155-3p and miR-155-5p acts as a switch to determine the final autophagy-related outcome under hypoxic environment (Wan et al. 2014).

4.5.4 Angiogenesis

Autophagic activity is extremely important for angiogenesis and some miRNAs are shown to regulate autophagy-induced angiogenesis in tumor vascularization. In endothelial progenitor cells, miR-195 is reported to inhibit autophagy by targeting GABARAP like 1 (GABARAPL1) protein, and knocking down miR-195 led to stimulation of autophagy and promoted angiogenesis and cell growth under hypoxia (Mo et al. 2016). Inhibition of autophagy by the addition of 3-MA blocked all the above responses indicating the role of autophagy in tumor growth and angiogenesis (Mo et al. 2016). In prostate cancer cells, miR-212 was found to be downregulated and it is shown that miR-212 is able to inhibit autophagy by directly targeting the autophagy activator SIRT1 (Ramalinga et al. 2015). Overexpressing miR-212 led to the suppression of angiogenesis and the death of cancer cells (Ramalinga et al. 2015). Another study has shown that inhibition of miR-130a led to autophagy induction and initiated cell death in endothelial progenitor cells (Xu et al. 2014). The level of miR-1273g-3p is also elucidated to be increased upon glucose level fluctuations that induced autophagy and inhibited angiogenesis in cancer cells (Guo et al. 2016).

4.5.5 Autophagy Regulating miRNAs as Biomarkers in Cancer

Some of the miRNAs involved in the regulation of autophagy have been reported as potential biomarkers of cancer. A recent study by Fan et al. 2020 unraveled the link between miR-1246 expression and sensitivity of irradiation of non-small cell lung cancer (NSCLC) cells. They had shown that the expression of miR-1246 promoted the resistance of NSCLC cells toward irradiation through inhibition of mTOR and activation of autophagy. Furthermore, the study also established the prospective role of miR-1246 to act as a biomarker for predicting the efficacy of radiotherapy in NSCLC patients and a probable target for radiotherapy sensitization.

Another recent study (Guo et al. 2019) demonstrates the effect of miR-384 on tumor progression and autophagy in NSCLC cells. Based on a previous study (Fan et al. 2017), the authors of this paper found that the expression of miR-384 was significantly downregulated in NSCLC cells and reasoned that miR-384 can be a potential therapeutic target. They also found elevated expression of Collagen α-1 (X) chain (COL10A1) and co-related inverse relationship between miR-384 and COL10A1. Overexpression of miR-384 or inhibition in COL10A1 expression led to inhibition in NSCLC cell proliferation and tumor growth through autophagy induction. Overall, this study concluded that miR-384 can promote apoptosis and autophagy in NSCLC cells by downregulating COL10A1 and this potential can be used as a biomarker for the prediction of NSCLC (Guo et al. 2019).

A recent study demonstrated the effect of miR-1251-5p on autophagy and its consequence on tumor progression in ovarian cancer cells (Shao et al. 2019). Autophagy, being a complex process, is capable of either promoting or inhibiting tumorigenesis depending on the tissue and context (Chen and Debnath 2010; Kroemer et al. 2010; Yun and Lee 2018). In the above-mentioned study, miR-1251-5p mimics increased cell cycle progression and cell proliferation whereas overexpression of tumor-binding cofactor C (TBCC) increased the expression of p62 and α/β-tubulin along with inhibition of the expression of CDK4 and LC3BII which resulted in suppression of cell growth and autophagy. In ovarian cancer cells, effects on TBCC were rescued by miR-1251-5p and promoted tumor growth. Since miR-1251-5p can directly target TBCC and enhances autophagy leading to the promotion of carcinogenesis in ovarian cancer, it can be used as a biomarker to know the severity of the disease.

In breast cancer tissues, expression of miR-205 and miR-342 levels were found to be low (Savad et al. 2012) whereas the high expression of miR-155 and miR-493 is correlated as a better recovery of patients from cancer (Gasparini et al. 2014) and suppression of miR-30e and miR-27a is associated with worsening of the disease (Gasparini et al. 2014). In ovarian cancer, a decrease in miR-152 level is correlated with cisplatin resistance (He et al. 2015) and miR-29b expression is associated with recovery (Dai et al. 2014). A decrease in the expression of miR-212 in sera and tumor tissue of patients is used as a diagnosis for prostate cancer (Ramalinga et al. 2015).

4.5.6 Role of Autophagy Regulating miRNA in Tumor Therapy

Upon identification of miRNAs in 1993, its diagnostic application in different diseases has been demonstrated. miRNA plays a central role in controlling tumor suppression in cancer. It was found by Calin et al. that downregulation or deletion of miR-16-1 and miR-15a in 13q14 leads to the progression of CLL (Calin et al. 2002). Expression of miR-16-1 and miR-15a in CLL decreases the expression of B-cell lymphoma 2 (BCL2) that has been shown to be induced by heat shock and cell stress and prevent apoptosis of tumor cells, thereby giving credence to the fact that miRNAs can be used in cancer therapy (Tsujimoto 1989; Cimmino et al. 2005). Takamizawa et al. have found that in A549 lung adenocarcinoma cells, overexpression of miRNA *let-7* inhibits the growth of cancer cells in vivo (Takamizawa et al. 2004). In epithelial ovarian cancer (EOC) cells, miR-199a regulates IKKβ expression, and inhibits the proliferation of these cells (Chen et al. 2008). Downregulation of miR-221 by isoflavone, bio response formulated 3,3-′-diindolylmethane (BR-DIM) and difluorinated curcumin (CDF), inhibits proliferation of pancreatic cancer cells (Sarkar et al. 2013). In colorectal cancer (CRC), upregulation of miR-324-5p suppresses the proliferation of colorectal tumor cells and its invasion by targeting embryonic lethal abnormal vision-like protein 1 (ELAVL1, Gu et al. 2019). miR-331-3p is reported to reduce the expression of erythroblastic oncogene B-2 (ERBB-2) in prostate cancer (PCa, Epis et al. 2009). ERBB-2 associated with androgen receptor signaling that promotes cell proliferation proteins known to prevent apoptosis of tumor cells which was induced by heat shock and chemotherapeutics (Vernimmen et al. 2003). miR-199b-5b also suppresses tumor progression in breast cancer by inhibiting angiogenesis because miR-199b-5p treated mice showed a reduction in tumor size and the number of blood vessels in tumors and nearby tissues (Lin et al. 2019). miR-524 also affects tumor growth and angiogenesis by inhibiting the expression of angiopoietin-2 (He et al. 2014). Role of different miRNAs like miR-320, and miR-29b have been elucidated to suppress angiogenesis by inhibiting expression of neuropilin1 and Akt3 respectively (Wu et al. 2014; Li et al. 2017). In breast cancer cells, overexpression of miR-340 is elucidated to downregulate the level of ROCK1 and inhibits invasion, migration, and proliferation of tumor cells (Maskey et al. 2017). In hepatocellular carcinoma overexpression on miR-145 downregulates the ROCK1 expression and inhibits cell proliferation (Ding et al. 2016).

4.5.6.1 miRNA Regulates Autophagy and Inhibits Tumor Progression

Autophagy plays a dual role in cancer progression and suppression. Several studies say that autophagy promotes tumor survival by supplying nutrients to the stressed cancer cells (White and DiPaola 2009). In colorectal cancer stem cells, aberrant expression of miR-140-5p inhibits growth and interferes with autophagy through *son of a mother against decapentaplegic* (Smad 2) and ATG12 inhibition (Zhai et al. 2015). miR-502 has been reported to inhibit autophagy by suppressing Rab1 in colon cancer cells. It is also known to inhibit cancer cell growth and arrest cell cycle that impedes tumor progression (Zhai et al. 2013). In small cell lung cancer cells, overexpression of miR-143 decreases autophagy by targeting ATG2B leading to inhibition of cancer cell proliferation (Wei et al. 2015). Overexpression of miR-193b is reported to increase autophagy in oesophageal cancer cells and non-apoptotic cell death (Nyhan et al. 2016). On the contrary, miR-9-3p is known to inhibit autophagy and also decreases the expression of BCL-2 leading to apoptosis in medullary thyroid cancer cells (Gundara et al. 2015). Downregulation of autophagy was also manifested by miR-17-5p through a reduction in Beclin1 expression in paclitaxel resistance cancer cells (Chatterjee et al. 2014). Heterogenous unclear ribonucleoproteinA1 (hnRNP A1) that prevents apoptosis of cancerous cells by increasing connective tissue growth factor (CTGF) and CyclinD1 has been elucidated to be degraded through the autophagic pathway by miR-18a over-expression in colon cancer cells (Fujiya et al. 2014).

4.5.6.2 miRNA Inhibits Chemoresistance Tumor Cell Growth by Modulating Autophagy

In osteosarcoma cells, chemotherapy-induced autophagy imparts chemoresistance whereas miR-101 block the activation of autophagy in chemoresistance osteosarcoma cells and enhance their sensitivity to chemotherapy (Chang et al. 2014). In gastric cancer cells, autophagy is shown to help the tumor cells in survival against different drugs by converting them to chemoresistance cells. Overexpression of miR-23b-3p in multidrug resistance tumor cells, is reported to inhibit autophagy by targeting ATG12 and high-mobility group protein B2 (HMGB2) that increases sensitivity to different drugs (An et al. 2015). In hepatocarcinoma cells, autophagy prevents the apoptosis of cancerous cells induced by drug-like cisplatin and helps in promoting cell growth, whereas miR-101 overexpression has been reported to inhibit autophagy and induce apoptosis in cancerous cells (Xu et al. 2013). Similarly, in 5-fluorouracil treated colorectal cancer cells, miR-22 inhibits autophagy and promotes apoptosis for efficiently killing the tumor cells (Zhang et al. 2015a). HMGB1 has been shown to be targeted by miR-22 to inhibit autophagy that prevents cell proliferation, migration, and invasion of osteocarcinoma cells (Guo et al. 2014).

4.5.6.3 miRNA Inhibits Radioresistance Tumor Cells Growth by Modulating Autophagy

Autophagy helps the cancerous cells to survive from radiation treatment and makes them resistant to radiation. In breast cancer cells, autophagy induced radiation resistance is inhibited by miR-200 and makes the cells sensitized to radiation treatment (Sun et al. 2015b). Hypoxia also induced autophagy to produce radioresistance prostate cancer cells but overexpression of miR-124 and miR-144 decreases hypoxia-mediated autophagy and converts radioresistance tumor cells to radiosensitive cells (Gu et al. 2016).

4.6 Conclusion

miRNAs play a key role in regulating different biological processes. It has been predicted that around 60% of all the protein-coding genes are regulated by miRNA. As discussed in the review earlier, miRNAs directly or indirectly modulate autophagy in different cancer types under different physiological conditions and in response to different types of stress signals. Either a single miRNA can regulate the expression of different autophagy-related proteins/pathways or many different miRNAs can also control a single important autophagy-related protein and pathways. The same stress signal or stimuli can modulate the expression of different miRNAs in different cancer cell types. These miRNAs are generally termed as oncomirs and tumor suppressors. Most of these miRNAs are reported to regulate various autophagy-related genes. Autophagy plays an important role during cancer progression and spread. Dysregulation or aberrant expression of autophagy regulating miRNAs is involved in the development of different cancers. Due to the ability of autophagy regulating miRNAs to control cancer progression, they are being used in cancer treatment as they sensitize cells to chemotherapy and radiation therapy. They can also be used as cancer biomarkers to accurately predict the diagnosis of disease and to check the patient's response to the treatment. Thus, miRNA manipulations by using antagomirs, mimics, gene therapy, gene delivery, or other strategies can be used for cancer treatment. Comprehensive knowledge of miRNAs and related networks might contribute to the efforts involving autophagy modulation as an innovative treatment approach.

Acknowledgments Department of Science and Technology (DST), Govt. of India (EMR/2016/000048, EEQ/2016/000205 and DST/ INSPIRE /Faculty award/ 2014/DST/ INSPIRE /04/2014/01662) is acknowledged by R.D. for the generous financial support to his laboratory. R.D. is also thankful to the Ministry of Human Resource and Development (MHRD), Govt. of India for intramural support. The research fellowship by DST to A.B. and MHRD to A.M., A.K. and L.N. is deeply acknowledged. D.M. is thankful to DST for Kishore Vaigyanik Protsahan Yojana (KVPY) fellowship.
Declaration of competing interest: The authors declare that they have no known competing financial interests that could have appeared to influence the work reported in this paper.

References

An Y, Zhang Z, Shang Y, Jiang X, Dong J, Yu P, Nie Y, Zhao Q (2015) miR-23b-3p regulates the chemoresistance of gastric cancer cells by targeting ATG12 and HMGB2. Cell Death Dis 6: e1766

Bao L, Lv L, Feng J, Chen Y, Wang X, Han S, Zhao H (2016) miR-487b-5p regulates temozolomide resistance of lung cancer cells through LAMP2-medicated autophagy. DNA Cell Biol 35:385–392

Bartel DP (2004) MicroRNAs: genomics, biogenesis, mechanism, and function. Cell 116:281–297

Behura A, Mishra A, Chugh S, Mawatwal S, Kumar A, Manna D, Mishra A, Singh R, Dhiman R (2019) ESAT-6 modulates Calcimycin-induced autophagy through microRNA-30a in mycobacteria infected macrophages. J Infect 79:139–152

Broughton JP, Lovci MT, Huang JL, Yeo GW, Pasquinelli AE (2016) Pairing beyond the seed supports microRNA targeting specificity. Mol Cell 64:320–333

Calin GA, Dumitru CD, Shimizu M, Bichi R, Zupo S, Noch E, Aldler H, Rattan S, Keating M, Rai K (2002) Frequent deletions and down-regulation of micro-RNA genes miR15 and miR16 at 13q14 in chronic lymphocytic leukemia. Proc Natl Acad Sci 99:15524–15529

Cecconi F, Levine B (2008) The role of autophagy in mammalian development: cell makeover rather than cell death. Dev Cell 15:344–357

Chang Y, Yan W, He X, Zhang L, Li C, Huang H, Nace G, Geller DA, Lin J, Tsung A (2012) miR-375 inhibits autophagy and reduces viability of hepatocellular carcinoma cells under hypoxic conditions. Gastroenterology 143:177–187.e8

Chang Z, Huo L, Li K, Wu Y, Hu Z (2014) Blocked autophagy by miR-101 enhances osteosarcoma cell chemosensitivity in vitro. Sci World J 2014:794756

Chatterjee A, Chattopadhyay D, Chakrabarti G (2014) miR-17-5p downregulation contributes to paclitaxel resistance of lung cancer cells through altering beclin1 expression. PLoS One 9: e95716

Chen N, Debnath J (2010) Autophagy and tumorigenesis. FEBS Lett 584:1427–1435

Chen R, Alvero A, Silasi D, Kelly M, Fest S, Visintin I, Leiser A, Schwartz P, Rutherford T, Mor G (2008) Regulation of IKKβ by miR-199a affects NF-κB activity in ovarian cancer cells. Oncogene 27:4712–4723

Chen K, Fan W, Wang X, Ke X, Wu G, Hu C (2012a) MicroRNA-101 mediates the suppressive effect of laminar shear stress on mTOR expression in vascular endothelial cells. Biochem Biophys Res Commun 427:138–142

Chen Y, Liersch R, Detmar M (2012b) The miR-290-295 cluster suppresses autophagic cell death of melanoma cells. Sci Rep 2:808

Chen S, Li P, Li J, Wang Y, Du Y, Chen X, Zang W, Wang H, Chu H, Zhao G (2015) MiR-144 inhibits proliferation and induces apoptosis and autophagy in lung cancer cells by targeting TIGAR. Cell Physiol Biochem 35:997–1007

Choi SH, Lyu SY, Park WB (2004) Mistletoe lectin induces apoptosis and telomerase inhibition in human A253 cancer cells through dephosphorylation of Akt. Arch Pharm Res 27:68

Cimmino A, Calin GA, Fabbri M, Iorio MV, Ferracin M, Shimizu M, Wojcik SE, Aqeilan RI, Zupo S, Dono M (2005) miR-15 and miR-16 induce apoptosis by targeting BCL2. Proc Natl Acad Sci 102:13944–13949

Clotaire DZJ, Zhang B, Wei N, Gao R, Zhao F, Wang Y, Lei M, Huang W (2016) MiR-26b inhibits autophagy by targeting ULK2 in prostate cancer cells. Biochem Biophys Res Commun 472:194–200

Comincini S, Allavena G, Palumbo S, Morini M, Durando F, Angeletti F, Pirtoli L, Miracco C (2013) microRNA-17 regulates the expression of ATG7 and modulates the autophagy process, improving the sensitivity to temozolomide and low-dose ionizing radiation treatments in human glioblastoma cells. Cancer Biol Ther 14:574–586

Dai F, Zhang Y, Chen Y (2014) Involvement of miR-29b signaling in the sensitivity to chemotherapy in patients with ovarian carcinoma. Hum Pathol 45:1285–1293

Demirci MDS, Bağcı C, Allmer J (2016) Differential expression of toxoplasma gondii microRNAs in murine and human hosts. In: Non-coding RNAs and inter-kingdom communication. Springer, Cham, pp 143–159

Deretic V, Levine B (2009) Autophagy, immunity, and microbial adaptations. Cell Host Microbe 5:527–549

Ding W, Tan H, Zhao C, Li X, Li Z, Jiang C, Zhang Y, Wang L (2016) MiR-145 suppresses cell proliferation and motility by inhibiting ROCK1 in hepatocellular carcinoma. Tumor Biol 37:6255–6260

Epis MR, Giles KM, Barker A, Kendrick TS, Leedman PJ (2009) miR-331-3p regulates ERBB-2 expression and androgen receptor signaling in prostate cancer. J Biol Chem 284:24696–24704

Fader CM, Sánchez DG, Mestre MB, Colombo MI (2009) TI-VAMP/VAMP7 and VAMP3/ cellubrevin: two v-SNARE proteins involved in specific steps of the autophagy/multivesicular body pathways. Biochim Biophys Acta Mol Cell Res 1793:1901–1916

Fan N, Zhang J, Cheng C, Zhang X, Feng J, Kong R (2017) MicroRNA-384 represses the growth and invasion of non-small-cell lung cancer by targeting astrocyte elevated gene-1/Wnt signaling. Biomed Pharmacother 95:1331–1337

Fan L, Wang J, Cao Q, Ding X, Li B (2020) Aberrant miR-1246 expression promotes radioresistance in non-small cell lung cancer: a potential prognostic biomarker and radiotherapy sensitization target. Am J Cancer Res 10:314

Fang Y, Xue JL, Shen Q, Chen J, Tian L (2012) MicroRNA-7 inhibits tumor growth and metastasis by targeting the phosphoinositide 3-kinase/Akt pathway in hepatocellular carcinoma. Hepatology 55:1852–1862

Feng L, Ma Y, Sun J, Shen Q, Liu L, Lu H, Wang F, Yue Y, Li J, Zhang S (2014) YY1-MIR372-SQSTM1 regulatory axis in autophagy. Autophagy 10:1442–1453

Fornari F, Milazzo M, Chieco P, Negrini M, Calin GA, Grazi GL, Pollutri D, Croce CM, Bolondi L, Gramantieri L (2010) MiR-199a-3p regulates mTOR and c-Met to influence the doxorubicin sensitivity of human hepatocarcinoma cells. Cancer Res 70:5184–5193

Frankel LB, Lund AH (2012) MicroRNA regulation of autophagy. Carcinogenesis 33:2018–2025

Frankel LB, Wen J, Lees M, Høyer-Hansen M, Farkas T, Krogh A, Jäättelä M, Lund AH (2011) microRNA-101 is a potent inhibitor of autophagy. EMBO J 30:4628–4641

Fu G, Brkić J, Hayder H, Peng C (2013) MicroRNAs in human placental development and pregnancy complications. Int J Mol Sci 14:5519–5544

Fujita N, Hayashi-Nishino M, Fukumoto H, Omori H, Yamamoto A, Noda T, Yoshimori T (2008) An Atg4B mutant hampers the lipidation of LC3 paralogues and causes defects in autophagosome closure. Mol Biol Cell 19:4651–4659

Fujiya M, Konishi H, Kamel MM, Ueno N, Inaba Y, Moriichi K, Tanabe H, Ikuta K, Ohtake T, Kohgo Y (2014) microRNA-18a induces apoptosis in colon cancer cells via the autophagolysosomal degradation of oncogenic heterogeneous nuclear ribonucleoprotein A1. Oncogene 33:4847–4856

Gasparini P, Cascione L, Fassan M, Lovat F, Guler G, Balci S, Irkkan C, Morrison C, Croce CM, Shapiro CL (2014) microRNA expression profiling identifies a four microRNA signature as a novel diagnostic and prognostic biomarker in triple negative breast cancers. Oncotarget 5:1174

Ge Y-Y, Shi Q, Zheng Z-Y, Gong J, Zeng C, Yang J, Zhuang S-M (2014) MicroRNA-100 promotes the autophagy of hepatocellular carcinoma cells by inhibiting the expression of mTOR and IGF-1R. Oncotarget 5:6218

Gibbings D, Mostowy S, Jay F, Schwab Y, Cossart P, Voinnet O (2012) Selective autophagy degrades DICER and AGO2 and regulates miRNA activity. Nat Cell Biol 14:1314–1321

Gozuacik D, Akkoc Y, Ozturk DG, Kocak M (2017) Autophagy-regulating microRNAs and cancer. Front Oncol 7:65

Gregory RI, Shiekhattar R (2005) MicroRNA biogenesis and cancer. Cancer Res 65:3509–3512

Grund S, Diederichs S (2010) microRNA biogenesis and its impact on RNA interference. In: RNA technologies and their applications. Springer, Berlin, pp 325–354

Gu H, Liu M, Ding C, Wang X, Wang R, Wu X, Fan R (2016) Hypoxia-responsive miR-124 and miR-144 reduce hypoxia-induced autophagy and enhance radiosensitivity of prostate cancer cells via suppressing PIM 1. Cancer Med 5:1174–1182

Gu C, Zhang M, Sun W, Dong C (2019) Upregulation of miR-324-5p inhibits proliferation and invasion of colorectal cancer cells by targeting ELAVL1. Oncol Res 27:515–524

Gundara JS, Zhao J, Gill AJ, Lee JC, Delbridge L, Robinson BG, McLean C, Serpell J, Sidhu SB (2015) Noncoding RNA blockade of autophagy is therapeutic in medullary thyroid cancer. Cancer Med 4:174–182

Guo S, Bai R, Liu W, Zhao A, Zhao Z, Wang Y, Wang Y, Zhao W, Wang W (2014) miR-22 inhibits osteosarcoma cell proliferation and migration by targeting HMGB1 and inhibiting HMGB1-mediated autophagy. Tumor Biol 35:7025–7034

Guo X, Xue H, Guo X, Gao X, Xu S, Yan S, Han X, Li T, Shen J, Li G (2015) MiR224-3p inhibits hypoxia-induced autophagy by targeting autophagy-related genes in human glioblastoma cells. Oncotarget 6:41620

Guo J, Sang Y, Yin T, Wang B, Yang W, Li X, Li H, Kang Y (2016) miR-1273g-3p participates in acute glucose fluctuation-induced autophagy, dysfunction, and proliferation attenuation in human umbilical vein endothelial cells. Am J Physiol Endocrinol Metab 310:E734–E743

Guo Q, Zheng M, Xu Y, Wang N, Zhao W (2019) MiR-384 induces apoptosis and autophagy of non-small cell lung cancer cells through the negative regulation of collagen α-1 (X) chain gene. Biosci Rep 39:BSR20181523

Ha M, Kim VN (2014) Regulation of microRNA biogenesis. Nat Rev Mol Cell Biol 15:509

Hall DP, Cost NG, Hegde S, Kellner E, Mikhaylova O, Stratton Y, Ehmer B, Abplanalp WA, Pandey R, Biesiada J (2014) TRPM3 and miR-204 establish a regulatory circuit that controls oncogenic autophagy in clear cell renal cell carcinoma. Cancer Cell 26:738–753

Hayes J, Peruzzi PP, Lawler S (2014) MicroRNAs in cancer: biomarkers, functions and therapy. Trends Mol Med 20:460–469

He C, Klionsky DJ (2009) Regulation mechanisms and signaling pathways of autophagy. Annu Rev Genet 43:67–93

He T, Qi F, Jia L, Wang S, Song N, Guo L, Fu Y, Luo Y (2014) MicroRNA-542-3p inhibits tumour angiogenesis by targeting angiopoietin-2. J Pathol 232:499–508

He J, Yu J-J, Xu Q, Wang L, Zheng JZ, Liu L-Z, Jiang B-H (2015) Downregulation of ATG14 by EGR1-MIR152 sensitizes ovarian cancer cells to cisplatin-induced apoptosis by inhibiting cytoprotective autophagy. Autophagy 11:373–384

Huang W (2017) MicroRNAs: biomarkers, diagnostics, and therapeutics. In: Bioinformatics in MicroRNA research. Springer, New York, pp 57–67

Huang Y, Chuang AY, Ratovitski EA (2011) Phospho-ΔNp63α/miR-885-3p axis in tumor cell life and cell death upon cisplatin exposure. Cell Cycle 10:3938–3947

Huang Y, Guerrero-Preston R, Ratovitski EA (2012) Phospho-ΔNp63α-dependent regulation of autophagic signaling through transcription and micro-RNA modulation. Cell Cycle 11:1247–1259

Hwang H-W, Wentzel EA, Mendell JT (2007) A hexanucleotide element directs microRNA nuclear import. Science 315:97–100

Jegga AG, Schneider L, Ouyang X, Zhang J (2011) Systems biology of the autophagy-lysosomal pathway. Autophagy 7:477–489

Jing Z, Han W, Sui X, Xie J, Pan H (2015) Interaction of autophagy with microRNAs and their potential therapeutic implications in human cancers. Cancer Lett 356:332–338

Kaminskyy V, Zhivotovsky B (2012) Proteases in autophagy. Biochim Biophys Acta Prot Proteomics 1824:44–50

Khvorova A, Reynolds A, Jayasena SD (2003) Functional siRNAs and miRNAs exhibit strand bias. Cell 115:209–216

Kim DH, Sætrom P, Snøve O, Rossi JJ (2008) MicroRNA-directed transcriptional gene silencing in mammalian cells. Proc Natl Acad Sci 105:16230–16235

Kim VN, Han J, Siomi MC (2009) Biogenesis of small RNAs in animals. Nat Rev Mol Cell Biol 10:126–139

Korkmaz G, Le Sage C, Tekirdag KA, Agami R, Gozuacik D (2012) miR-376b controls starvation and mTOR inhibition-related autophagy by targeting ATG4C and BECN1. Autophagy 8:165–176

Korkmaz G, Tekirdag KA, Ozturk DG, Kosar A, Sezerman OU, Gozuacik D (2013) MIR376A is a regulator of starvation-induced autophagy. PLoS One 8:e82556

Kovaleva V, Mora R, Park YJ, Plass C, Chiramel AI, Bartenschlager R, Döhner H, Stilgenbauer S, Pscherer A, Lichter P (2012) miRNA-130a targets ATG2B and DICER1 to inhibit autophagy and trigger killing of chronic lymphocytic leukemia cells. Cancer Res 72:1763–1772

Krishnan P, Damaraju S (2018) The challenges and opportunities in the clinical application of noncoding RNAs: the road map for miRNAs and piRNAs in cancer diagnostics and prognostics. Int J Genomics 2018:5848046

Kroemer G, Mariño G, Levine B (2010) Autophagy and the integrated stress response. Mol Cell 40:280–293

Li Y, Cai B, Shen L, Dong Y, Lu Q, Sun S, Liu S, Ma S, Ma PX, Chen J (2017) MiRNA-29b suppresses tumor growth through simultaneously inhibiting angiogenesis and tumorigenesis by targeting Akt3. Cancer Lett 397:111–119

Liang H, Zhang J, Zen K, Zhang C-Y, Chen X (2013) Nuclear microRNAs and their unconventional role in regulating non-coding RNAs. Protein Cell 4:325–330

Lin X, Qiu W, Xiao Y, Ma J, Xu F, Zhang K, Gao Y, Chen Q, Li Y, Li H (2019) MiR-199b-5p suppresses tumor angiogenesis mediated by vascular endothelial cells in breast cancer by targeting ALK1. Front Genet 10:1397

Liu S, Wang H, Mu J, Wang H, Peng Y, Li Q, Mao D, Guo L (2020) MiRNA-211 triggers an autophagy-dependent apoptosis in cervical cancer cells: regulation of Bcl-2. Naunyn Schmiedeberg's Arch Pharmacol 393:359–370

Ma Y, Yang H-Z, Dong B-J, Zou H-B, Zhou Y, Kong X-M, Huang Y-R (2014) Biphasic regulation of autophagy by miR-96 in prostate cancer cells under hypoxia. Oncotarget 5:9169

MacRae IJ, Ma E, Zhou M, Robinson CV, Doudna JA (2008) In vitro reconstitution of the human RISC-loading complex. Proc Natl Acad Sci 105:512–517

Maskey N, Li D, Xu H, Song H, Wu C, Hua K, Song J, Fang L (2017) MicroRNA-340 inhibits invasion and metastasis by downregulating ROCK1 in breast cancer cells. Oncol Lett 14:2261–2267

Mikhaylova O, Stratton Y, Hall D, Kellner E, Ehmer B, Drew AF, Gallo CA, Plas DR, Biesiada J, Meller J (2012) VHL-regulated MiR-204 suppresses tumor growth through inhibition of LC3B-mediated autophagy in renal clear cell carcinoma. Cancer Cell 21:532–546

Mizushima N (2019) The ATG conjugation systems in autophagy. Curr Opin Cell Biol 63:1–10

Mizushima N, Komatsu M (2011) Autophagy: renovation of cells and tissues. Cell 147:728–741

Mizushima N, Levine B, Cuervo AM, Klionsky DJ (2008) Autophagy fights disease through cellular self-digestion. Nature 451:1069–1075

Mo J, Zhang D, Yang R (2016) MicroRNA-195 regulates proliferation, migration, angiogenesis and autophagy of endothelial progenitor cells by targeting GABARAPL1. Biosci Rep 36:e00396

Nyhan MJ, O'Donovan TR, Boersma AW, Wiemer EA, McKenna SL (2016) MiR-193b promotes autophagy and non-apoptotic cell death in oesophageal cancer cells. BMC Cancer 16:101

O'Brien J, Hayder H, Zayed Y, Peng C (2018) Overview of microRNA biogenesis, mechanisms of actions, and circulation. Front Endocrinol 9:402

Okada C, Yamashita E, Lee SJ, Shibata S, Katahira J, Nakagawa A, Yoneda Y, Tsukihara T (2009) A high-resolution structure of the pre-microRNA nuclear export machinery. Science 326:1275–1279

Park JH, Shin C (2014) MicroRNA-directed cleavage of targets: mechanism and experimental approaches. BMB Rep 47:417

Paul P, Chakraborty A, Sarkar D, Langthasa M, Rahman M, Bari M, Singha RS, Malakar AK, Chakraborty S (2018) Interplay between miRNAs and human diseases. J Cell Physiol 233:2007–2018

Pennati M, Lopergolo A, Profumo V, De Cesare M, Sbarra S, Valdagni R, Zaffaroni N, Gandellini P, Folini M (2014) miR-205 impairs the autophagic flux and enhances cisplatin cytotoxicity in castration-resistant prostate cancer cells. Biochem Pharmacol 87:579–597

Ramalinga M, Roy A, Srivastava A, Bhattarai A, Harish V, Suy S, Collins S, Kumar D (2015) MicroRNA-212 negatively regulates starvation induced autophagy in prostate cancer cells by inhibiting SIRT1 and is a modulator of angiogenesis and cellular senescence. Oncotarget 6:34446

Ravikumar B, Imarisio S, Sarkar S, O'Kane CJ, Rubinsztein DC (2008) Rab5 modulates aggregation and toxicity of mutant huntingtin through macroautophagy in cell and fly models of Huntington disease. J Cell Sci 121:1649–1660

Reggiori F, Ungermann C (2017) Autophagosome maturation and fusion. J Mol Biol 429:486–496

Saçar MD, Bağcı C, Allmer J (2014) Computational prediction of microRNAs from toxoplasma gondii potentially regulating the hosts' gene expression. Genomics Proteomics Bioinformatics 12:228–238

Salmanidis M, Pillman K, Goodall G, Bracken C (2014) Direct transcriptional regulation by nuclear microRNAs. Int J Biochem Cell Biol 54:304–311

Sarkar S, Dubaybo H, Ali S, Goncalves P, Kollepara SL, Sethi S, Philip PA, Li Y (2013) Down-regulation of miR-221 inhibits proliferation of pancreatic cancer cells through up-regulation of PTEN, p27kip1, p57kip2, and PUMA. Am J Cancer Res 3:465

Savad S, Mehdipour P, Miryounesi M, Shirkoohi R, Fereidooni F, Mansouri F, Modarressi MH (2012) Expression analysis of MiR-21, MiR-205, and MiR-342 in breast cancer in Iran. Asian Pac J Cancer Prev 13:873–877

Schneider JL, Cuervo AM (2014) Autophagy and human disease: emerging themes. Curr Opin Genet Dev 26:16–23

Seitz H, Zamore PD (2006) Rethinking the microprocessor. Cell 125:827–829

Shao Y, Liu X, Meng J, Zhang X, Ma Z, Yang G (2019) MicroRNA-1251-5p promotes carcinogenesis and autophagy via targeting the tumor suppressor TBCC in ovarian cancer cells. Mol Ther 27:1653–1664

Shi G, Shi J, Liu K, Liu N, Wang Y, Fu Z, Ding J, Jia L, Yuan W (2013) Increased miR-195 aggravates neuropathic pain by inhibiting autophagy following peripheral nerve injury. Glia 61:504–512

Shomron N, Levy C (2009) MicroRNA-biogenesis and pre-mRNA splicing crosstalk. BioMed Res Int 2009:594678

Songane M, Kleinnijenhuis J, Netea MG, van Crevel R (2012) The role of autophagy in host defence against Mycobacterium tuberculosis infection. Tuberculosis 92:388–396

Sun K-T, Chen MY, Tu M-G, Wang I-K, Chang S-S, Li C-Y (2015a) MicroRNA-20a regulates autophagy related protein-ATG16L1 in hypoxia-induced osteoclast differentiation. Bone 73:145–153

Sun Q, Liu T, Yuan Y, Guo Z, Xie G, Du S, Lin X, Xu Z, Liu M, Wang W (2015b) MiR-200c inhibits autophagy and enhances radiosensitivity in breast cancer cells by targeting UBQLN1. Int J Cancer 136:1003–1012

Suzuki H, Osawa T, Fujioka Y, Noda NN (2017) Structural biology of the core autophagy machinery. Curr Opin Struct Biol 43:10–17

Takamizawa J, Konishi H, Yanagisawa K, Tomida S, Osada H, Endoh H, Harano T, Yatabe Y, Nagino M, Nimura Y (2004) Reduced expression of the let-7 microRNAs in human lung cancers in association with shortened postoperative survival. Cancer Res 64:3753–3756

Tan S, Shi H, Ba M, Lin S, Tang H, Zeng X, Zhang X (2016) miR-409-3p sensitizes colon cancer cells to oxaliplatin by inhibiting Beclin-1-mediated autophagy. Int J Mol Med 37:1030–1038

Taniguchi K, Sugito N, Kumazaki M, Shinohara H, Yamada N, Nakagawa Y, Ito Y, Otsuki Y, Uno B, Uchiyama K (2015) MicroRNA-124 inhibits cancer cell growth through PTB1/PKM1/PKM2 feedback cascade in colorectal cancer. Cancer Lett 363:17–27

Tao J, Liu W, Shang G, Zheng Y, Huang J, Lin R, Chen L (2015) MiR-207/352 regulate lysosomal-associated membrane proteins and enzymes following ischemic stroke. Neuroscience 305:1–14

Tekirdag KA, Akkoc Y, Kosar A, Gozuacik D (2016) MIR376 family and cancer. Histol Histopathol 31:841–855

Tomasetti M, Monaco F, Manzella N, Rohlena J, Rohlenova K, Staffolani S, Gaetani S, Ciarapica V, Amati M, Bracci M (2016) MicroRNA-126 induces autophagy by altering cell metabolism in malignant mesothelioma. Oncotarget 7:36338

Tsujimoto Y (1989) Stress-resistance conferred by high level of bcl-2 alpha protein in human B lymphoblastoid cell. Oncogene 4:1331–1336

Tüfekci KU, Öner MG, Meuwissen RLJ, Genç Ş (2014) The role of microRNAs in human diseases. In: miRNomics: MicroRNA biology and computational analysis. Eds. Malik Yousef and Jens Allmer. Springer, New York, pp 33–50

Velikkakath AKG, Nishimura T, Oita E, Ishihara N, Mizushima N (2012) Mammalian Atg2 proteins are essential for autophagosome formation and important for regulation of size and distribution of lipid droplets. Mol Biol Cell 23:896–909

Vernimmen D, Guéders M, Pisvin S, Delvenne P, Winkler R (2003) Different mechanisms are implicated in ERBB2 gene overexpression in breast and in other cancers. Br J Cancer 89:899–906

Wampfler J, Federzoni EA, Torbett BE, Fey MF, Tschan MP (2015) Low DICER1 expression is associated with attenuated neutrophil differentiation and autophagy of NB4 APL cells. J Leukoc Biol 98:357–363

Wan G, Xie W, Liu Z, Xu W, Lao Y, Huang N, Cui K, Liao M, He J, Jiang Y (2014) Hypoxia-induced MIR155 is a potent autophagy inducer by targeting multiple players in the MTOR pathway. Autophagy 10:70–79

Wang Y, Liu J, Liu C, Naji A, Stoffers DA (2013) MicroRNA-7 regulates the mTOR pathway and proliferation in adult pancreatic β-cells. Diabetes 62:887–895

Wang Z, Wang N, Liu P, Chen Q, Situ H, Xie T, Zhang J, Peng C, Lin Y, Chen J (2014) MicroRNA-25 regulates chemoresistance-associated autophagy in breast cancer cells, a process modulated by the natural autophagy inducer isoliquiritigenin. Oncotarget 5:7013

Wang B, Zhong Y, Huang D, Li J (2016a) Macrophage autophagy regulated by miR-384-5p-mediated control of Beclin-1 plays a role in the development of atherosclerosis. Am J Transl Res 8:606

Wang J, Chen J, Sen S (2016b) MicroRNA as biomarkers and diagnostics. J Cell Physiol 231:25–30

Wei Y, Li L, Wang D, Zhang C-Y, Zen K (2014) Importin 8 regulates the transport of mature microRNAs into the cell nucleus. J Biol Chem 289:10270–10275

Wei J, Ma Z, Li Y, Zhao B, Wang D, Jin Y, Jin Y (2015) miR-143 inhibits cell proliferation by targeting autophagy-related 2B in non-small cell lung cancer H1299 cells. Mol Med Rep 11:571–576

White E, DiPaola RS (2009) The double-edged sword of autophagy modulation in cancer. Clin Cancer Res 15:5308–5316

Winter J, Jung S, Keller S, Gregory RI, Diederichs S (2009) Many roads to maturity: microRNA biogenesis pathways and their regulation. Nat Cell Biol 11(3):228–234

Wu H, Wang F, Hu S, Yin C, Li X, Zhao S, Wang J, Yan X (2012) MiR-20a and miR-106b negatively regulate autophagy induced by leucine deprivation via suppression of ULK1 expression in C2C12 myoblasts. Cell Signal 24:2179–2186

Wu D, Huang H-J, He C-N, Wang K-Y (2013) MicroRNA-199a-3p regulates endometrial cancer cell proliferation by targeting mammalian target of rapamycin (mTOR). Int J Gynecol Cancer 23:1191–1197

Wu Y-Y, Chen Y-L, Jao Y-C, Hsieh I-S, Chang K-C, Hong T-M (2014) miR-320 regulates tumor angiogenesis driven by vascular endothelial cells in oral cancer by silencing neuropilin 1. Angiogenesis 17:247–260

Xiao J, Zhu X, He B, Zhang Y, Kang B, Wang Z, Ni X (2011) MiR-204 regulates cardiomyocyte autophagy induced by ischemia-reperfusion through LC3-II. J Biomed Sci 18:35

Xu N, Zhang J, Shen C, Luo Y, Xia L, Xue F, Xia Q (2012) Cisplatin-induced downregulation of miR-199a-5p increases drug resistance by activating autophagy in HCC cell. Biochem Biophys Res Commun 423(4):826–831

Xu Y, An Y, Wang Y, Zhang C, Zhang H, Huang C, Jiang H, Wang X, Li X (2013) miR-101 inhibits autophagy and enhances cisplatin-induced apoptosis in hepatocellular carcinoma cells. Oncol Rep 29:2019–2024

Xu Q, Meng S, Liu B, Li MQ, Li Y, Fang L, Li YG (2014) MicroRNA-130a regulates autophagy of endothelial progenitor cells through Runx3. Clin Exp Pharmacol Physiol 41:351–357

Xu R, Liu S, Chen H, Lao L (2016) MicroRNA-30a downregulation contributes to chemoresistance of osteosarcoma cells through activating Beclin-1-mediated autophagy. Oncol Rep 35:1757–1763

Xue H, Yuan G, Guo X, Liu Q, Zhang J, Gao X, Guo X, Xu S, Li T, Shao Q (2016) A novel tumor-promoting mechanism of IL6 and the therapeutic efficacy of tocilizumab: hypoxia-induced IL6 is a potent autophagy initiator in glioblastoma via the p-STAT3-MIR155-3p-CREBRF pathway. Autophagy 12:1129–1152

Yang X, Zhong X, Tanyi JL, Shen J, Xu C, Gao P, Zheng TM, DeMichele A, Zhang L (2013) mir-30d Regulates multiple genes in the autophagy pathway and impairs autophagy process in human cancer cells. Biochem Biophys Res Commun 431:617–622

Yi R, Qin Y, Macara IG, Cullen BR (2003) Exportin-5 mediates the nuclear export of pre-microRNAs and short hairpin RNAs. Genes Dev 17:3011–3016

Yi H, Liang B, Jia J, Liang N, Xu H, Ju G, Ma S, Liu X (2013) Differential roles of miR-199a-5p in radiation-induced autophagy in breast cancer cells. FEBS Lett 587:436–443

Yu Y, Yang L, Zhao M, Zhu S, Kang R, Vernon P, Tang D, Cao L (2012) Targeting microRNA-30a-mediated autophagy enhances imatinib activity against human chronic myeloid leukemia cells. Leukemia 26:1752–1760

Yun CW, Lee SH (2018) The roles of autophagy in cancer. Int J Mol Sci 19:3466

Zeng Y, Cullen BR (2004) Structural requirements for pre-microRNA binding and nuclear export by Exportin 5. Nucleic Acids Res 32:4776–4785

Zeng Y, Huo G, Mo Y, Wang W, Chen H (2015) MIR137 regulates starvation-induced autophagy by targeting ATG7. J Mol Neurosci 56:815–821

Zhai H, Song B, Xu X, Zhu W, Ju J (2013) Inhibition of autophagy and tumor growth in colon cancer by miR-502. Oncogene 32:1570–1579

Zhai H, Fesler A, Ba Y, Wu S, Ju J (2015) Inhibition of colorectal cancer stem cell survival and invasive potential by hsa-miR-140-5p mediated suppression of Smad2 and autophagy. Oncotarget 6:19735

Zhang H, Tang J, Li C, Kong J, Wang J, Wu Y, Xu E, Lai M (2015a) MiR-22 regulates 5-FU sensitivity by inhibiting autophagy and promoting apoptosis in colorectal cancer cells. Cancer Lett 356:781–790

Zhang X, Shi H, Lin S, Ba M, Cui S (2015b) MicroRNA-216a enhances the radiosensitivity of pancreatic cancer cells by inhibiting beclin-1-mediated autophagy. Oncol Rep 34:1557–1564

Zhang J-A, Zhou B-R, Xu Y, Chen X, Liu J, Gozali M, Wu D, Z-q Y, Luo D (2016) MiR-23a-depressed autophagy is a participant in PUVA-and UVB-induced premature senescence. Oncotarget 7:37420

Zhu H, Wu H, Liu X, Li B, Chen Y, Ren X, Liu C-G, Yang J-M (2009) Regulation of autophagy by a beclin 1-targeted microRNA, miR-30a, in cancer cells. Autophagy 5:816–823

Zou Z, Wu L, Ding H, Wang Y, Zhang Y, Chen X, Chen X, Zhang C-Y, Zhang Q, Zen K (2012) MicroRNA-30a sensitizes tumor cells to cis-platinum via suppressing beclin 1-mediated autophagy. J Biol Chem 287:4148–4156

Exploring the Metabolic Implications of Autophagy Modulation in Tumor Microenvironment

Subhadip Mukhopadhyay

Abstract

Autophagy is a process that is involved in the bulk recycling of cellular materials to replenish cellular demands during crisis. However, in cancer cells, the process of autophagy is dual-faced. Generally, during the initial stages of tumorigenesis, autophagy inhibits proliferation, but when the cancer cells face extreme nutrient deprivation and other metabolic stress then autophagy protects them to metastasize. Advances in clinical findings have identified many targets involved in the autophagic pathway of cancer. This led to the development of trials using protective autophagy inhibitor and lethal autophagy inducer in order to trigger cell death dependent on autophagy. Interestingly, the existence of solely autophagy-dependent death of the cancer cells is yet to be established in patients. Nevertheless, the present concept of autophagy-dependent cell death is highly debatable from the point of therapy. Specificity and efficacy issues of autophagy-dependent cell death remain largely uncertain. This brings us to focus on the controversies and lacunas in the understanding of cell death and autophagy. The tumor microenvironment (TME) is regarded as a bed of metabolic cascade that ensures proper refueling of the nutrient-deprived core facing extremes of pH, starvation, hypoxia, immunogenic intrusion, and therapeutic insults. The novelty of this chapter lies in its comprehensive outlook to highlight how autophagy modulation alters TME and its significance in rewiring the metabolism.

S. Mukhopadhyay (✉)
Department of Life Sciences, National Institute of Technology Rourkela, Rourkela, Odisha, India

Present Address: Department of Radiation Oncology, NYU Medical School, Laura and Isaac Perlmutter Cancer Center, New York, NY, USA
e-mail: subhadip.mukhopadhyay@nyulangone.org

Keywords

Autophagy · Apoptosis · Metabolism · Glycolysis · Amino acids · Fats · Therapy · Tumor microenvironment · Immune system

5.1 Introduction

Tumors achieve their metabolic rewiring due to the fueling from their immediate microenvironment that makes them dynamic pseudoorgans (Lyssiotis and Kimmelman 2017). Tumor microenvironment (TME) represents an ecosystem of cells of heterogeneous lineage which is mainly comprised of cancer cells, stromal cells, interstitial fluids, immune cells, cytokines, chemokines, tumor-associated macrophages (TAM), cancer-associated fibroblasts (CAFs), and endothelial cells. TME provides the tumor a niche that makes them metabolically independent and breaks free off the homeostatic control of the normal cells thereby making them behave like a self-regulating organ(oid) like structure (Fig. 5.1). In order to

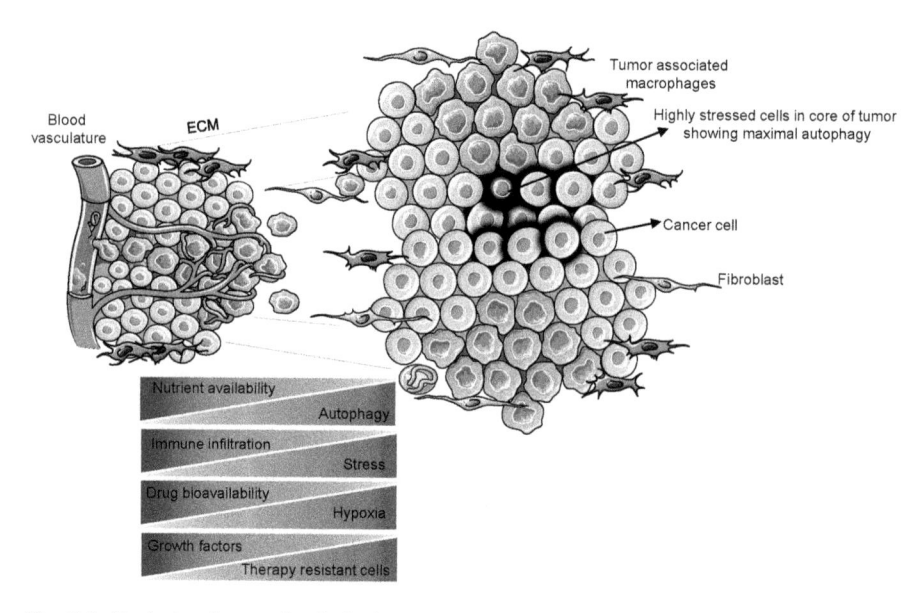

Fig. 5.1 Exploring the arm that fuels the tumor microenvironment: ecosystem of TME comprises of heterogeneous cellular components from extracellular matrix (ECM) comprising cancer-associated fibroblasts, stromal cells, to various immune cells (T, NK, B cells, macrophages), and fats. TME creates a shield-like structure that is impervious to nutrient uptake, drug bioavailability, or access to growth factors, resulting in massive autophagy increase in the core where there is high stress due to starvation and hypoxia, accumulation of metabolic wastes, thereby resulting in the production of therapy-resistant cancer cells. This figure was created using Servier Medical Art templates, which are licensed under a Creative Commons Attribution 3.0 Unported License; https://smart.servier.com

understand the metabolic rewiring in a tumor, it becomes imperative to understand the phenomena of autophagy by virtue of which it can beat all stress response.

The term autophagy comes from the Greek meaning for "*self-eating*" which encompasses a highly evolutionarily conserved catabolic process digesting the cell's own machinery to bypass a stress condition. Generally, the mechanism of autophagy remains conserved from yeast to higher mammals and involves the development of a crescent-shaped double-membrane structure called phagophore or isolation membrane, sequestering damaged part of the stressed cell *en masse* and gives rise to a ring-shaped "*autophagosome*". This autophagosome fuses with a lysosome to give rise to "*autolysosome*" where the acidic milieu degrades the cellular cargo into recyclable metabolites which are absorbed into the cytosolic pool for further cellular nourishment. Till now 41 *Atg* (*Au*tophagy-related) genes have been identified to regulate the process of autophagy, and most of the genes have been identified using yeast as a model organism. This phenomenon remains holistically conserved across different eukaryotes.

5.2 Fundamentals of Autophagy and its Duality in TME

Briefly speaking, autophagy can be categorized broadly into three types – macroautophagy, microautophagy, and chaperone-mediated autophagy. In this chapter, we will be dealing with macroautophagy which will be mentioned henceforth simply as autophagy. Before we open the topic about the relevance of autophagy in the tumor microenvironment, let us understand how the mechanism of autophagy function and why there is so much debate regarding this. Stress due to various stimuli like a response to therapy, increasing population of cancer cells, hypoxic tumor core region, alteration of pH, temperature, nutrient starvation, and pressure from infiltrating immune cells signal envisages inactivation of mTOR (mammalian Target Of Rapamycin) triggering the de novo synthesis of a double-layered isolation membrane that leads to nucleation of the initiation step of autophagy. Stress evokes a natural autophagic response in a cell through the class III phosphatidylinositol-3-kinase, hVps34 (human vesicular protein sorting 34) complex. Binding of Beclin1 with hVps34 is involved in rapid catalysis of PI3P (phosphatidylinositol-3-phosphate) production. PI3P, Beclin1, and ATG14 mostly remains tethered together as a specialized pre-autophagic structure called omagesome (named as such due to the Ω shaped structure) giving rise to the initial development of autophagosome. The presence of PI3P phosphatases like PTEN, JUMPY plays an important regulatory signal in the development of budding autophagic vesicle structure (Mukhopadhyay et al. 2016).

In the mammalian system, the initiation of autophagy is quite complex and involves the ULK1/2 complex which comprises FIP200, ATG 13, and ATG101 proteins. After initiation, the second major phase of autophagy is elongation, which is taken care of by the two ubiquitination-like systems LC3 and ATG5-ATG12 processing which gives rise to a mature double-membrane autophagosome structure. Finally, the autophagosome docks and fuses with the lysosome thereby forming

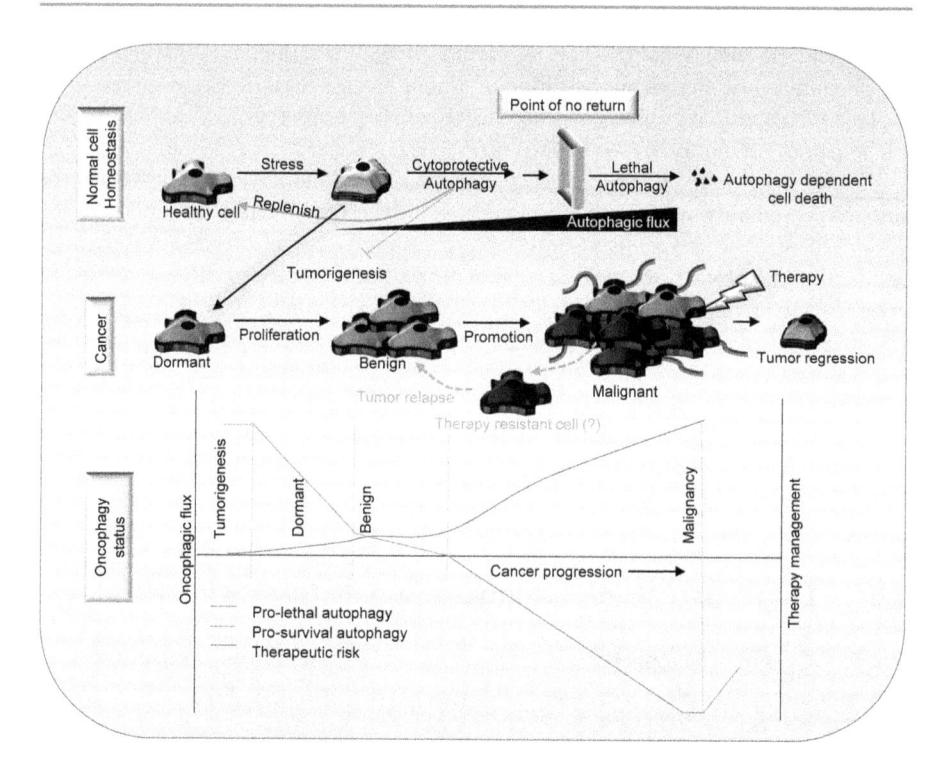

Fig. 5.2 Schematic representation of the dynamics of autophagy. Healthy cells exhibiting normal homeostasis undergoes cytoprotective autophagy during stress, however, when this continues for a protracted period beyond a point of no return then it becomes lethal and causes autophagy-dependent cell death. A cell undergoing tumorigenesis face extreme metabolic inhibition by cytoprotective autophagy, however, once the malignancy sets in then it serves as a promoter of tumor proliferation. A characteristic duality of autophagy which is different from the normal cellular autophagy is defined as oncophagy and the status of its progression according to the stages of the tumor is shown

autolysosome which help to inundate the cytoplasm with nutrients. Laura et al. showed that survivability of the disseminated solitary dormant breast cancer and metastatic tumor recurrence rely on triggering autophagy-mediated through ATG7 (Vera-Ramirez et al. 2018). The topic of autophagy is a highly complicated process with different and diverse nature at every turn of the tumorigenesis process. Any sort of stress implicated in a living cell provides an impetus to evoke an autophagy response. However, the dynamics of autophagy is not operated by a simple "on/off" switch in context to a cell that is undergoing tumorigenesis to develop into a cancer cell (Fig. 5.2).

Initially, autophagy behaves in a cytoprotective manner to reduce the stress and tumorigenic consequences in a cell. But once the benign tumor metastasizes to full-blown malignancy this cytoprotective mechanism shifts into a lethal phenotype by supporting and/or fueling the growing tumor cells to ensure their rapid proliferation and hijacks the metabolic fueling from the body.

Work in this field deserves special attention because of the janus role of autophagy in cancer. In a healthy cell, the cytoprotective role of autophagy prevents the buildup of excess stress-mediated damage and tumorigenesis. Protracted stress-mediated autophagy in these cells commits the cell to cross the "autophagic point of no return" an elusive concept that commits the cell to crossover an irreversible autophagy barricade that pushes the cytoprotective phenomena to a lethal makeshift (as elaborated in Fig. 5.2, Loos et al. 2011) and undergoes autophagy-dependent cell death (type II programmed cell death). Therefore, for a cancer cell, it becomes crucial to understand the autophagic point of no return to target them without making them therapeutically latent to pose risks of cancer recurrence. Autophagic modulation of TME helps the cancer cells strive for austere conditions like hypoxia, extreme acidosis, and accumulation of biological wastes in the interstitial spaces along with shielding the infiltrating chemotherapy.

Emerging research has focused on the vital role of TME that is capable of modulating autophagy and vice-versa. The immediate TME faces a rush of a diverse cellular niche. How the tumor environment can modulate the autophagy quotient during therapy to act against or in favor of the clinician is a hot field of research in the present time and will be discussed in detail in the later sections. There exists a lot of lacunae between medical research and patient therapy. This work tries to highlight these points with a focus on the major paradoxes and paradigms of the molecular mechanism at work behind the autophagic dynamics that modulate the TME inside a cancer cell.

5.3 Autophagic Regulation of Different Metabolites in TME

5.3.1 Autophagic Contributions of Glycolytic Intermediates

Tumor cells become attuned to dysregulated glucose metabolism leading to disease recurrence and cancer cell survival benefit under acidic pH stress by glycolysis to mediate autophagic stress through the PI3K/AKT/mTOR pathway that prolongs cancer cell survival by resisting apoptosis (Lue et al. 2017; Wojtkowiak et al. 2012). Low glucose in TME triggers energy sensor AMP-activated protein kinase (AMPK) mediated autophagy induction which becomes responsible for the degradation of damaged or long-lived components to replenish nutrients to compensate the stress (Williams et al. 2009). On the other hand, high glucose TME accelerated the SREBP1-autophagy axis to play a crucial role in pancreatic ductal adenocarcinoma (PDAC) progression (Zhou et al. 2019). Karsli-Uzunbas et al. showed that conditionally deleted ATG7 is expendable for short-term survival, but help to avoid the lethal hypoglycemic condition, cachexia during starvation highlighting a new role of autophagy in glucose homeostasis, and lung tumor maintenance (Karsli-Uzunbas et al. 2014). Chemotherapeutic insults in autophagy inhibited condition under glucose restricted condition led to the upregulation of period circadian clock 2 (PER2), a tumor suppressor protein in the colorectal cancer cell (Schroll et al. 2017). Interestingly, GPx1 degradation caused by glucose deprivation triggered

ROS/AMPK signaling mediated autophagy in PDAC. Also, GPx1 may modulate glycolysis inhibition in pancreatic cancer under the glucose-starved situation while GPx1 overexpression and autophagy inhibition resulted in caspase-dependent apoptosis (Meng et al. 2018).

Upregulation of another glycolytic enzyme hexokinase (HKII) was found to be involved in escalating autophagy in the hypoxic TME of tongue squamous cell carcinoma (Chen et al. 2018). In a stressful TME, miRNA-7 inhibits autophagy by upregulation of LKB1-AMPK-mTOR signaling, thereby diminishing the supply of intracellular glucose pool available to fuel glycolysis (Gu et al. 2017). Glucose shortage mediated metabolic catastrophe resulted in autophagy-dependent BCR/Abl protein degradation in chronic myeloid leukemia cells, which are resistant to tyrosine kinase inhibitor-based therapy (Bono et al. 2016). βIII-Tubulin was known to decrease the reliance of cells on glycolytic metabolism, priming them to cope with low glucose stress in TME and protect from the endoplasmic reticulum (ER) by co-immunoprecipitating with GRP78 in non-small cell lung cancer (NSCLC) (Parker et al. 2016). GRP78 expression in limited glucose condition, led to dysregulation of pyruvate kinase M2 (PKM2) that triggered mitochondrial pyruvate dehydrogenase A (PDHA) and B (PDHB), causing a metabolic transition from glycolysis to the TCA cycle (Li et al. 2015). Glucose limitation in TME led to GRP78 mediated autophagic activation that led to degradation of IKKβ, which instigated inactivation of the NF–κB pathway and consequently altered the expression of PKM2, GLUT1, and HIF-1α.

Another glycolytic metabolite lactate helps the viability of glucose deprived melanoma TME in hypoxic conditions by repressing autophagy (Matsuo et al. 2019). In this regard tumor suppressor gene ANKDD1A reduces the half-life of HIF1α through an increase of FIH1; downregulation of glucose uptake and lactate production impedes autophagy along with triggering apoptosis in hypoxic glioblastoma multiforme (GBM) TME (Feng et al. 2019). Hypoxic TME of PDAC is involved in autophagy-mediated degradation of stromal lumican involved in regulating cancer progression (Sarcar et al. 2019).

Acidification of hypoxic and glucose deprived tumor cores by exogenous lactate supplementation is reported to prevent cell death by inhibiting autophagy in B16 melanoma cells (Matsuo and Sadzuka 2018). On the other hand, depletion of MCT4 on NK cells executes a compensatory metabolic rewiring by inducing autophagy to minimize the acidic extracellular breast tumor microenvironment that is involved with the export of lactate (Long et al. 2018). Following the phenomenon of reverse Warburg effect, cancer cells utilize lactate as an active metabolite and shuttle to TME to inflict metabolic reprogramming. In this regard, 2-methoxyestradiol, can inverse L-lactate-induced metabolic reprogramming in osteosarcoma 143B tumor cells (Gorska-Ponikowska et al. 2017). Phosphoproteomics screening established that mitochondrial Akt-PDK1 signaling alters tumor metabolism toward glycolysis by inhibiting autophagy, apoptosis, and support proliferation of hypoxic tumors (Chae et al. 2016).

5.3.2 Autophagy Contributions of Amino Acids

In this section, we briefly report about the major contributions of amino acids that are under the autophagic regulatory control network in TME. Among the amino acids, the glutamine plays a crucial role to fuel the tricarboxylic acid (TCA) cycle and maintain the pool size of antioxidants and ammonia.

Glutaminolysis-derived ammonia diffuses into the interstitial space of TME and is involved in regulating anti autophagy protein TIGAR (Mariño and Kroemer 2010) in MCF7 cells. Coculturing tamoxifen-resistant cancer cell with stromal fibroblasts drives catabolism of tumor stroma along with the anabolic activity of tumor cell (Albanese et al. 2011). Witkiewicz et al. showed that screening of lethal breast TME supported "Autophagic Tumor Stroma Model of Cancer Metabolism" and identified that loss of stromal caveolin-1 status in breast cancers leads to autophagy-mediated recycling of nutrients like lactate, ketones, and glutamine to feed anabolic events in cancer that led to metastasis and poor clinical prognosis (Witkiewicz et al. 2011). Interestingly, from the perspective of autophagy, it was reported that SNAT7 is a critical primary lysosomal glutamine exporter essential for the growth and proliferation of cancer (Verdon et al. 2017). Besides this, the relevance of arginine in TME was elucidated by work from Beth Levine's lab where it was shown that autophagic impairment leads to the secretion of arginase and subsequent degradation of arginine was recognized as a metabolic vulnerability in cancer (Poillet-Perez et al. 2018).

On the other hand, preconditioning of primary human renal proximal tubular epithelial cells without tryptophan led to enhanced survivability in hypoxic conditions by triggering autophagy (Eleftheriadis et al. 2017). Exemplary findings established that the autophagic modulation in pancreatic stellate cells controls alanine secretion which feeds the PDAC in an austere tumor microenvironment to promote growth (Sousa et al. 2016).

5.3.3 Autophagy Contributions of Lipids

Vital research is being carried out to understand how the TME is being inundated with lipid molecules which in turn like a vicious cycle promote more fat-derived energy to fuel the tumor cell. Wen et al. co-cultured adipocytes with colon cancer cells to establish the mechanism by virtue of which adipocyte-derived free fatty acids are released to the cancer cells, thereby inducing autophagy as a result of AMPK activation (Wen et al. 2017). SIRT3 mediated autophagy was reported to adipocyte differentiation, and lipophagy mediated through increased activity of pyruvate dehydrogenase under high salt conditions which may result in an influx of lipid droplets in TME (Wei et al. 2019; Zhang et al. 2020).

5.4 Autophagic Regulation of Immune System in TME

An ecosystem in a TME comprises a plethora of immunomodulatory signals secreted by the cancer cells, CAFs, stromal cells, fibroblasts, macrophages, and T cells that modify each other to define the metabolic plasticity of the tumor. Tumor-associated macrophages (TAMs) create an environment conducive to facilitate the progression of the tumor. Tumor cell-released autophagosomes (TRAPs) stimulate IL-10-producing B cells and inhibit neutrophil production to repress antitumor immunogenic signals. Moreover, within TME, TRAPs are also responsible for converting TAMs into immunosuppressive M2 like macrophage secreting PD-L1, IL10. Inhibition of autophagy by Beclin1 silencing led to a decrease of TRAPs and enhanced T cell activation (Wen et al. 2018). The rush of antitumor response combined with translocation of calreticulin (CRT), extracellular release of ATP, danger-associated molecular patterns (DAMPs), high mobility group box 1 (HMGB1), and stimulation of type I interferon (IFN) led to a form of cell death which is regarded as immunogenic cell death (ICD). Targeting ICD by a combination of autophagy and chemotherapy is a potential approach to improve the prognosis of cancer patients with a long-term immune memory response to protect against the possible chance of tumor recurrence (Wang et al. 2018). For instance, Sigma1 utilizes autophagy to eradicate the functional PD-L1 from the cell surface to regulate the tumor immune microenvironment (Maher et al. 2018).

There is debate over the topic of mechanism that autophagy utilizes to trigger anticancer immunity while there are reports that show autophagy disarm anticancer immunity mediated by cytotoxic T cells and natural killer (NK) cells (Li et al. 2017). Autophagy-dependent regulation of TGF-β in myeloid cells leads to M2 macrophage accumulation in TME and is involved in controlling metastasis along with epithelial-mesenchymal transition (EMT) of tumor cells (Jinushi et al. 2017). The motility of aggressive metastatic cancer cell in a TME cannot be attributed due to a single factor, rather a hypoxic tumor core due to limited vascularization creating a hypoxic region which gives rise to mesenchymal-like carcinoma cells that exhibit high EMT and acquire stem cell-like propensity (Pietilä et al. 2016). Hypoxia-induced autophagy eliminates pro-apoptotic NK-derived serine protease GZMB/granzyme B, thereby blocking NK-mediated target cell apoptosis in breast tumor (Viry et al. 2014). Interestingly, chemotherapy dependent p53 activation helps in granzyme B and NK cells mediated breast tumor killing through induction of autophagy (Chollat-Namy et al. 2019).

Recent work confirmed that hypoxic TME activates autophagy as well as suppresses the immune surveillance of melanoma by NK cells through modulation of Cx43-mediated intercellular communications (Tittarelli et al. 2015). Although autophagy can be expendable for chemotherapy-induced cell death, but it is critical for its immunogenicity (Michaud et al. 2011). Tumor cells with functional autophagy responded to chemotherapy by attracting dendritic cells and T lymphocytes into TME. Inhibition of autophagy suppressed the discharge of ATP from dying cancel cell. Besides it will be interesting to figure out how autophagy can be used as a tool in a framework where cancer immunoediting integrates the immune system's dual host-protective and tumor-promoting roles (Schreiber et al. 2011).

On the other hand inhibition of autophagy, as clinically identified by loss of LC3, HMGB1 staining followed by characteristic immune infiltration, and metastasis-free survival was identified to be a driver of tumor progression due to their adverse role in anticancer immunosurveillance (Ladoire et al. 2016). On a similar note, Atg5-mutant KRas (G12D)-driven lung cancer was reduced by depletion of CD25+ Treg cells thereby demonstrating that autophagy accelerates tumor progression; however, it represses early oncogenesis, highlighting a role between autophagy and regulatory T cell controlled anticancer immunity (Rao et al. 2014).

Gene profiling studies in early tumors of FIP200 conditional knockout mice saw T cells infiltration and CXCL10 secretion in TME (Wei et al. 2011). Interestingly, proinflammatory cytokine IL-1β secretion via an unconventional export pathway depending on Atg5 and involving Golgi reassembly stacking protein (GRASP) paralogues, GRASP55 (GORASP2) and Rab8a (Dupont et al. 2011). Moreover, elimination of LC3B and Beclin 1 was showed to accompany an increase in ROS, along with heightened stimulation of caspase-1 and secretion of interleukin 1β (IL-1β) and IL-18 (Nakahira et al. 2011). Autophagy blockade led to M2 to M1 repolarization of TAM to increased sensitivity of laryngeal cancer cells to cisplatin in mice (Guo et al. 2019).

Autophagic involvement in TME often presents a puzzle in that the infiltrating lymphocytes with a TME are dysfunctional in situ, however, they exhibit stem cell-like properties with massive metastatic potential. It was reported that increased abundance of extracellular potassium restricts T cell effector activity by restraining nutrient uptake, thus evoking autophagy and decline of histone acetylation at effector and exhaustion loci. This in turn produces CD8+ T cells with enhanced in vivo persistence, multipotency, and tumor clearance (Vodnala et al. 2019).

Besides this, the nonconventional role of autophagy was recently highlighted by Cunha et al., where they showed antitumor effects of LC3-associated phagocytosis (LAP) machinery impairment require tumor-infiltrating T cells, dependent upon STING and type I interferon response. Furthermore, they also showed that autophagy induction by myeloid cells in TME suppress T lymphocytes by effecting LAP (Cunha et al. 2018).

5.5 Concluding Remarks

The most important aspect of the knowledge of autophagic relevance holds a clue to modify the TME to determine the route to sensitization of a tumor (Mukhopadhyay et al. 2016). The conventional therapeutic regime in clinical trials focusing on autophagy, makes use of CQ/HCQ-based autophagy inhibition followed by different drugs like cisplatin, paclitaxel, FOLFOX, or radiation. Considering the duality of autophagy (Panda et al. 2015) and the intersecting cross-talking pathway shared with apoptosis (Mukhopadhyay et al. 2014) it becomes important to understand the autophagic point of no return.

But the biggest challenge in this regard of utilizing autophagy inhibition/induction-based medicine is that we still do not have any standard drug for clinical use that is extremely potent and specific towards tumor cells. Most of the drugs result in a holistic modulation of the tumor and all the neighboring cells so it becomes difficult for the effector killer cells or the therapy to intrude the thick TME and reach the core of the highly aggressive cancer cell that have already achieved high autophagic potential as an anti-therapeutic defense response. Subsequent to this effect it becomes a very big problem to estimate the accurate dose of therapy. It remains a clinical dilemma, to hit the right target without inducing side effects (Yang and Klionsky 2020). Moreover, there is also variation in stage-specific autophagy level that adds complexity to its dual character. So, it becomes imperative to search for any cancer stage-specific genetic event that can be assured to be a good modulatory marker of TME. An in-depth investigation revealing the molecular mechanism of the potential clinical drug on the interplay of TME's autophagy and its metabolic signatures should be carried out to deliver a coherent approach for tackling this Achilles' heel of cancer.

References

Albanese C, Machado FS, Tanowitz HB (2011) Glutamine and the tumor microenvironment: understanding the mechanisms that fuel cancer progression. Cancer Biol Ther 12 (12):1098–1100. https://doi.org/10.4161/cbt.12.12.18856. Epub 2011 Dec 15

Bono S, Lulli M, D'Agostino VG, Di Gesualdo F, Loffredo R, Cipolleschi MG, Provenzani A, Rovida E, Dello Sbarba P (2016) Different BCR/Abl protein suppression patterns as a converging trait of chronic myeloid leukemia cell adaptation to energy restriction. Oncotarget 7 (51):84810–84825. https://doi.org/10.18632/oncotarget.13319

Chae YC, Vaira V, Caino MC, Tang HY, Seo JH, Kossenkov AV, Ottobrini L, Martelli C, Lucignani G, Bertolini I, Locatelli M, Bryant KG, Ghosh JC, Lisanti S, Ku B, Bosari S, Languino LR, Speicher DW, Altieri DC (2016) Mitochondrial Akt regulation of hypoxic tumor reprogramming. Cancer Cell 30(2):257–272. https://doi.org/10.1016/j.ccell.2016.07.004

Chen G, Zhang Y, Liang J, Li W, Zhu Y, Zhang M, Wang C, Hou J (2018) Deregulation of hexokinase II is associated with glycolysis, autophagy, and the epithelial-mesenchymal transition in tongue squamous cell carcinoma under hypoxia. Biomed Res Int 2018:8480762. https://doi.org/10.1155/2018/8480762. eCollection 2018

Chollat-Namy M, Ben Safta-Saadoun T, Haferssas D, Meurice G, Chouaib S, Thiery J (2019) The pharmalogical reactivation of p53 function improves breast tumor cell lysis by granzyme B and NK cells through induction of autophagy. Cell Death Dis 10(10):695. https://doi.org/10.1038/s41419-019-1950-1

Cunha LD, Yang M, Carter R, Guy C, Harris L, Crawford JC, Quarato G, Boada-Romero E, Kalkavan H, Johnson MDL, Natarajan S, Turnis ME, Finkelstein D, Opferman JT, Gawad C, Green DR (2018) LC3-associated phagocytosis in myeloid cells promotes tumor immune tolerance. Cell 175(2):429–441.e16. https://doi.org/10.1016/j.cell.2018.08.061. Epub 2018 Sep 20

Dupont N, Jiang S, Pilli M, Ornatowski W, Bhattacharya D, Deretic V (2011) Autophagy-based unconventional secretory pathway for extracellular delivery of IL-1β. EMBO J 30 (23):4701–4711. https://doi.org/10.1038/emboj.2011.398

Eleftheriadis T, Pissas G, Sounidaki M, Antoniadis N, Antoniadi G, Liakopoulos V, Stefanidis I (2017) Preconditioning of primary human renal proximal tubular epithelial cells without tryptophan increases survival under hypoxia by inducing autophagy. Int Urol Nephrol 49 (7):1297–1307. https://doi.org/10.1007/s11255-017-1596-9. Epub 2017 Apr 17

Feng J, Zhang Y, She X, Sun Y, Fan L, Ren X, Fu H, Liu C, Li P, Zhao C, Liu Q, Liu Q, Li G, Wu M (2019) Hypermethylated gene ANKDD1A is a candidate tumor suppressor that interacts with FIH1 and decreases HIF1α stability to inhibit cell autophagy in the glioblastoma multiforme hypoxia microenvironment. Oncogene 38(1):103–119. https://doi.org/10.1038/s41388-018-0423-9. Epub 2018 Aug 6

Gorska-Ponikowska M, Kuban-Jankowska A, Daca A, Nussberger S (2017) 2-Methoxyestradiol reverses the pro-carcinogenic effect of l-lactate in osteosarcoma 143B cells. Cancer Genomics Proteomics 14(6):483–493. Review

Gu DN, Jiang MJ, Mei Z, Dai JJ, Dai CY, Fang C, Huang Q, Tian L (2017) microRNA-7 impairs autophagy-derived pools of glucose to suppress pancreatic cancer progression. Cancer Lett 400:69–78. https://doi.org/10.1016/j.canlet.2017.04.020. Epub 2017 Apr 25

Guo Y, Feng Y, Cui X, Wang Q, Pan X (2019) Autophagy inhibition induces the repolarisation of tumour-associated macrophages and enhances chemosensitivity of laryngeal cancer cells to cisplatin in mice. Cancer Immunol Immunother 68(12):1909–1920. https://doi.org/10.1007/s00262-019-02415-8. Epub 2019 Oct 22

Jinushi M, Morita T, Xu Z, Kinoshita I, Dosaka-Akita H, Yagita H, Kawakami Y (2017) Autophagy-dependent regulation of tumor metastasis by myeloid cells. PLoS One 12(7): e0179357. https://doi.org/10.1371/journal.pone.0179357. eCollection 2017

Karsli-Uzunbas G, Guo JY, Price S, Teng X, Laddha SV, Khor S, Kalaany NY, Jacks T, Chan CS, Rabinowitz JD, White E (2014) Autophagy is required for glucose homeostasis and lung tumor maintenance. Cancer Discov 4(8):914–927. https://doi.org/10.1158/2159-8290.CD-14-0363. Epub 2014 May 29

Ladoire S, Enot D, Senovilla L, Ghiringhelli F, Poirier-Colame V, Chaba K, Semeraro M, Chaix M, Penault-Llorca F, Arnould L, Poillot ML, Arveux P, Delaloge S, Andre F, Zitvogel L, Kroemer G (2016) The presence of LC3B puncta and HMGB1 expression in malignant cells correlate with the immune infiltrate in breast cancer. Autophagy 12(5):864–875. https://doi.org/10.1080/15548627.2016.1154244. Epub 2016 Mar 16

Li Z, Wang Y, Newton IP, Zhang L, Ji P, Li Z (2015) GRP78 is implicated in the modulation of tumor aerobic glycolysis by promoting autophagic degradation of KKβ. Cell Signal 27 (6):1237–1245. https://doi.org/10.1016/j.cellsig.2015.02.030. Epub 2015 Mar 5

Li YY, Feun LG, Thongkum A, Tu CH, Chen SM, Wangpaichitr M, Wu C, Kuo MT, Savaraj N (2017) Autophagic mechanism in anti-cancer immunity: its pros and cons for cancer therapy. Int J Mol Sci 18(6):E1297. Review. https://doi.org/10.3390/ijms18061297

Long Y, Gao Z, Hu X, Xiang F, Wu Z, Zhang J, Han X, Yin L, Qin J, Lan L, Yin F, Wang Y (2018) Downregulation of MCT4 for lactate exchange promotes the cytotoxicity of NK cells in breast carcinoma. Cancer Med 7(9):4690–4700. https://doi.org/10.1002/cam4.1713. Epub 2018 Jul 26

Loos B, Genade S, Ellis B, Lochner A, Engelbrecht AM (2011) At the core of survival: autophagy delays the onset of both apoptotic and necrotic cell death in a model of ischemic cell injury. Exp Cell Res 317(10):1437–1453. https://doi.org/10.1016/j.yexcr.2011.03.011. Epub 2011 Mar 21

Lue HW, Podolak J, Kolahi K, Cheng L, Rao S, Garg D, Xue CH, Rantala JK, Tyner JW, Thornburg KL, Martinez-Acevedo A, Liu JJ, Amling CL, Truillet C, Louie SM, Anderson KE, Evans MJ, O'Donnell VB, Nomura DK, Drake JM, Ritz A, Thomas GV (2017) Metabolic reprogramming ensures cancer cell survival despite oncogenic signaling blockade. Genes Dev 31(20):2067–2084. https://doi.org/10.1101/gad.305292.117

Lyssiotis CA, Kimmelman AC (2017) Metabolic interactions in the tumor microenvironment. Trends Cell Biol 27(11):863–875. Review. https://doi.org/10.1016/j.tcb.2017.06.003. Epub 2017 Jul 19

Maher CM, Thomas JD, Haas DA, Longen CG, Oyer HM, Tong JY, Kim FJ (2018) Small-molecule Sigma1 modulator induces autophagic degradation of PD-L1. Mol Cancer Res 16 (2):243–255. https://doi.org/10.1158/1541-7786.MCR-17-0166. Epub 2017 Nov 8

Mariño G, Kroemer G (2010) Ammonia: a diffusible factor released by proliferating cells that induces autophagy. Sci Signal 3(124):pe19. Review. https://doi.org/10.1126/scisignal.3124pe19

Matsuo T, Sadzuka Y (2018) Extracellular acidification by lactic acid suppresses glucose deprivation-induced cell death and autophagy in B16 melanoma cells. Biochem Biophys Res Commun 496(4):1357–1361. https://doi.org/10.1016/j.bbrc.2018.02.022. Epub 2018 Feb 5

Matsuo T, Daishaku S, Sadzuka Y (2019) Lactic acid promotes cell survival by blocking autophagy of b16f10 mouse melanoma cells under glucose deprivation and hypoxic conditions. Biol Pharm Bull 42(5):837–839. https://doi.org/10.1248/bpb.b18-00919

Meng Q, Xu J, Liang C, Liu J, Hua J, Zhang Y, Ni Q, Shi S, Yu X (2018) GPx1 is involved in the induction of protective autophagy in pancreatic cancer cells in response to glucose deprivation. Cell Death Dis 9(12):1187. https://doi.org/10.1038/s41419-018-1244-z

Michaud M, Martins I, Sukkurwala AQ, Adjemian S, Ma Y, Pellegatti P, Shen S, Kepp O, Scoazec M, Mignot G, Rello-Varona S, Tailler M, Menger L, Vacchelli E, Galluzzi L, Ghiringhelli F, di Virgilio F, Zitvogel L, Kroemer G (2011) Autophagy-dependent anticancer immune responses induced by chemotherapeutic agents in mice. Science 334(6062):1573–1577. https://doi.org/10.1126/science.1208347

Mukhopadhyay S, Panda PK, Sinha N, Das DN, Bhutia SK (2014) Autophagy and apoptosis: where do they meet? Apoptosis 19(4):555–566. Review. https://doi.org/10.1007/s10495-014-0967-2

Mukhopadhyay S, Sinha N, Das DN, Panda PK, Naik PP, Bhutia SK (2016) Clinical relevance of autophagic therapy in cancer: investigating the current trends, challenges, and future prospects. Crit Rev Clin Lab Sci 53(4):228–252. Review. https://doi.org/10.3109/10408363.2015.1135103. Epub 2016 Feb 16

Nakahira K, Haspel JA, Rathinam VA, Lee SJ, Dolinay T, Lam HC, Englert JA, Rabinovitch M, Cernadas M, Kim HP, Fitzgerald KA, Ryter SW, Choi AM (2011) Autophagy proteins regulate innate immune responses by inhibiting the release of mitochondrial DNA mediated by the NALP3 inflammasome. Nat Immunol 12(3):222–230. https://doi.org/10.1038/ni.1980. Epub 2010 Dec 12

Panda PK, Mukhopadhyay S, Das DN, Sinha N, Naik PP, Bhutia SK (2015) Mechanism of autophagic regulation in carcinogenesis and cancer therapeutics. Semin Cell Dev Biol 39:43–55. Review. https://doi.org/10.1016/j.semcdb.2015.02.013. Epub 2015 Feb 25

Parker AL, Turner N, McCarroll JA, Kavallaris M (2016) βIII-Tubulin alters glucose metabolism and stress response signaling to promote cell survival and proliferation in glucose-starved non-small cell lung cancer cells. Carcinogenesis 37(8):787–798. https://doi.org/10.1093/carcin/bgw058

Pietilä M, Ivaska J, Mani SA (2016) Whom to blame for metastasis, the epithelial-mesenchymal transition or the tumor microenvironment? Cancer Lett 380(1):359–368. Review. https://doi.org/10.1016/j.canlet.2015.12.033. Epub 2016 Jan 11

Poillet-Perez L, Xie X, Zhan L, Yang Y, Sharp DW, Hu ZS, Su X, Maganti A, Jiang C, Lu W, Zheng H, Bosenberg MW, Mehnert JM, Guo JY, Lattime E, Rabinowitz JD, White E (2018) Autophagy maintains tumour growth through circulating arginine. Nature 563(7732):569–573. https://doi.org/10.1038/s41586-018-0697-7. Epub 2018 Nov 14

Rao S, Tortola L, Perlot T, Wirnsberger G, Novatchkova M, Nitsch R, Sykacek P, Frank L, Schramek D, Komnenovic V, Sigl V, Aumayr K, Schmauss G, Fellner N, Handschuh S, Glösmann M, Pasierbek P, Schlederer M, Resch GP, Ma Y, Yang H, Popper H, Kenner L, Kroemer G, Penninger JM (2014) A dual role for autophagy in a murine model of lung cancer. Nat Commun 5:3056. https://doi.org/10.1038/ncomms4056

Sarcar B, Li X, Fleming JB (2019) Hypoxia-induced autophagy degrades stromal lumican into tumor microenvironment of pancreatic ductal adenocarcinoma: a mini-review. J Cancer Treat Diagn 3(1):22–27. https://doi.org/10.29245/2578-2967/2019/1.1165. Epub 2019 Feb 22

Schreiber RD, Old LJ, Smyth MJ (2011) Cancer immunoediting: integrating immunity's roles in cancer suppression and promotion. Science 331(6024):1565–1570. Review. https://doi.org/10.1126/science.1203486

Schroll MM, LaBonia GJ, Ludwig KR, Hummon AB (2017) Glucose restriction combined with autophagy inhibition and chemotherapy in hct 116 spheroids decreases cell clonogenicity and viability regulated by tumor suppressor genes. J Proteome Res 16(8):3009–3018. https://doi.org/10.1021/acs.jproteome.7b00293. Epub 2017 Jul 3

Sousa CM, Biancur DE, Wang X, Halbrook CJ, Sherman MH, Zhang L, Kremer D, Hwang RF, Witkiewicz AK, Ying H, Asara JM, Evans RM, Cantley LC, Lyssiotis CA, Kimmelman AC (2016) Pancreatic stellate cells support tumour metabolism through autophagic alanine secretion. Nature 536(7617):479–483; 540(7631):150. https://doi.org/10.1038/nature19084. Epub 2016 Aug 10

Tittarelli A, Janji B, Van Moer K, Noman MZ, Chouaib S (2015) The selective degradation of synaptic Connexin 43 protein by hypoxia-induced autophagy impairs natural killer cell-mediated tumor cell killing. J Biol Chem 290(39):23670–23679. https://doi.org/10.1074/jbc.M115.651547. Epub 2015 Jul 28

Vera-Ramirez L, Vodnala SK, Nini R, Hunter KW, Green JE (2018) Autophagy promotes the survival of dormant breast cancer cells and metastatic tumour recurrence. Nat Commun 9 (1):1944. https://doi.org/10.1038/s41467-018-04070-6

Verdon Q, Boonen M, Ribes C, Jadot M, Gasnier B, Sagné C (2017) SNAT7 is the primary lysosomal glutamine exporter required for extracellular protein-dependent growth of cancer cells. Proc Natl Acad Sci U S A 114(18):E3602–E3611. https://doi.org/10.1073/pnas.1617066114. Epub 2017 Apr 17

Viry E, Baginska J, Berchem G, Noman MZ, Medves S, Chouaib S, Janji B (2014) Autophagic degradation of GZMB/granzyme B: a new mechanism of hypoxic tumor cell escape from natural killer cell-mediated lysis. Autophagy 10(1):173–175. https://doi.org/10.4161/auto.26924. Epub 2013 Nov 15

Vodnala SK, Eil R, Kishton RJ, Sukumar M, Yamamoto TN, Ha NH, Lee PH, Shin M, Patel SJ, Yu Z, Palmer DC, Kruhlak MJ, Liu X, Locasale JW, Huang J, Roychoudhuri R, Finkel T, Klebanoff CA, Restifo NP (2019) T cell stemness and dysfunction in tumors are triggered by a common mechanism. Science 363(6434):eaau0135. https://doi.org/10.1126/science.aau0135

Wang YJ, Fletcher R, Yu J, Zhang L (2018) Immunogenic effects of chemotherapy-induced tumor cell death. Genes Dis 5(3):194–203. Review. https://doi.org/10.1016/j.gendis.2018.05.003. eCollection 2018 Sep

Wei H, Wei S, Gan B, Peng X, Zou W, Guan JL (2011) Suppression of autophagy by FIP200 deletion inhibits mammary tumorigenesis. Genes Dev 25(14):1510–1527. https://doi.org/10.1101/gad.2051011

Wei T, Huang G, Liu P, Gao J, Huang C, Sun M, Shen W (2019) Sirtuin 3-mediated pyruvate dehydrogenase activity determines brown adipocytes phenotype under high-salt conditions. Cell Death Dis 10(8):614. https://doi.org/10.1038/s41419-019-1834-4

Wen YA, Xing X, Harris JW, Zaytseva YY, Mitov MI, Napier DL, Weiss HL, Mark Evers B, Gao T (2017) Adipocytes activate mitochondrial fatty acid oxidation and autophagy to promote tumor growth in colon cancer. Cell Death Dis 8(2):e2593. https://doi.org/10.1038/cddis.2017.21

Wen ZF, Liu H, Gao R, Zhou M, Ma J, Zhang Y, Zhao J, Chen Y, Zhang T, Huang F, Pan N, Zhang J, Fox BA, Hu HM, Wang LX (2018) Tumor cell-released autophagosomes (TRAPs) promote immunosuppression through induction of M2-like macrophages with increased expression of PD-L1. J Immunother Cancer 6(1):151. https://doi.org/10.1186/s40425-018-0452-5

Williams T, Forsberg LJ, Viollet B, Brenman JE (2009) Basal autophagy induction without AMP-activated protein kinase under low glucose conditions. Autophagy 5(8):1155–1165. Epub 2009 Nov 16

Witkiewicz AK, Kline J, Queenan M, Brody JR, Tsirigos A, Bilal E, Pavlides S, Ertel A, Sotgia F, Lisanti MP (2011) Molecular profiling of a lethal tumor microenvironment, as defined by stromal caveolin-1 status in breast cancers. Cell Cycle 10(11):1794–1809. Epub 2011 Jun 1

Wojtkowiak JW, Rothberg JM, Kumar V, Schramm KJ, Haller E, Proemsey JB, Lloyd MC, Sloane BF, Gillies RJ (2012) Chronic autophagy is a cellular adaptation to tumor acidic pH microenvironments. Cancer Res 72(16):3938–3947. https://doi.org/10.1158/0008-5472.CAN-11-3881. Epub 2012 Jun 19

Yang Y, Klionsky DJ (2020) Autophagy and disease: unanswered questions. Cell Death Differ 27:858. Review. https://doi.org/10.1038/s41418-019-0480-9 [Epub ahead of print]

Zhang T, Liu J, Tong Q, Lin L (2020) SIRT3 Acts as a Positive Autophagy Regulator to Promote Lipid Mobilization in Adipocytes via Activating AMPK. Int J Mol Sci 21(2):372. https://doi.org/10.3390/ijms21020372

Zhou C, Qian W, Li J, Ma J, Chen X, Jiang Z, Cheng L, Duan W, Wang Z, Wu Z, Ma Q, Li X (2019) High glucose microenvironment accelerates tumor growth via SREBP1-autophagy axis in pancreatic cancer. J Exp Clin Cancer Res 38(1):302. https://doi.org/10.1186/s13046-019-1288-7

Mitophagy and Reverse Warburg Effect: Metabolic Compartmentalization of Tumor Microenvironment

6

Prajna Paramita Naik

Abstract

'The Warburg effect' is one of the aberrant glucose metabolism pathways in cancer cells that generate malignant phenotypes and promotes cancer progression. However, in the year 2009, a novel model called 'two-compartment metabolic coupling' model or 'the reverse Warburg effect' was proposed where the tumor stromal plays a crucial role in the process of tumor progression. Based on this new model, the present review summarizes the autophagic stroma model of cancer and multiple compartment model of tumor metabolism. Cancer-associated fibroblast cells in tumor microenvironment undergo aerobic glycolysis (the reverse Warburg effect) just like the cancer cells. Such a phenomenon is possible only due to the forced activation of glycolysis by decreasing the mitochondrial mass and/or generating dysfunctional mitochondria. The tumor stroma is often found with autophagic and mitophagic activities as evidenced by the higher expression of autophagic and mitophagic signature molecules. Moreover, caveolin-1 and hypoxia-inducible factor-1α play a fundamental role in governing the mitophagy-mediated occurrence of 'reverse Warburg effect'. To the surprise, cancer stem cell also follows the same strategy to exploit the tumor stroma in order to derive high energy fuels for its survival and proliferation. Such parasitic energy-coupling between the cancer cell and cancer-associated fibroblasts makes the fibroblasts a metabolic slave. The metabolic coupling is the result of the paracrine regulation where oxidative stress generated in adjacent fibroblasts by the reactive oxygen species (ROS) produced by cancer cells along with the up-regulation of the oncometabolite transport process through various transporters. This review also discusses the paradigm shift from 'the Warburg effect' to 'the reverse Warburg effect'. It also describes the pivotal role of mitophagy in triggering the 'the reverse Warburg effect'.

P. P. Naik (✉)
P.G. Department of Zoology, Vikram Deb (Auto) College, Jeypore, Odisha, India

Keywords

Mitophagy · Reverse Warburg effect · Metabolic plasticity · Tumor microenvironment

6.1 Introduction

Recently, tumor microenvironment (TME) has gained greater attention due to its roles as an essential contributing factor in the progression of cancer as well as its association with poor clinical outcomes. Moreover, tumors are very often regarded as organs owing to their distinct vasculature wherein cancer cells are protected by a protumor microenvironment (Egeblad et al. 2010). It is now obvious that tumor is not a pure homogeneous population in vivo. Rather, the cells in tumor are set in 'cancer cell nests', which together constitute tumor microenvironment. TME comprises (1) extracellular matrix (ECM), (2) cancer cells, (3) cancer stem cells (CSCs), (4) infiltrating immune cells [B lymphocytes and T lymphocytes, eosinophil, neurophill, basophil, natural killer (NK) cells, mast cells, antigen presenting cells (APC) (dendritic cells and macrophages) and tumor-associated macrophages (TAMs)], (5) Stromal cells [fibroblast cells, myofibroblasts and cancer-associated fibroblasts (CAFs)], and (6) angiogenic endothelial cells [Tumour associated endothelial (TAE) cells] along with their precursors (pericytes) (Friedl and Alexander 2011; Hanahan and Coussens 2012). The TME is a discrete and dynamic domain that guarantees well-defined functional attributes leading to the formation of a suitable habitat that protects cancer cells from various genetic and epigenetic insults and gives a favourable environment for growth and development. Many documents report the occurrence of a desmoplastic 'reactive stroma' encompassing CAFs and myofibroblast-like cells that provides a protumor microenvironment (Zhang et al. 2013). In this regard, it is believed that stromal fibroblasts manipulate the onco-metabolic processes and vice versa (Avagliano et al. 2018; Zhang et al. 2013). Interestingly, reports claim that fibroblasts cells when co-cultured with cancer cells lose their mitochondria. On the contrary, the cancer cells showed an increased mitochondrial mass. Such behavioural aspects of fibroblasts and cancer cells depict the host–parasite relationship where cancer cells act as 'parasites' and stromal fibroblast cells as 'host' (Ko et al. 2011; Martinez-Outschoorn et al. 2011b; Zhang et al. 2013). Similar to the previous reports on infectious 'intracellular' parasites which employs oxidative stress and autophagy to yield host-derived recycled nutrients, the cancer cells also behave as 'extracellular' parasites. Cancer cells exert oxidative stress which acts as a 'weapon' to induce autophagic elimination of mitochondria, by mitophagy to obtain nutrients from neighbouring stromal cells and are compelled to perform aerobic glycolysis to generate energy-rich metabolites (e.g. lactate and ketones) to 'feed' nearby cancer cells (Ko et al. 2011; Whitaker-Menezes et al. 2011a). This paradigm is referred to as 'The Autophagic Tumor Stroma Model of Cancer Metabolism' (Lisanti et al. 2010; Pavlides et al. 2010; Whitaker-Menezes et al. 2011a). Such a strategy employed by cancer cells simply via inducing oxidative stress wherever they go allow them to seed anywhere and offers decreased dependency on blood vessels for food supply during metastasis. This model advocates the quintessential role of autophagy and mitophagy in the alterations of stromal catabolism to promote anabolic growth of cancer cells in order to promote survival advantage during cancer progression (Pavlides et al. 2012). Hence, antioxidants and autophagy inhibitors that have potential in uncoupling such parasitic metabolic relationships inside the tumor microenvironment would provide novel insights into cancer therapeutics.

6.2 Cancer Metabolism: Pasteur Effect, Inverted Pasteur Effect, Warburg Effect and Reverse Warburg Effect

In the 18th century, Antoine Lavoisier reported that to release energy, living organisms slowly burn the metabolic fuels by consuming oxygen. Later on, Louis Pasteur postulated the *'Pasteur effect'*, which proposes that 'fermentation is an alternate form of life and that fermentation is suppressed by respiration'. Six decades later, Warburg proposed that augmented glucose fermentation and diminished respiration is the chief ground of carcinogenesis. It was considered as a counter to the *'Pasteur effect'* and popularly known as *'aerobic glycolysis'*. However, another scientist cotemporary to Warburg named Crabtree demonstrated that aerobic glycolysis is used as an energy source during pathological overgrowths. Moreover, glucose uptake and glycolytic activity were shown to have a negative effect on oxygen consumption collectively known as *'inverted Pasteur effect'* or *'Crabtree effect'* (Vadlakonda et al. 2013). Intriguingly, it is the metabolic responses of tumor cells which permit them to thrive and establish in a particular microenvironment. Such metabolic adaptation noted in cancer cells controls the therapeutic responses. Here, tumor progression and metastasis are dependent on the metabolic adaptation of both cancer and non-cancerous cells present in the vicinity of tissue or organ. Moreover, the metabolism involves both the intracellular network that distributes and offers organic compounds, and the extracellular organic and signalling molecules that facilitate intercellular signalling, which in turn regulates the metabolic functioning of a cell. It is well known that the metabolic interactions among the tumor cells and the stromal cells provide survival advantages where the stromal cells facilitates metabolic substrates supplementation in the form of glutamine, lactate, and fatty acids.

6.2.1 Warburg Effect

Initially, the German chemist and physician Otto Warburg and colleagues in the 1920s performed experiments on the lactate production and oxygen consumption in tissues derived from liver carcinoma of rat (Warburg 1925). To their surprise, they found strikingly different glucose metabolism between normal and cancer cells. Moreover, cancer cells were found to be more reliant on glycolysis despite oxygen availability. This led to the discovery of 'Warburg effect' which referes to the phenomenon of preferred aerobic glycolysis and enhanced lactate production despite the availability of sufficient oxygen in the vicinity. This hypothesis of Warburg proposes that usually cancer cells produce energy by non-oxidative glucose break down, i.e., 'Glycolysis'; whereas normal cells produce ATP through oxidative phosphorylation (OXPHOS) (Warburg 1956). Warburg effect is documented in many cancers including cancer of lungs, colorectal cancer, glioblastoma and breast. DeBerardinis et al. have experimentally proven the occurrence of Warburg effect by studying the cancer cells incubated with 10 mM C-13-labelled glucose under oxygenated conditions (DeBerardinis et al. 2007). Elevated levels of glycolytic metabolites were observed during the metabolomic analysis when 4 mM glucose was perfused to cancer cells prior to experiments suggesting the occurrence Warburg phenomenon (Fantin et al. 2006). Only 2 ATP per molecule of glucose is produced through glycolysis under the anaerobic condition, which is much lower than OXPHOS (i.e. 30 or 32 ATP per glucose molecule). This indicates that in comparison with aerobic glycolysis near about 15 times more glucose is needed to be anaerobically catabolized to generate the same amount of energy. As a result, tumor cells require more glucose, and to make that happen, there is a ten times faster uptake of glucose in tumor cells than the normal cells. It has been suggested that a lower yield but a higher rate of ATP production provides selective advantage to cells competing for limited and shared energy resources. Moreover, cancer cells compete with stromal cells and other cells in the tumor microenvironment due to limited availability of glucose (Chang et al. 2015; Pfeiffer et al. 2001; Slavov et al. 2014). Moreover, cancer cells prefer anaerobic glycolysis for production of ATP because of limited O_2 exposure and hypoxic conditions (Bartrons and Caro 2007). It has been found that high glucose levels in the culture media considerably reduces mitochondrial respiration and vice versa (Gohil et al. 2010; Marroquin et al. 2007). Under high (25 mM) and low (1 mM) glucose conditions, the cancer cells were cultured to investigate oxygen consumption rates (OCRs; mitochondrial respiration) and extracellular acidification rates (ECAR; glycolysis). It was noticed that upon culture under high glucose conditions, cancer cells showed either high OCR-low ECAR or low OCR-high ECAR, or high/moderate OCR-high/moderate ECAR. However, under low glucose conditions, the cancer cells showed high–moderate OCR with very little ECAR as other substrates are used for cellular ATP production (Potter et al. 2016). Moreover, Warburg effect is not consistent as seen in a study with rat hepatoma carried out by Weinhouse. According to this study, the slow-growing cells were more oxidative, whereas the more proliferative cells were more glycolytic. There occurs a dynamic interplay between oxidative and glycolytic states called as metabolic flexibility or metabolic plasticity (Jose et al. 2011; Obre and Rossignol 2015). Such metabolic flexibility is dependent on the environmental conditions and cancer-associated mutations (Astuti et al. 2001; Baysal et al. 2000; Dang et al. 2009; Yan et al. 2009). This kind of dynamic interplay of cancer cells is accompanied with mitochondrial dysfunction. Moreover, studies showed that an increased glycolytic rate is the consequence of decreased mitochondrial mass in cancer cells (Gogvadze et al. 2010).

Warburg effect has been explained in many cancer types and their role in cancer is proposed to be associated with transcriptional and post-translational related metabolic changes. One of the transcription factors, hypoxia-inducible factor 1 (HIF-1) up-regulates expression of glycolytic enzymes, glucose transporters, and pyruvate dehydrogenase kinases (PDKs). Up-regulation of PDKs phosphorlyates and deactivates mitochondrial pyruvate dehydrogenase (PDH) complex and tricarboxylic acid (TCA) cycle (Semenza 2010). Many transcription regulators like alpha estrogen-related receptor (ERR) and MYC are associated with Warburg effect in the same manner (Yeung et al. 2008). Increased expression of MYC in tumors is proposed to be linked with an enhanced glycolytic rate and pathophysiology of metabolic modifications. MYC overexpression leads to high uptake of fluorodeoxyglucose (FDG) in human breast cancer (Palaskas et al. 2011). Again, MYC enhances the Warburg effect by elevating glucose flux and preventing the entry of pyruvate into the TCA cycle. Moreover, MYC overexpression is reported to enhance the activity of the glycolytic enzyme, transmembrane transport of glucose, glutamine transporters and glutaminase-1 activity (Dang et al. 2008; Gao et al. 2009; Nilsson et al. 2012; Osthus et al. 2000; Shim et al. 1997). The orphan nuclear receptor, ERR regulates oxidative metabolism, and mitochondrial biogenesis along with augmented glucose metabolism (Villena and Kralli 2008). Similarly, tumor suppressor protein p53 can lower the glycolysis rate by enhancing the enzymatic activity of fructose-2,6-bisphosphatase and thereby, increase the oxidative phosphorylation process. Warburg effect is also associated with a diminished level of expression of p53 in cancer cells linked with increased glycolysis (Bensaad et al. 2006; Maddocks and Vousden 2011). p53 is also shown to promotes OXPHOS by elevating cytochrome c oxidase and loss of expression of p53 in cancer cells therefore can induce the Warburg effect (Matoba et al. 2006). Moreover, the Warburg effect is also investigated for its association with post-translational regulation in cancer metabolism. Oncogenic phosphorylation events on metabolic enzymes promote aerobic glycolysis. It has been found that hexokinase (HK) and phosphofructokinase-2 (PFK-2) phosphorylation by AKT; downstream of PI3K activation facilitates glucose transporter (GLUT) expression and its localisation to the plasma membrane (Robey and Hay 2009). Studies on various cancer models have depicted the relationship of glycolysis and post-transcriptional modification of the M2 isoform of pyruvate kinase (PKM2). Post-translational modifications of PKM2 like the K305 acetylation decreases its enzymatic activity and modifies the glycolytic pathway. Moreover, it leads to increased degradation of such enzymes by activating chaperone-mediated autophagy. The expression of Y105F mutant of PKM2 in tumor cells was shown to have reduced lactate production and increased oxygen consumption which was consequently found to induce tumor xenograft development (Hitosugi et al. 2009). Again, the induction of PI3K/AKT pathway led to an increase in the phosphorylation of PFK-2, and HK and enhanced glucose influx; subsequently up-regulating the glycolytic pathway (Høyer-Hansen and Jäättelä 2007; Zheng et al. 2011). As mentioned earlier, high demand for glucose is an important feature of cancer and in tumors, there is enhanced relative uptake of FDG, or fluorodeoxyglucose F 18 (18F-FDG). However, it has been found that 18F-FDG uptake is considerably high in hypoxic cancer cells than normoxic ones. Moreover, 18F-FDG uptake in the normoxic cancer cells is typically low and is similar to cells of stromal or necrotic regions leading to the question whether the Warburg effect actually applies to normoxic cancer cells or not. It has also been reported that there is an augmented increase in 18F-FDG uptake in hypoxic cancer cells than the normoxic ones in both in vivo and in vitro culture studies. Hence, cancer cells are supposed to have increased demand for glucose in the absence of oxygen which is logically explained by the Pasteur effect (Zhang et al. 2015).

6.2.2 The Reverse Warburg Effect

Previously, it was believed that the Warburg effect is a phenomenon performed only by cancer cells. However, human skin keloid fibroblasts were also shown to produce energy in the form of ATP mostly via glycolysis. Similarly, hypoxic microenvironments in tumors and keloids led to the activation of the same phenomenon (Vincent et al. 2008). Firstly, Pavlides et al. (2009) described that hydrogen peroxide (H_2O_2) released by cancer cells could induce oxidative stress in CAFs resulting in the loss of mitochondrial function leading to a metabolic switch from OXPHOS to glycolysis. Subsequent to the glycolytic switch, the lactate production by CAF accelerates (Zong et al. 2016). Lactates are transported to extracellular space via monocarboxylate transporter 4 (MCT4) and taken up by the cancer cells by MCT1 for its use in oxidative metabolism (Martinez-Outschoorn et al. 2011a, 2013). Furthermore, many co-culture systems involving fibroblast cells and cancer cells were experimented in this regard and it was observed that epithelial cancer cells are potentially capable of inducing the Warburg effect in stromal fibroblasts (Martinez-Outschoorn et al. 2011a). This phenomenon is popularly regarded as 'reverse Warburg effect' (RWE) (Fig. 6.1) (Jiang et al. 2019; Pavlides et al. 2009). Basically, the hypoxic and nutritionally challenged tumor microenvironment exploits CAFs as 'metabolic slaves' (Roy and Bera 2016). Interestingly, it is to be noted down that in the current situation stromal fibroblasts cells and not the cancer cells are undertaking the Warburg effect. The reverse Warburg effect was proposed to explain metabolic flexibility as mentioned earlier. To understand it better, it can be explained in two steps. In the first step, the cancer cells educate CAFs to boost aerobic glycolysis which leads to the enhanced production of energy-rich fuels (e.g. pyruvate, ketone bodies, fatty acids and lactate). Furthermore, in the second step, these energy-rich fuels produced by CAFs are utilized in mitochondrial OXPHOS by the cancer cells. Particularly, the lactates are converted to pyruvate by lactate dehydrogenase-B (LDH-B) enzymes (Martinez-Outschoorn et al. 2010b). It is also found that in normoxic microenvironment, the oxidative tumor performs OXPHOS to leave behind glucose for its utilization by glycolytic cancer cells in the hypoxic microenvironment (Bonuccelli et al. 2010; Feron 2009; Porporato et al. 2011; Sandulache et al. 2011; Sonveaux et al. 2008; Wilde et al. 2017). In other words enhanced lactate production is utilized in mitochondrial OXPHOS in cancer cells for energy production. This is also accompanied by diminished expression of caveolin-1 (Cav-1) in stromal cells. According to report, the Cav-1 is an important structural protein that plays an essential role in endocytosis, vesicular transport, and other

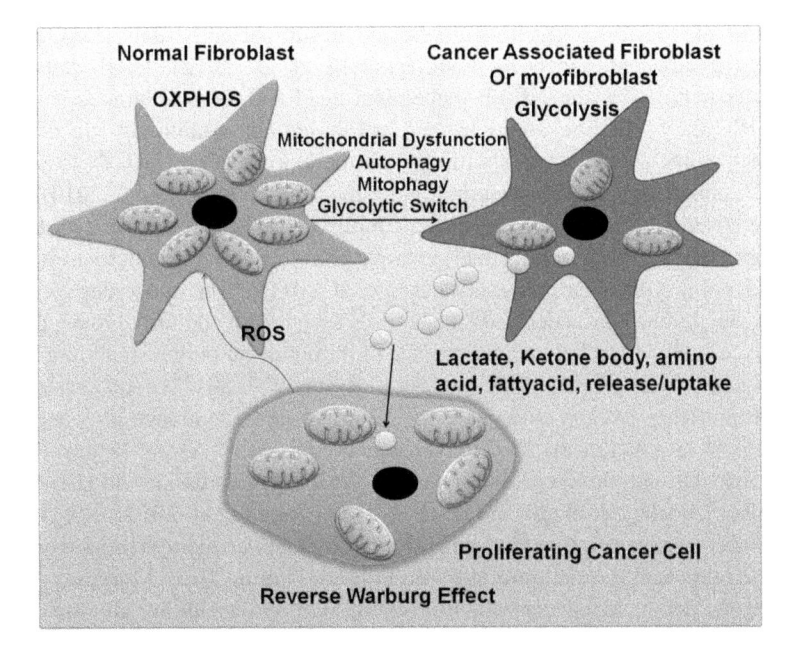

Fig. 6.1 Schematic representation of reverse Warburg effect. Diagram shows the mitochondrial dysfunction in fibroblast cells mediated by the ROS released by cancer cells leading to the up-regulation of aerobic glycolysis. The release of lactate and other high energy fuels by fibroblasts are then utilised by cancer cells for metabolic processes like OXPHOS

signalling pathways. It is also reported to aggravate oxidative stress mediated mitochondrial dysfunctions in CAFs (Sotgia et al. 2009; Witkiewicz et al. 2009, 2010). Furthermore, it is noticed that there is an elevated expression of mono-carboxylate transporters (MCTs) which performs the role of 'energy transfer device' (e.g., lactate) between CAFs and cancer cells through 'lactate shuttle' (Choi et al. 2013; Cirri and Chiarugi 2012; Rae et al. 2009; Whitaker-Menezes et al. 2011b; Witkiewicz et al. 2012). The reverse Warburg effect is basically the co-existence of metabolic alterations in both stromal cells and cancer cells depending on the demand of energy. On one hand, H_2O_2 secreted by cancer cells forms an oxidative microenvironment and up-regulate MCT1 to mediate enhance uptake of lactate. On the other hand, the stromal cells react to oxidative stress generated by H_2O_2 by mediating HIF-1-induced autophagic flux. This may lead to at least two levels of consequences. Firstly, it can mediate the degradation of Cav-1 by the autophagic process leading to tumor progression. Secondly, it can induce the transactivation of glycolytic enzymes by HIF-1 and up-regulation of MCT4 that mediates the efflux of lactate. Moreover, high expression of MCT4 in stromal cells is associated with a poor overall survival rate than low stromal Cav-1 status (Galluzzi et al. 2012). Again, loss of Cav-1 in stromal cells up-regulates the expression of glycolytic enzyme pyruvate kinase M2 (PKM2) and glycolysis. Cav-1 null stromal cells are also demonstrated to have acclerated lactate dehydrogenase-A (LDH-A) activity. Moreover, according to

documents an increased glycolysis pathway in stromal cells fuels the OXPHOS pathway in adjacent cancer cells (Capparelli et al. 2012). CAF cells when co-cultured with RAS-and NF-κB–dependent head and neck squamous cell carcinoma (HNSCC) cell line could trigger metabolic reprogramming via oxidative stress that brings about lactate shuttling with the help of MCT1/MCT4 to promote metabolic coupling between the tumor and tumor stroma (Curry et al. 2014). Along with the stromal and tumoral metabolic coupling, the lactate shuttle maintains an acid–base balance by inhibiting the generation of a fatal acidic microenvironment (Lee and Yoon 2015; Martinez-Outschoorn et al. 2011b). The tumor-derived lactate and autophagic fibroblasts derived lactate together perform additional roles. Lactates when taken up by endothelial cells via MCT1 stimulate autocrine signalling by NF-kB/IL-8 pathway to promote angiogenesis (Vegran et al. 2011). Lactate release via MCT4 from the breast and colon cancer cells are reported to induce IL-8-dependent angiogenesis (Azuma et al. 2007; Polet and Feron 2013). Again, lactate can also induce vascular endothelial growth factor-A (VEGFA) expression via HIF-1α activation (De Saedeleer et al. 2012; Lee et al. 2015; Sanità et al. 2014). The 'biofuel' lactate-induced generation of IL-8 and VEGFA together encourage pro-survival and pro-angiogenic activities in tumor growth (Polet and Feron 2013). Moreover, to fulfil the energetic demands, cancer cells use several nutrients including glucose, lactate, and glutamine. The schematic diagram of the molecular regulation of the reverse Warburg effect is represented in Fig. 6.2.

6.2.3 Two Compartment Model of Tumor Metabolism

A compartment-specific role of autophagy in tumor metabolism was proposed to explain the metabolic paradigm (Fig. 6.3). As discussed earlier, this model describes that occurance of autophagy, mitochondrial dysfunction, and mitophagy in tumor stroma results in the recycling of nutrients and provides chemical high-energy 'fuels,' and building blocks. This triggers the anabolic growth of tumors by inducing oxidative mitochondrial metabolism and autophagy mediated resistance in cancer cells. This stromal-epithelial metabolic coupling is popularly termed as the 'two-compartment tumor metabolism' (Martinez-Outschoorn et al. 2012; Salem et al. 2012). This hypothesis is stringently verified by experimenting on two genetic variants involving the fibroblasts with constitutive autophagic activity in addition to mitochondrial dysfunction and autophagy-resistant cancer cells with enhanced mitochondrial activity. Golgi phosphoprotein 3 (GOLPH3) overexpressing autophagy resistant cells are shown to display mitochondrial biogenesis. However, the damage-regulated autophagy modulator (DRAM) and liver kinase B1 (LKB1) overexpressing cells stimulated AMP-kinase activation and autophagy in CAFs (Salem et al. 2012). This autophagic fibroblasts exhibited mitochondrial dysfunction, increased glycolysis and generation of mitochondrial fuels. Both types of cells, that is, autophagic fibroblasts and autophagy resistant cancer cells promoted tumor growth where CAFs displayed glycolysis and cancer cell showed increased OXPHOS. In breast cancer, the activation of GPER/cAMP/

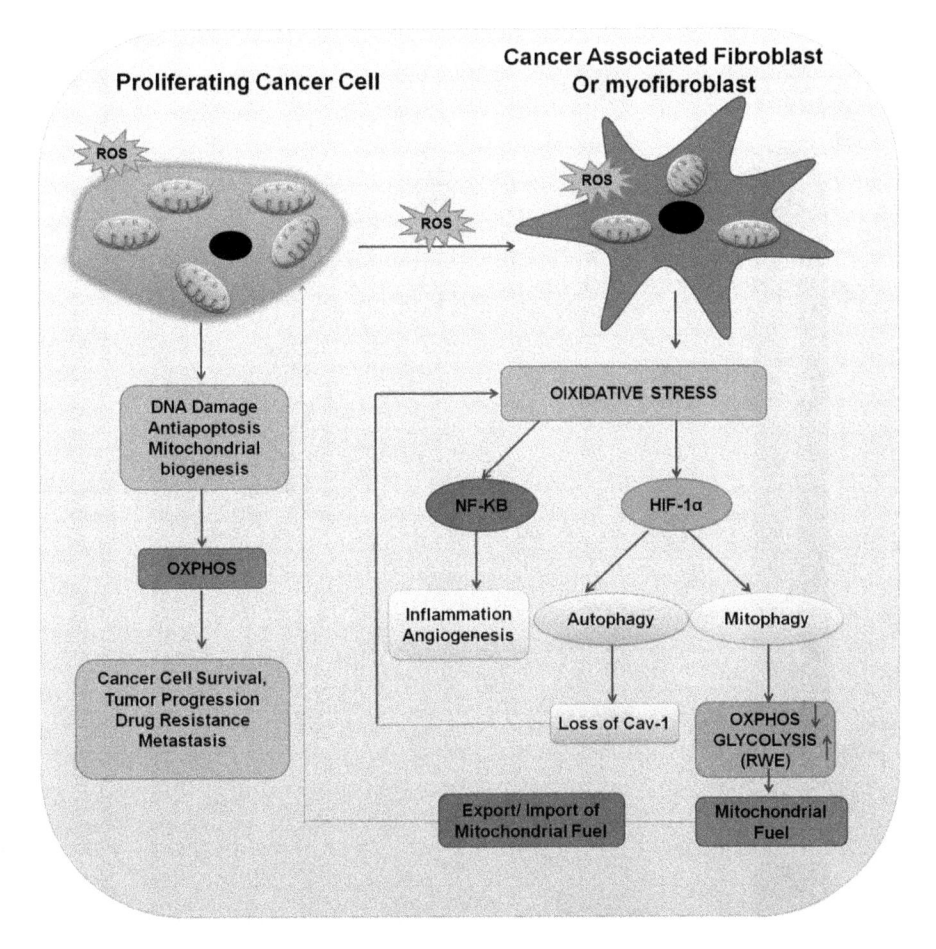

Fig. 6.2 Molecular mechanism of reverse Warburg effect. Oxidative stress generated in CAFs by ROS produced from cancer cells causes activation of autophagy/mitophagy and loss of Cav-1. Again, the stimulation of glycolysis occurs to release lactate which is then transported from CAFs to cancer cells to carry out OXPHOS

PKA/CREB and PI3K/AKT/mTORC1 signalling pathways in CAFs stimulate the aerobic glycolysis switch to secrete pyruvate and lactate for fuelling the OXPHOS in cancer cells (Yu et al. 2017). Moreover, 'Warburg-like' cancer metabolism and DNA damage response in tumor microenvironment share a strong association by activating the downstream signalling of DNA damage/repair target gene DRAM (Salem et al. 2013; Sotgia et al. 2013; Yang et al. 2016b). Cancer cells usually take advantage such resulting metabolites from the altered metabolism operating in CAFs. Moreover, higher expression of GLUT1 is seen in the fibroblasts cells when co-cultured with prostate cancer cells (Kihira et al. 2011; Sanita et al. 2014; Sun et al. 2014). Moreover, CAFs also exports lactate through MCT4 (Andersen et al. 2015; Sanità et al. 2014). Intriguingly, the prostate cancer cells on the other

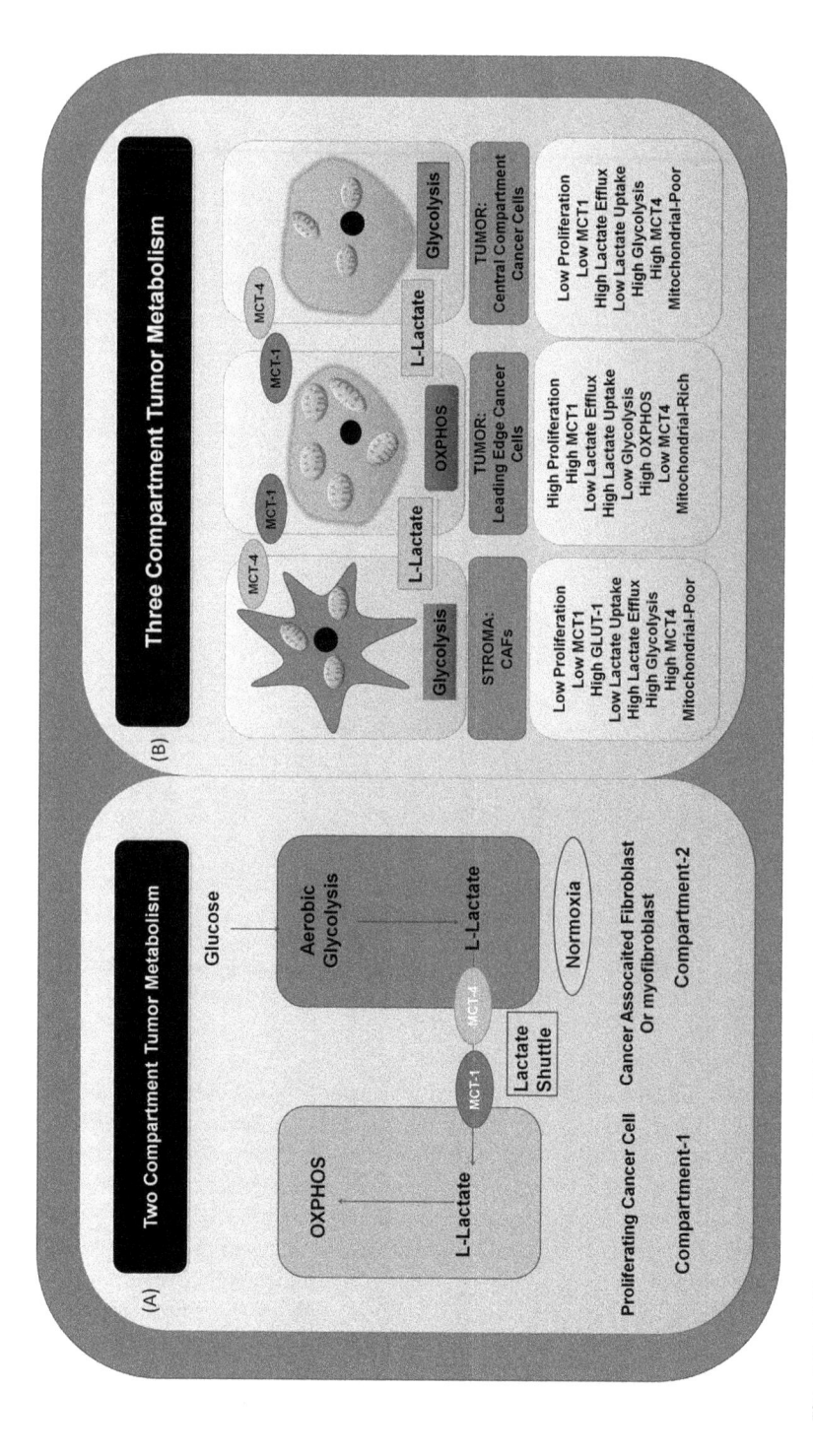

Fig. 6.3 (a) Two compartment and (b) three compartment tumor metabolism. The two-compartment tumor metabolism involves the metabolic coupling between cancer and cancer-associated fibroblast cells. The three-compartment tumor metabolism involves the cancer cells at edge, cancer cells at core and cancer-associated fibroblast cells

hand showed decreased GLUT1 expression and increased lactate influx via MCT1 (Fiaschi et al. 2012). Similar findings are also reported in breast cancer cells (Johnson et al. 2017; Le Floch et al. 2011; Witkiewicz et al. 2012). Furthermore, in another co-culture system with pancreatic cancer cells, MCT1 inhibition in CAFs is shown to decrease the expression pyruvate kinase 2 (PK2) along with the glucose import and lactate secretion (Giannoni et al. 2015). Consistent with the reverse Warburg effect, metastatic breast cancer cells could potentially amplify OXPHOS whereas the adjacent stromal cells were reported to lack detectable mitochondria and perform glycolysis. This was proved by double labelling experiments with the molecular marker for glycolysis (MCT4) and OXPHOS (TOMM20 or COX). This experimet discovered the presence of at least two distinct metabolic compartments that lie side-by-side both in primary tumors and their metastases (Sotgia et al. 2012). Again, there is little information about lipid metabolism in tumor microenvironment. Breast cancer cells in response to CAFs-conditioned media led to the overexpression of fatty acid transporter 1 (FATP1) and accumulation of lipid in cancer cells. Here, FATP1 negotiates the symbiosis of lipid metabolism between breast cancer cells and CAFs (Lopes-Coelho et al. 2018). CAFs are also shown to transport lysophospholipids (lsyo-PLs) directly to the pancreatic cancer cells via lipid droplets. Moreover, fibroblasts release lipids to the neighbouring cancers cells through microvesicles in melanoma and prostate cancer (Lopes-Coelho et al. 2018; Santi et al. 2015). Similar to lipid transport, glutamine release by CAFs and uptake by cancer cells show another aspect of metabolic coupling in tumor stroma. In ovarian cancer, glutamine metabolism in CAFs is reported to promote tumor growth with an increase in glutamine transporter SLC6A14 in cancer cells when co-cultured with CAFs. Thus, co-targeting the glutaminase in cancer and glutamine synthetase in CAFs will provide new insight into cancer therapeutics (Ko et al. 2011; Yang et al. 2016a).

6.2.4 Three Compartment Model of Tumor Metabolism

Reports also propose the presence of a three compartments model of tumor metabolism in cancer (Fig. 6.3). According to this model, the first metabolic compartment comprising cancer cells in periphery depends on OXPHOS. Whereas the second metabolic compartment consisting of cells occupying the deeper layer of tumors is often found to have more glycolytic (aerobic or anaerobic) activity. The third metabolic compartment comprising of fibroblast cells in tumor stroma undergoes aerobic glycolysis. Interestingly, this model is experimentally shown through elevated MCT4 expression in tumor stroma and deeper tumor to facilitate the release of biofuels. However, MCT1 expression level was found to be higher in the leading tumor edge. The OXPHOS execution in the leading tumor edge was confirmed by functional LDH-B and mitochondrial metabolism marker translocase of outer mitochondrial membrane 20 (TOMM20) (Curry et al. 2014). High oxidative stress (MCT4$^+$) is an important feature in cancer tissues as well as tumor stromal cells with higher tumor stage. High oxidative stress is also a marker for cancer-associated fibroblasts and a key hallmark of cancer tissues which render them the ability to exploit the adjacent proliferating and mitochondrial-rich cancer cells. Two of the metabolic compartment that are the 'non-proliferating' populations of cells (Ki-67$^-$/MCT4$^+$) supply high-energy mitochondrial 'fuels' to the 'proliferative' cancer cells to catabolize and derive energy thereby play an essential role in determining the clinical outcome of the disease. In normal mucosa of head and neck cancers also, there are evidence of the presence of three-compartment metabolism. The normal basal stem cells are the proliferative (Ki-67$^+$), mitochondrial-rich (TOMM20$^+$/COX$^+$) and can have the ability to uptake mitochondrial fuel like L-lactate and ketone bodies (MCT1$^+$). It can be inferred that OXPHOS is a common characteristic of both normal stem cells and proliferating cancer cells (Fig. 6.4) (Curry et al. 2013). In head and neck cancer, similarly, a population of highly proliferative epithelial cancer cells with high mitochondrial content and ability for mitochondrial fuels import (Ki-67$^+$/TOMM20$^+$/COX$^+$/MCT1$^+$) were identified. Such proliferating cells are found to be surrounded by the non-proliferating epithelial tumor as well as stromal cells (Ki-67$^-$) that are deficient in mitochondria (TOMM20$^-$/COX$^-$/MCT1$^-$) and are displaying oxidative stress and glycolysis (MCT4$^+$). For simpler understanding it can be stated that 'Three compartment tumor metabolism' comprises (1) non-proliferative and mitochondrial-poor cancer cells (Ki-67$^-$/TOMM20$^-$/COX$^-$/MCT1$^-$); (2) proliferative and mitochondrial-rich cancer cells (Ki-67$^+$/TOMM20$^+$/COX$^+$/MCT1$^+$) and (3) non-proliferative and mitochondrial-poor stromal cells (Ki-67$^-$/TOMM20$^-$/COX$^-$/MCT1$^-$). The non-proliferative cancer along with the stromal cells offeres metabolites for OXPHOS to be operated in proliferating cancer cells (Bagordakis et al. 2016). Rapidly proliferating and poorly differentiated stem-cell-like HNSCC cancer cells have higher level of OXPHOS activity. Recently, Curry et al. documented a potential relationship of cancer stemness with lactate/ketone uptake and mitochondrial metabolism in head and neck cancer. Moreover, the three-compartment metabolism in HNSCC tumors involves the (1) hyper-proliferative (Ki-67$^+$), mitochondrial-rich (TOMM20$^+$/COX$^+$) and mitochondrial fuels import abled (MCT1$^+$) poorly differentiated cancer cells (CSCs) that undergo OXPHOS. Contrary to the proliferating cancer cells, the stromal cells and well-differentiated cancer cells are mitochondrial-poor, glycolytic and non-proliferative. The non-proliferating compartments of tumor are MCT4$^+$, that is, with characteristics like oxidative stress and mitochondrial dysfunction that can generate and export of L-lactate and ketone body (Curry et al. 2013) (Fig. 6.4). However, another category of cells, that are, cancer stem cells (CSCs) are too an important regulator of TME and tumorigenesis. Intriguingly, CSCs are documented to rely more on OXPHOS for their energy production. However, the CSCs also seem to be metabolically plastic i.e., they can exhibit both glycolytic/and oxidative phenotype (combined phenotype) depending on the demand. The quiescent and non-proliferative CSCs are OXPHOS phenotype with high mitochondrial mass whereas the proliferative CSCs show combined phenotype (Chae and Kim, 2018). Therefore, it is obvious that there is a multicompartmental metabolism in the tumor depending on the micro-environmental condition and demand for proliferation (Fig. 6.4).

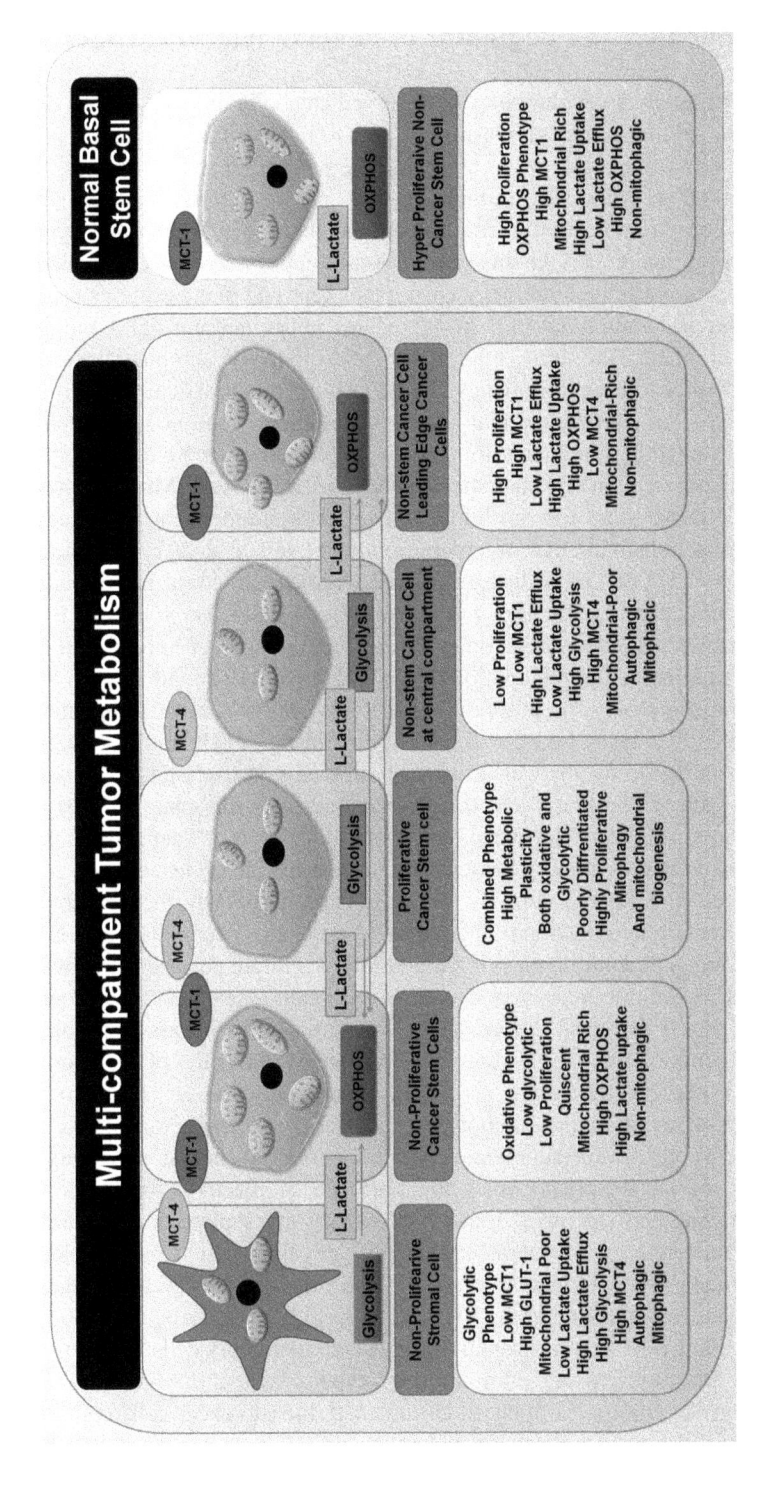

Fig. 6.4 Multi-compartment tumor metabolism. Figure displays the multiple compartments available in tumor stroma for metabolic coupling which comprises of nonstem cancer cells of edge, nonstem cancer cells of the core, non-proliferative cancer stem cells, proliferative cancer stem cells and cancer-associated fibroblasts

6.3 Mitophagy: The Regulator of Reverse Warburg Effect

Autophagy is basically an evolutionarily conserved catabolic process where a cell is programmed to self digest its cytoplasmic content to release metabolites and generate ATP during nutrient starvation, hypoxia, chemo/radio-therapeutic stress, and oncogene activation. Moreover, autophagy also provides a survival advantage where cancer cells are protected in response to metabolic deprivation and hypoxia during tumor progression. Unlike this bulk or non-selective autophagy, the selective autophagy or cargo-specific autophagy encourages removal of superfluous or damaged organelles and long-lived protein aggregates under nutrient-rich conditions. Lemasters et al. coined the term "mitophagy" to explain the autophagolysosomal degradation of mitochondria. Mitophagy maintains mitochondrial quality control and homeostasis in cells during normal cyto-physiological condition. Autophagic removal of dysfunctional mitochondria and metabolic turnover by mitophagy can promote cellular protection and chemoresistance in many cancers. Many reports are available to describe about the pro-tumor role of Parkin-dependent mitophagy as they regulate metabolism in tumor microenvironment (Naik et al. 2019). Occurrence of glycolysis and OXPHOS is directly related with the structural and functional dynamics of mitochondria. Healthy mitochondria can easily carry out OXPHOS while the dysfunctional ones cannot do so. Therefore, they must be eliminated from the cells via mitophagy. The elimination of dysfunctional mitochondria leads to the decrease in mitochondrial content forcing the cells to opt for alternative bioenergetics pathway like glycolysis. In order to exploit the fibroblast cells to release energy rich fuels, cancer cells must instigate mitophagy in CAFs According to Lisanti et al. stromal cells, such as fibroblasts lose their mitochondria by mitophagy to carry out the reverse Warburg Effect (Martinez-Outschoorn et al. 2010a). There is also translational evidence to support mitophagic tumor stroma of cancer metabolism. It was found that in the co-culture system of cancer cells and fibroblasts, the later showed remarkable alteration in mitochondrial content. Moreover, cancer cells are shown to display very low mitochondrial mass under homotypic culture conditions. However, when cultured with fibroblasts, there occurs a significant increase in the mitochondrial mass in cancer cells and a decrease in the mitochondrial mass in fibroblasts (Martinez-Outschoorn et al. 2010a, d). Mitophagy has to play a critical responsibilty in the generation of a glycolytic phenotype in cancer (Fig. 6.5). It is also important to mention that homotypic cancer cells when administered with lactate showed a considerable augmentation in mitochondrial content, indicating that administration of lactate phenocopies the presence of reactive fibroblasts with activated mitophagy. A unilateral transfer of energy takes place from glycolytic stromal fibroblast cells to oxidative cancer cells developing a parasitic relationship among them. Upon oxidative stress via the release of ROS by the cancer cells, Cav-1 is degraded (Sung et al. 2018). Interestingly, a study by Castello-Cros et al. showed that Cav-1 loss in stromal fibroblasts in mammary cells leads to overexpression of plasminogen activator inhibitor type 1 and 2 (PAI-1/2). PAI-1/2 overexpressing fibroblast cells upon co-culture with breast cancer cells could promote tumor growth and metastasis by inducing OXPHOS in nearby cancer cells.

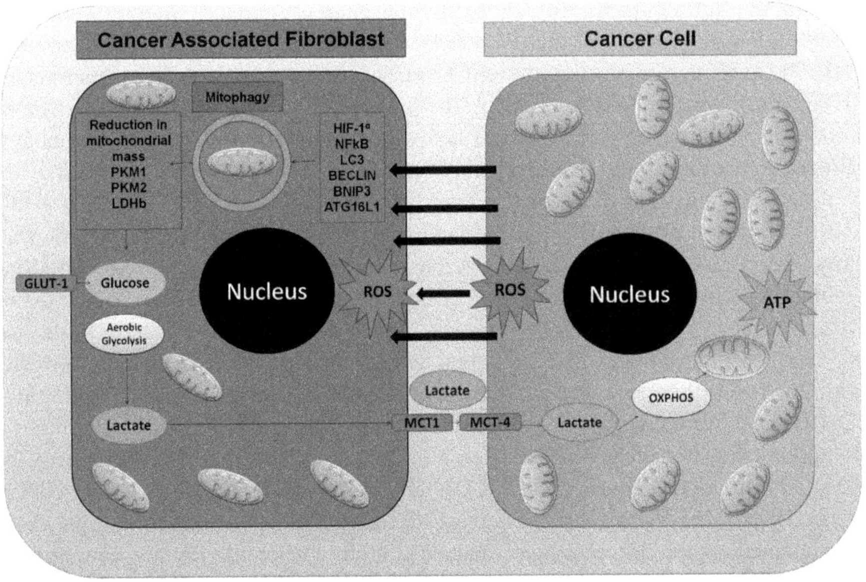

6 Mitophagy and Reverse Warburg Effect: Metabolic Compartmentalization... 131

Fig. 6.5 Mitophagy regulating the reverse Warburg effect. Release of ROS to nearby CAFs causes mitochondrial dysfunction and activation HIF-1α, NF-kB, DRAM, LC3, BNIP3, BNIP3L, ATG16L1 and so on which further leads to the activation of autophagy as well as mitophagy. Mitophagy activation leads to the reduction of mitochondrial mass and OXPHOS. This also causes the stimulation of glycolysis in CAFs by the up-regulation of PKM1/2 and LDH-B expression. The lactate produced by aerobic glycolysis is then transported from CAF by MCT4 to cancer cells via the MCT1 to be used for OXPHOS

Moreover, it led to the up-regulation of activated fibroblasts markers such as calponin, vimentin, and fibronectin (Muda 2011). Subsequently, autophagy/mitophagy gets initiated in activated fibroblasts as evidenced from the overexpression of Beclin-1 and LAMP-1/2. Nextly, the ROS released by autophagic fibroblasts are reported to promote genomic instability in the cancer cells in the vicinity, thereby leading to the stimulation of further oncogenic mutations to support cancer cell proliferation. It can be summarized that the autophagic stroma model of cancer proposes the provocation of oxidative stress mediated mitochondrial dysfunction leading to the activation of autophagy/mitophagy process that finally helps in tumor invasion and metastasis. According to reports, as mitophagy occurs in the stromal fibroblast cells, they are compelled to perform aerobic glycolysis, resulting in the formation of excess lactate and/or ketones. One intriguing study involving the stable over-expression of autophagy and mitophagy gene BCL2/adenovirus E1B 19 kDa protein-interacting protein 3 (BNIP3), Cathepsin B (CTSB) and Autophagy-related protein 16-1 (ATG16L1) in telomerse-immortalised human fibroblasts (hTERT-BJ1) showed mitochondrial dysfunction, and constitutive autophagic/mitophagic features ensuing aerobic glycolysis to produce L-lactate and ketone bodies. Moreover, it was found that hypoxia-triggered break down of Cav-1 leads to the up-regulation of autophagy/mitophagy signature proteins such as microtubule-associated light chain 3 (LC3), ATG16L, BNIP3 and BCL2/adenovirus E1B 19 kDa protein-interacting protein 3-like (BNIP3L) (Martinez-Outschoorn et al. 2010c). Furthermore, knockdown of Cav-1 through si-RNA approach in stromal fibroblasts could enhance the level of lysosomal signature proteins and mitophagy markers. In another study, mammary fat pads of Cav-1 $^{(-/-)}$ null mice showed overexpression of autophagy/mitophagy markers like LC3 and BNIP3L (Qian et al. 2019; Thompson et al. 2012). Additionally, Cav-1 knockdown in fibroblasts was shown to promote ROS production, oxidative stress generation, and mitochondrial dysfunction which further led to the acceleration of autophagy/mitophagy. Moreover, in breast cancer patients, proteins like BNIP3L, PKM2 and LDH-B are expressed in high amount in Cav-1 deficient tumor stromal compartment (Lisanti et al. 2010). Furthermore, the expression of aerobic glycolysis marker PKM2, and LDH-B and mitophagy marker BNIP3L in CAFs are reported as effective biomarkers for the identification of high-risk cancer patients (Chiavarina et al. 2011; Martinez-Outschoorn et al. 2010d; Salem et al. 2012; Sung et al. 2020). Again, PKM1 has the ability to increase the glycolytic potential of stromal cells accompanied with the enhanced lactate output, whereas PKM2 can potentially induce the NF-kB-mediated autophagy induction and enhances the ketone body output. Such induction of oxidative stress-induced autophagy/mitophagy in the tumor stromal compartment offers a strategic mean to the cancer cells so that they can directly 'feed off' the nutrients, chemical building blocks, and energy-rich metabolites released by stromal fibroblasts (Chiavarina et al. 2011; Guido et al. 2012). This parasitic relationship and metabolic dependency also emphasizes a worthwhile solution to the 'autophagy paradox' in cancer aetiology and chemotherapy. Another document supporting this hypothesis reported the presence of a cytokine-mediated cross-talk between CAFs and cancer cells and a remarkable exchange of metabolites is noted among them. In this context, hypoxia-induced HIF-1α, cytokines, active ROS, and ammonia released by cancer cells in addition to the limited nutrient status in the tumor microenvironment are found to activate mitophagy and glycolysis in CAFs that culminates in the metabolic coupling through release of metabolites. As a result, anabolism is stimulated in cancer cells with downregulation of autophagy with consequent tumor growth (Sanita et al. 2014). The autophagy/mitophagy induction in hTERT (human telomerase reverse transcriptase)-immortalised fibroblasts as seen from the up-regulation of Beclin1, LAMP1, Cathepsin B, and BNIP3 is supposed to be mediated by DRAM genes. The DRAM overexpression in fibroblasts is also shown to downregulate the expression of Cav-1 and mitochondrial OXPHOS complexes indicating the onset of mitochondrial dysfunction and autophagy/mitophagy (Guido et al. 2012). Moreover, AMP–kinase activation also indicates the metabolic dysfunction in CAFs (Roy and Bera 2016). Moreover, under the up-regulated BNIP3 condition and DRAM overexpression, there was a decline in OXPHOS complexes I, III and IV suggesting the instigation of mitophagy onset (Liu et al. 2014; Salem et al. 2012). Again, any dysfunction in mitochondria triggers autophagy/mitophagy in CAFs that subsequently promotes reverse Warburg effect by agravating HIF-1α, and NF-kB. It is also accompanied with the activation of antioxidant defence by encouraging the up-regulation of antioxidant enzymes (peroxiredoxin-1) and antiapoptotic proteins (TIGAR) (Martinez-Outschoorn et al. 2010c).

6.4 Conclusion and Future Perspective

Altered metabolism always provides the means to cancer cells to meet the need for unrestricted proliferation. Metabolic reprogramming of TME is highly necessary for tumor initiation and progression. Additionally, tumors are often seen consisting of a metabolically heterogeneous population where different cell types with different metabotypes coexist and collaborate to assure cancer progression. In TME, the CAFs represent a crucial cell type governing the metabolic crosstalk between cell types. Cancer cells have also developed the potential to use a variety of fuel sources. Tumor cells are shown to have increased aerobic as well as mitochondrial metabolism for ATP production, redox balance in various tumor cell types. Such metabolic alteration can be targeted for therapeutic intervention. The metabolic slavery of CAFs in TME can be a prospective target in this aspect. Strategies to inactivate CAFs myofibroblastic phenotype and disruption of metabolic crosstalk between tumor cells and CAFs might decrease the aggressiveness of tumor. In this regard, human patients are now experimented for various early-phase clinical trials (Kishton and Rathmell 2015; Ross and Critchlow 2014; Vander Heiden 2011). However, there are two most important concerns to be taken care of while adopting this approach. The first one is the metabolic plasticity adopted by cancer cells for allowing them to undergo rapid metabolic rewiring as a compensatory response. Secondly, the chance of developing potential toxicity in rapidly proliferating normal cells by targeting fundamental metabolic pathways. However, targeting metabolic pathways with anti-metabolites has been employed for a long time and serves as a successful treatment modality in multiple cancer types. Undoubtedly, insights into new metabolic models would lead to the development of novel biomarkers and parallel therapies which in turn would facilitate the discovery of personalised cancer medicine. As the oxidative stress and autophagy/mitophagy play a central role in this process, novel powerful antioxidants, autophagy/mitophagy inhibitors need to be developed to mitigate cancer.

References

Andersen S, Solstad Ø, Moi L et al (2015) Organized metabolic crime in prostate cancer: the coexpression of MCT1 in tumor and MCT4 in stroma is an independent prognosticator for biochemical failure. Urol Oncol 33:338.e9–338.e3.38E17. https://doi.org/10.1016/j.urolonc.2015.05.013

Astuti D, Latif F, Dallol A, Dahia PL, Douglas F, George E, Sköldberg F, Husebye ES, Eng C, Maher ER (2001) Gene mutations in the succinate dehydrogenase subunit SDHB cause susceptibility to familial pheochromocytoma and to familial paraganglioma. Am J Hum Genet 69:49–54

Avagliano A, Granato G, Ruocco MR, Romano V, Belviso I, Carfora A, Montagnani S, Arcucci A (2018) Metabolic Reprogramming of Cancer Associated Fibroblasts: The Slavery of Stromal Fibroblasts. Biomed Res Int 2018:6075403

Azuma M, Shi M, Danenberg KD, Gardner H, Barrett C, Jacques CJ, Sherod A, Iqbal S, El-Khoueiry A, Yang D (2007) Serum lactate dehydrogenase levels and glycolysis significantly correlate with tumor VEGFA and VEGFR expression in metastatic CRC patients. Pharmacogenomics 8:1705–1713

Bagordakis E, Sawazaki-Calone I, Macedo CC, Carnielli CM, de Oliveira CE, Rodrigues PC, Rangel AL, Dos Santos JN, Risteli J, Graner E, Salo T, Paes Leme AF, Coletta RD (2016) Secretome profiling of oral squamous cell carcinoma-associated fibroblasts reveals organization and disassembly of extracellular matrix and collagen metabolic process signatures. Tumour Biol 37:9045–9057

Bartrons R, Caro J (2007) Hypoxia, glucose metabolism and the Warburg's effect. J Bioenerg Biomembr 39:223–229

Baysal BE, Ferrell RE, Willett-Brozick JE, Lawrence EC, Myssiorek D, Bosch A, van der Mey A, Taschner PE, Rubinstein WS, Myers EN (2000) Mutations in SDHD, a mitochondrial complex II gene, in hereditary paraganglioma. Science 287:848–851

Bensaad K, Tsuruta A, Selak MA, Vidal MNC, Nakano K, Bartrons R, Gottlieb E, Vousden KH (2006) TIGAR, a p53-inducible regulator of glycolysis and apoptosis. Cell 126:107–120

Bonuccelli G, Tsirigos A, Whitaker-Menezes D, Pavlides S, Pestell RG, Chiavarina B, Frank PG, Flomenberg N, Howell A, Martinez-Outschoorn UE, Sotgia F, Lisanti MP (2010) Ketones and lactate "fuel" tumor growth and metastasis: Evidence that epithelial cancer cells use oxidative mitochondrial metabolism. Cell Cycle 9:3506–3514

Capparelli C, Guido C, Whitaker-Menezes D, Bonuccelli G, Balliet R, Pestell TG, Goldberg AF, Pestell RG, Howell A, Sneddon S (2012) Autophagy and senescence in cancer-associated fibroblasts metabolically supports tumor growth and metastasis, via glycolysis and ketone production. Cell Cycle 11:2285–2302

Chae YC, Kim JH (2018) Cancer stem cell metabolism: target for cancer therapy. BMB Rep 51(7):319–326

Chang C-H, Qiu J, O'Sullivan D, Buck MD, Noguchi T, Curtis JD, Chen Q, Gindin M, Gubin MM, Van Der Windt GJ (2015) Metabolic competition in the tumor microenvironment is a driver of cancer progression. Cell 162:1229–1241

Chiavarina B, Whitaker-Menezes D, Martinez-Outschoorn UE, Witkiewicz AK, Birbe R, Howell A, Pestell RG, Smith J, Daniel R, Sotgia F, Lisanti MP (2011) Pyruvate kinase expression (PKM1 and PKM2) in cancer-associated fibroblasts drives stromal nutrient production and tumor growth. Cancer Biol Ther 12:1101–1113

Choi J, Kim DH, Jung WH, Koo JS (2013) Metabolic interaction between cancer cells and stromal cells according to breast cancer molecular subtype. Breast Cancer Res 15:R78

Cirri P, Chiarugi P (2012) Cancer-associated-fibroblasts and tumour cells: a diabolic liaison driving cancer progression. Cancer Metastasis Rev 31:195–208

Curry JM, Tuluc M, Whitaker-Menezes D, Ames JA, Anantharaman A, Butera A, Leiby B, Cognetti DM, Sotgia F, Lisanti MP, Martinez-Outschoorn UE (2013) Cancer metabolism, stemness and tumor recurrence: MCT1 and MCT4 are functional biomarkers of metabolic symbiosis in head and neck cancer. Cell Cycle 12:1371–1384

Curry JM, Sprandio J, Cognetti D, Luginbuhl A, Bar-ad V, Pribitkin E, Tuluc M (2014) Tumor microenvironment in head and neck squamous cell carcinoma. Semin Oncol 41:217–234

Dang CV, J-w K, Gao P, Yustein J (2008) The interplay between MYC and HIF in cancer. Nat Rev Cancer 8:51–56

Dang L, White DW, Gross S, Bennett BD, Bittinger MA, Driggers EM, Fantin VR, Jang HG, Jin S, Keenan MC (2009) Cancer-associated IDH1 mutations produce 2-hydroxyglutarate. Nature 462:739–744

De Saedeleer C, Copetti T, Porporato P, Verrax J, Feron O (2012) Lactate activates HIF-1 in oxidative but not in Warburg-phenotype human. PLoS One 7:e46571

DeBerardinis RJ, Mancuso A, Daikhin E, Nissim I, Yudkoff M, Wehrli S, Thompson CB (2007) Beyond aerobic glycolysis: transformed cells can engage in glutamine metabolism that exceeds the requirement for protein and nucleotide synthesis. Proc Natl Acad Sci 104:19345–19350

Egeblad M, Nakasone ES, Werb Z (2010) Tumors as organs: complex tissues tha interface with the entire organism. Dev Cell 18:884–901

Fantin VR, St-Pierre J, Leder P (2006) Attenuation of LDH-A expression uncovers a link between glycolysis, mitochondrial physiology, and tumor maintenance. Cancer Cell 9:425–434

Feron O (2009) Pyruvate into lactate and back: from the Warburg effect to symbiotic energy fuel exchange in cancer cells. Radiother Oncol 92:329–333

Fiaschi T, Marini A, Giannoni E, Taddei ML, Gandellini P, De Donatis A, Lanciotti M, Serni S, Cirri P, Chiarugi P (2012) Reciprocal metabolic reprogramming through lactate shuttle coordinately influences tumor-stroma interplay. Cancer Res 72:5130–5140

Friedl P, Alexander S (2011) Cancer invasion and the microenvironment: plasticity and reciprocity. Cell 147:992–1009

Galluzzi L, Kepp O, Kroemer G (2012) Reverse Warburg: straight to cancer. Cell Cycle 11:1059

Gao P, Tchernyshyov I, Chang T-C, Lee Y-S, Kita K, Ochi T, Zeller KI, De Marzo AM, Van Eyk JE, Mendell JT (2009) c-Myc suppression of miR-23a/b enhances mitochondrial glutaminase expression and glutamine metabolism. Nature 458:762–765

Giannoni E, Taddei ML, Morandi A, Comito G, Calvani M, Bianchini F, Richichi B, Raugei G, Wong N, Tang D (2015) Targeting stromal-induced pyruvate kinase M2 nuclear translocation impairs oxphos and prostate cancer metastatic spread. Oncotarget 6:24061

Gogvadze V, Zhivotovsky B, Orrenius S (2010) The Warburg effect and mitochondrial stability in cancer cells. Mol Asp Med 31:60–74

Gohil VM, Sheth SA, Nilsson R, Wojtovich AP, Lee JH, Perocchi F, Chen W, Clish CB, Ayata C, Brookes PS (2010) Nutrient-sensitized screening for drugs that shift energy metabolism from mitochondrial respiration to glycolysis. Nat Biotechnol 28:249–255

Guido C, Whitaker-Menezes D, Lin Z, Pestell RG, Howell A, Zimmers TA, Casimiro MC, Aquila S, Ando S, Martinez-Outschoorn UE, Sotgia F, Lisanti MP (2012) Mitochondrial fission induces glycolytic reprogramming in cancer-associated myofibroblasts, driving stromal lactate production, and early tumor growth. Oncotarget 3:798–810

Hanahan D, Coussens LM (2012) Accessories to the crime: functions of cells recruited to the tumor microenvironment. Cancer Cell 21:309–322

Hitosugi T, Kang S, Vander Heiden MG, Chung TW, Elf S, Lythgoe K, Dong S, Lonial S, Wang X, Chen GZ, Xie J, Gu TL, Polakiewicz RD, Roesel JL, Boggon TJ, Khuri FR, Gilliland DG, Cantley LC, Kaufman J, Chen J (2009) Tyrosine phosphorylation inhibits PKM2 to promote the Warburg effect and tumor growth. Sci Signal 2:ra73

Høyer-Hansen M, Jäättelä M (2007) AMP-activated protein kinase: a universal regulator of autophagy? Autophagy 3:381–383

Jiang E, Xu Z, Wang M, Yan T, Huang C, Zhou X, Liu Q, Wang L, Chen Y, Wang H, Liu K, Shao Z, Shang Z (2019) Tumoral microvesicle-activated glycometabolic reprogramming in fibroblasts promotes the progression of oral squamous cell carcinoma. FASEB J 33:5690–5703

Johnson JM, Cotzia P, Fratamico R, Mikkilineni L, Chen J, Colombo D, Mollaee M, Whitaker-Menezes D, Domingo-Vidal M, Lin Z (2017) MCT1 in invasive ductal carcinoma: monocarboxylate metabolism and aggressive breast cancer. Front Cell Dev Biol 5:27

Jose C, Bellance N, Rossignol R (2011) Choosing between glycolysis and oxidative phosphorylation: a tumor's dilemma? Biochim Biophys Acta Bioenerg 1807:552–561

Kihira Y, Yamano N, Izawa-Ishizawa Y, Ishizawa K, Ikeda Y, Tsuchiya K, Tamaki T, Tomita S (2011) Basic fibroblast growth factor regulates glucose metabolism through glucose transporter 1 induced by hypoxia-inducible factor-1α in adipocytes. Int J Biochem Cell Biol 43:1602–1611

Kishton RJ, Rathmell JC (2015) Novel therapeutic targets of tumor metabolism. Cancer J 21:62–69

Ko YH, Lin Z, Flomenberg N, Pestell RG, Howell A, Sotgia F, Lisanti MP, Martinez-Outschoorn UE (2011) Glutamine fuels a vicious cycle of autophagy in the tumor stroma and oxidative mitochondrial metabolism in epithelial cancer cells: implications for preventing chemotherapy resistance. Cancer Biol Ther 12:1085–1097

Le Floch R, Chiche J, Marchiq I, Naiken T, Ilc K, Murray CM, Critchlow SE, Roux D, Simon M-P, Pouysségur J (2011) CD147 subunit of lactate/H+ symporters MCT1 and hypoxia-inducible MCT4 is critical for energetics and growth of glycolytic tumors. Proc Natl Acad Sci 108:16663–16668

Lee M, Yoon JH (2015) Metabolic interplay between glycolysis and mitochondrial oxidation: the reverse Warburg effect and its therapeutic implication. World J Biol Chem 6:148–161

Lee DC, Sohn HA, Park Z-Y, Oh S, Kang YK, Lee K-m, Kang M, Jang YJ, Yang S-J, Hong YK (2015) A lactate-induced response to hypoxia. Cell 161:595–609

Lisanti MP, Martinez-Outschoorn UE, Chiavarina B, Pavlides S, Whitaker-Menezes D, Tsirigos A, Witkiewicz AK, Lin Z, Balliet RM, Howell A, Sotgia F (2010) Understanding the "lethal" drivers of tumor-stroma co-evolution: emerging role(s) for hypoxia, oxidative stress and autophagy/mitophagy in the tumor micro-environment. Cancer Biol Ther 10:537–542

Liu K, Shi Y, Guo XH, Ouyang YB, Wang SS, Liu DJ, Wang AN, Li N, Chen DX (2014) Phosphorylated AKT inhibits the apoptosis induced by DRAM-mediated mitophagy in hepatocellular carcinoma by preventing the translocation of DRAM to mitochondria. Cell Death Dis 5: e1078

Lopes-Coelho F, Andre S, Felix A, Serpa J (2018) Breast cancer metabolic cross-talk: fibroblasts are hubs and breast cancer cells are gatherers of lipids. Mol Cell Endocrinol 462:93–106

Maddocks OD, Vousden KH (2011) Metabolic regulation by p53. J Mol Med 89:237–245

Marroquin LD, Hynes J, Dykens JA, Jamieson JD, Will Y (2007) Circumventing the Crabtree effect: replacing media glucose with galactose increases susceptibility of HepG2 cells to mitochondrial toxicants. Toxicol Sci 97:539–547

Martinez-Outschoorn UE, Balliet RM, Rivadeneira DB, Chiavarina B, Pavlides S, Wang C, Whitaker-Menezes D, Daumer KM, Lin Z, Witkiewicz AK, Flomenberg N, Howell A, Pestell RG, Knudsen ES, Sotgia F, Lisanti MP (2010a) Oxidative stress in cancer associated fibroblasts drives tumor-stroma co-evolution: a new paradigm for understanding tumor metabolism, the field effect and genomic instability in cancer cells. Cell Cycle 9:3256–3276

Martinez-Outschoorn UE, Pavlides S, Whitaker-Menezes D, Daumer KM, Milliman JN, Chiavarina B, Migneco G, Witkiewicz AK, Martinez-Cantarin MP, Flomenberg N, Howell A, Pestell RG, Lisanti MP, Sotgia F (2010b) Tumor cells induce the cancer associated fibroblast phenotype via caveolin-1 degradation: implications for breast cancer and DCIS therapy with autophagy inhibitors. Cell Cycle 9:2423–2433

Martinez-Outschoorn UE, Trimmer C, Lin Z, Whitaker-Menezes D, Chiavarina B, Zhou J, Wang C, Pavlides S, Martinez-Cantarin MP, Capozza F, Witkiewicz AK, Flomenberg N, Howell A, Pestell RG, Caro J, Lisanti MP, Sotgia F (2010c) Autophagy in cancer associated fibroblasts promotes tumor cell survival: role of hypoxia, HIF1 induction and NFkappaB activation in the tumor stromal microenvironment. Cell Cycle 9:3515–3533

Martinez-Outschoorn UE, Whitaker-Menezes D, Pavlides S, Chiavarina B, Bonuccelli G, Casey T, Tsirigos A, Migneco G, Witkiewicz A, Balliet R, Mercier I, Wang C, Flomenberg N, Howell A, Lin Z, Caro J, Pestell RG, Sotgia F, Lisanti MP (2010d) The autophagic tumor stroma model of cancer or "battery-operated tumor growth": a simple solution to the autophagy paradox. Cell Cycle 9:4297–4306

Martinez-Outschoorn UE, Lin Z, Trimmer C, Flomenberg N, Wang C, Pavlides S, Pestell RG, Howell A, Sotgia F, Lisanti MP (2011a) Cancer cells metabolically "fertilize" the tumor microenvironment with hydrogen peroxide, driving the Warburg effect: implications for PET imaging of human tumors. Cell Cycle 10:2504–2520

Martinez-Outschoorn UE, Pavlides S, Howell A, Pestell RG, Tanowitz HB, Sotgia F, Lisanti MP (2011b) Stromal–epithelial metabolic coupling in cancer: integrating autophagy and metabolism in the tumor microenvironment. Int J Biochem Cell Biol 43:1045–1051

Martinez-Outschoorn UE, Balliet RM, Lin Z, Whitaker-Menezes D, Howell A, Sotgia F, Lisanti MP (2012) Hereditary ovarian cancer and two-compartment tumor metabolism: epithelial loss of BRCA1 induces hydrogen peroxide production, driving oxidative stress and NFκB activation in the tumor stroma. Cell Cycle 11:4152–4166

Martinez-Outschoorn UE, Curry JM, Ko YH, Lin Z, Tuluc M, Cognetti D, Birbe RC, Pribitkin E, Bombonati A, Pestell RG, Howell A, Sotgia F, Lisanti MP (2013) Oncogenes and inflammation rewire host energy metabolism in the tumor microenvironment: RAS and NFkappaB target stromal MCT4. Cell Cycle 12:2580–2597

Matoba S, Kang J-G, Patino WD, Wragg A, Boehm M, Gavrilova O, Hurley PJ, Bunz F, Hwang PM (2006) p53 regulates mitochondrial respiration. Science 312:1650–1653

Muda M (2011) DUSPs strike again: Comment on: Kozarova A, et al. Cell Cycle 2011; 10: 1669–78. Cell Cycle 10:2827–2835

Naik PP, Birbrair A, Bhutia SK (2019) Mitophagy-driven metabolic switch reprograms stem cell fate. Cell Mol Life Sci 76:27–43

Nilsson LM, Forshell TZP, Rimpi S, Kreutzer C, Pretsch W, Bornkamm GW, Nilsson JA (2012) Mouse genetics suggests cell-context dependency for Myc-regulated metabolic enzymes during tumorigenesis. PLoS Genet 8:e1002573

Obre E, Rossignol R (2015) Emerging concepts in bioenergetics and cancer research: metabolic flexibility, coupling, symbiosis, switch, oxidative tumors, metabolic remodeling, signaling and bioenergetic therapy. Int J Biochem Cell Biol 59:167–181

Osthus RC, Shim H, Kim S, Li Q, Reddy R, Mukherjee M, Xu Y, Wonsey D, Lee LA, Dang CV (2000) Deregulation of glucose transporter 1 and glycolytic gene expression by c-Myc. J Biol Chem 275:21797–21800

Palaskas N, Larson SM, Schultz N, Komisopoulou E, Wong J, Rohle D, Campos C, Yannuzzi N, Osborne JR, Linkov I (2011) 18F-fluorodeoxy-glucose positron emission tomography marks MYC-overexpressing human basal-like breast cancers. Cancer Res 71:5164–5174

Pavlides S, Whitaker-Menezes D, Castello-Cros R, Flomenberg N, Witkiewicz AK, Frank PG, Casimiro MC, Wang C, Fortina P, Addya S, Pestell RG, Martinez-Outschoorn UE, Sotgia F, Lisanti MP (2009) The reverse Warburg effect: aerobic glycolysis in cancer associated fibroblasts and the tumor stroma. Cell Cycle 8:3984–4001

Pavlides S, Tsirigos A, Migneco G, Whitaker-Menezes D, Chiavarina B, Flomenberg N, Frank PG, Casimiro MC, Wang C, Pestell RG (2010) The autophagic tumor stroma model of cancer: Role of oxidative stress and ketone production in fueling tumor cell metabolism. Cell Cycle 9:3485–3505

Pavlides S, Vera I, Gandara R, Sneddon S, Pestell RG, Mercier I, Martinez-Outschoorn UE, Whitaker-Menezes D, Howell A, Sotgia F (2012) Warburg meets autophagy: cancer-associated fibroblasts accelerate tumor growth and metastasis via oxidative stress, mitophagy, and aerobic glycolysis. Antioxid Redox Signal 16:1264–1284

Pfeiffer T, Schuster S, Bonhoeffer S (2001) Cooperation and competition in the evolution of ATP-producing pathways. Science 292:504–507

Polet F, Feron O (2013) Endothelial cell metabolism and tumour angiogenesis: glucose and glutamine as essential fuels and lactate as the driving force. J Intern Med 273:156–165

Porporato PE, Dhup S, Dadhich RK, Copetti T, Sonveaux P (2011) Anticancer targets in the glycolytic metabolism of tumors: a comprehensive review. Front Pharmacol 2:49

Potter M, Newport E, Morten KJ (2016) The Warburg effect: 80 years on. Biochem Soc Trans 44:1499–1505

Qian XL, Pan YH, Huang QY, Shi YB, Huang QY, Hu ZZ, Xiong LX (2019) Caveolin-1: a multifaceted driver of breast cancer progression and its application in clinical treatment. Onco Targets Ther 12:1539–1552

Rae C, Nasrallah FA, Broer S (2009) Metabolic effects of blocking lactate transport in brain cortical tissue slices using an inhibitor specific to MCT1 and MCT2. Neurochem Res 34:1783–1791

Robey RB, Hay N (2009) Is Akt the "Warburg kinase"?-Akt-energy metabolism interactions and oncogenesis. Semin Cancer Biol 19:25–31

Ross SJ, Critchlow SE (2014) Emerging approaches to target tumor metabolism. Curr Opin Pharmacol 17:22–29

Roy A, Bera S (2016) CAF cellular glycolysis: linking cancer cells with the microenvironment. Tumour Biol 37:8503–8514

Salem AF, Whitaker-Menezes D, Lin Z, Martinez-Outschoorn UE, Tanowitz HB, Al-Zoubi MS, Howell A, Pestell RG, Sotgia F, Lisanti MP (2012) Twocompartment tumor metabolism: autophagy in the tumor microenvironment and oxidative mitochondrial metabolism (OXPHOS) in cancer cells. Cell Cycle 11:2545–2556

Salem AF, Al-Zoubi MS, Whitaker-Menezes D, Martinez-Outschoorn UE, Lamb R, Hulit J, Howell A, Gandara R, Sartini M, Galbiati F (2013) Cigarette smoke metabolically promotes cancer, via autophagy and premature aging in the host stromal microenvironment. Cell Cycle 12:818–825

Sandulache VC, Ow TJ, Pickering CR, Frederick MJ, Zhou G, Fokt I, Davis-Malesevich M, Priebe W, Myers JN (2011) Glucose, not glutamine, is the dominant energy source required for proliferation and survival of head and neck squamous carcinoma cells. Cancer 117:2926–2938

Sanita P, Capulli M, Teti A, Galatioto GP, Vicentini C, Chiarugi P, Bologna M, Angelucci A (2014) Tumor-stroma metabolic relationship based on lactate shuttle can sustain prostate cancer progression. BMC Cancer 14:154

Sanità P, Capulli M, Teti A, Galatioto GP, Vicentini C, Chiarugi P, Bologna M, Angelucci A (2014) Tumor-stroma metabolic relationship based on lactate shuttle can sustain prostate cancer progression. BMC Cancer 14:154

Santi A, Caselli A, Ranaldi F, Paoli P, Mugnaioni C, Michelucci E, Cirri P (2015) Cancer associated fibroblasts transfer lipids and proteins to cancer cells through cargo vesicles supporting tumor growth. Biochim Biophys Acta 1853:3211–3223

Semenza GL (2010) HIF-1: upstream and downstream of cancer metabolism. Curr Opin Genet Dev 20:51–56

Shim H, Dolde C, Lewis BC, Wu C-S, Dang G, Jungmann RA, Dalla-Favera R, Dang CV (1997) c-Myc transactivation of LDH-A: implications for tumor metabolism and growth. Proc Natl Acad Sci 94:6658–6663

Slavov N, Budnik BA, Schwab D, Airoldi EM, van Oudenaarden A (2014) Constant growth rate can be supported by decreasing energy flux and increasing aerobic glycolysis. Cell Rep 7:705–714

Sonveaux P, Vegran F, Schroeder T, Wergin MC, Verrax J, Rabbani ZN, De Saedeleer CJ, Kennedy KM, Diepart C, Jordan BF, Kelley MJ, Gallez B, Wahl ML, Feron O, Dewhirst MW (2008) Targeting lactate-fueled respiration selectively kills hypoxic tumor cells in mice. J Clin Invest 118:3930–3942

Sotgia F, Del Galdo F, Casimiro MC, Bonuccelli G, Mercier I, Whitaker-Menezes D, Daumer KM, Zhou J, Wang C, Katiyar S, Xu H, Bosco E, Quong AA, Aronow B, Witkiewicz AK, Minetti C, Frank PG, Jimenez SA, Knudsen ES, Pestell RG, Lisanti MP (2009) Caveolin-1−/− null mammary stromal fibroblasts share characteristics with human breast cancer-associated fibroblasts. Am J Pathol 174:746–761

Sotgia F, Whitaker-Menezes D, Martinez-Outschoorn UE, Flomenberg N, Birbe RC, Witkiewicz AK, Howell A, Philp NJ, Pestell RG, Lisanti MP (2012) Mitochondrial metabolism in cancer metastasis: visualizing tumor cell mitochondria and the "reverse Warburg effect" in positive lymph node tissue. Cell Cycle 11:1445–1454

Sotgia F, Martinez-Outschoorn UE, Lisanti MP (2013) Cancer metabolism: new validated targets for drug discovery. Oncotarget 4:1309

Sun P, Hu J-W, Xiong W-J, Mi J (2014) miR-186 regulates glycolysis through Glut1 during the formation of cancer-associated fibroblasts. Asian Pac J Cancer Prev 15:4245–4250

Sung JS, Kang S, Lee JH, Mun SG, Kim BG, Cho NH (2018) Integrin beta4-induced mitophagy promotes the lactate production of cancer-associated fibroblasts in breast cancer. In: AACR

Sung JS, Kang CW, Kang S, Jang Y, Chae YC, Kim BG, Cho NH (2020) ITGB4-mediated metabolic reprogramming of cancer-associated fibroblasts. Oncogene 39:664–676

Thompson DE, Siwicky MD, Moorehead RA (2012) Caveolin-1 expression is elevated in claudin-low mammary tumor cells. Cancer Cell Int 12:6

Vadlakonda L, Dash A, Pasupuleti M, Anil Kumar K, Reddanna P (2013) Did we get Pasteur, Warburg, and Crabtree on a right note? Front Oncol 3:186

Vander Heiden MG (2011) Targeting cancer metabolism: a therapeutic window opens. Nat Rev Drug Discov 10:671–684

Vegran F, Boidot R, Sonveaux P, Feron O (2011) 1004 ORAL lactate influx and efflux through monocarboxylate transporters bridge cancer cell metabolism and angiogenesis. Eur J Cancer 47: S98

Villena JA, Kralli A (2008) ERRalpha: a metabolic function for the oldest orphan. Trends Endocrinol Metab 19:269–276

Vincent AS, Phan TT, Mukhopadhyay A, Lim HY, Halliwell B, Wong KP (2008) Human skin keloid fibroblasts display bioenergetics of cancer cells. J Invest Dermatol 128:702–709

Warburg O (1925) The metabolism of carcinoma cells. J Cancer Res 9:148–163

Warburg O (1956) On the origin of cancer cells. Science 123:309–314

Whitaker-Menezes D, Martinez-Outschoorn UE, Flomenberg N, Birbe R, Witkiewicz AK, Howell A, Pavlides S, Tsirigos A, Ertel A, Pestell RG (2011a) Hyperactivation of oxidative mitochondrial metabolism in epithelial cancer cells in situ: visualizing the therapeutic effects of metformin in tumor tissue. Cell Cycle 10:4047–4064

Whitaker-Menezes D, Martinez-Outschoorn UE, Lin Z, Ertel A, Flomenberg N, Witkiewicz AK, Birbe RC, Howell A, Pavlides S, Gandara R, Pestell RG, Sotgia F, Philp NJ, Lisanti MP (2011b) Evidence for a stromal-epithelial "lactate shuttle" in human tumors: MCT4 is a marker of oxidative stress in cancer-associated fibroblasts. Cell Cycle 10:1772–1783

Wilde L, Roche M, Domingo-Vidal M, Tanson K, Philp N, Curry J, Martinez-Outschoorn U (2017) Metabolic coupling and the Reverse Warburg Effect in cancer: Implications for novel biomarker and anticancer agent development. Semin Oncol 44:198–203

Witkiewicz AK, Dasgupta A, Nguyen KH, Liu C, Kovatich AJ, Schwartz GF, Pestell RG, Sotgia F, Rui H, Lisanti MP (2009) Stromal caveolin-1 levels predict early DCIS progression to invasive breast cancer. Cancer Biol Ther 8:1071–1079

Witkiewicz AK, Dasgupta A, Sammons S, Er O, Potoczek MB, Guiles F, Sotgia F, Brody JR, Mitchell EP, Lisanti MP (2010) Loss of stromal caveolin-1 expression predicts poor clinical outcome in triple negative and basal-like breast cancers. Cancer Biol Ther 10:135–143

Witkiewicz AK, Whitaker-Menezes D, Dasgupta A, Philp NJ, Lin Z, Gandara R, Sneddon S, Martinez-Outschoorn UE, Sotgia F, Lisanti MP (2012) Using the "reverse Warburg effect" to identify high-risk breast cancer patients: stromal MCT4 predicts poor clinical outcome in triple-negative breast cancers. Cell Cycle 11:1108–1117

Yan H, Bigner DD, Velculescu V, Parsons DW (2009) Mutant metabolic enzymes are at the origin of gliomas. Cancer Res 69:9157–9159

Yang L, Achreja A, Yeung TL, Mangala LS, Jiang D, Han C, Baddour J, Marini JC, Ni J, Nakahara R, Wahlig S, Chiba L, Kim SH, Morse J, Pradeep S, Nagaraja AS, Haemmerle M, Kyunghee N, Derichsweiler M, Plackemeier T, Mercado-Uribe I, Lopez-Berestein G, Moss T, Ram PT, Liu J, Lu X, Mok SC, Sood AK, Nagrath D (2016a) Targeting stromal glutamine synthetase in tumors disrupts tumor microenvironment-regulated cancer cell growth. Cell Metab 24:685–700

Yang X, Xu X, Zhu J, Zhang S, Wu Y, Wu Y, Zhao K, Xing C, Cao J, Zhu H (2016b) miR-31 affects colorectal cancer cells by inhibiting autophagy in cancer-associated fibroblasts. Oncotarget 7:79617

Yeung S, Pan J, Lee M-H (2008) Roles of p53, MYC and HIF-1 in regulating glycolysis—the seventh hallmark of cancer. Cell Mol Life Sci 65:3981

Yu T, Yang G, Hou Y, Tang X, Wu C, Wu X, Guo L, Zhu Q, Luo H, Du Y (2017) Cytoplasmic GPER translocation in cancer-associated fibroblasts mediates cAMP/PKA/CREB/glycolytic axis to confer tumor cells with multidrug resistance. Oncogene 36:2131–2145

Zhang XH-F, Jin X, Malladi S, Zou Y, Wen YH, Brogi E, Smid M, Foekens JA, Massagué J (2013) Selection of bone metastasis seeds by mesenchymal signals in the primary tumor stroma. Cell 154:1060–1073

Zhang G, Li J, Wang X, Ma Y, Yin X, Wang F, Zheng H, Duan X, Postel GC, Li X-F (2015) The reverse Warburg effect and 18F-FDG uptake in non–small cell lung cancer A549 in mice: a pilot study. J Nucl Med 56:607–612

Zheng M, Wang Y-H, Wu X-N, Wu S-Q, Lu B-J, Dong M-Q, Zhang H, Sun P, Lin S-C, Guan K-L (2011) Inactivation of Rheb by PRAK-mediated phosphorylation is essential for energy-depletion-induced suppression of mTORC1. Nat Cell Biol 13:263–272

Zong WX, Rabinowitz JD, White E (2016) Mitochondria and cancer. Mol Cell 61:667–676

Mitochondrial Biogenesis, Mitophagy, and Mitophagic Cell Death in Cancer Regulation: A Comprehensive Review

Prakash Priyadarshi Praharaj, Bishnu Prasad Behera, Soumya Ranjan Mishra, Srimanta Patra, Kewal Kumar Mahapatra, Debasna Pritimanjari Panigrahi, Chandra Sekhar Bhol, and Sujit Kumar Bhutia

Abstract

Mitochondria are well known as the "energy factory" of the cell and are essential for better cell survival and programmed cell death and its quality control plays an imperative role in fine-tuning cellular and organismal homeostasis. For the betterment of mitochondrial homeostasis, a coordinated regulation is necessary between the genesis of functional and/or fresh mitochondria through mitochondrial biogenesis and the elimination of dysfunctional and/or damaged mitochondria through mitophagy. When mitochondria are formed in excess amounts, it has detrimental effects for several cell types, causing various pathophysiological conditions such as cancer. Moroever, mitophagy has evolved as a protective mechanism in various types of cancer cells upon mild to moderate stress controlling the size and quality of the functional mitochondrial pool. However, excessive activation without proper compensatory mitochondrial biogenesis may also give rise to malfunction of mitochondria and lead to mitophagic cell death. Similarly, triggering the biogenesis of new mitochondria has been advantageous, although the hyperactivity of which may lead to higher oxygen consumption and oxidative stress. Therefore, in this review we have discussed multiple conserving strategies, cancer cells developed to control mitochondrial biogenesis, mitophagy, and mitophagic cell death, promoting longevity and resistance toward extreme stress, and how their regulatory imbalance accelerates both survival and cell death during cancer progression.

P. P. Praharaj · B. P. Behera · S. R. Mishra · S. Patra · K. K. Mahapatra · D. P. Panigrahi · C. S. Bhol · S. K. Bhutia (✉)
Department of Life Science, National Institute of Technology Rourkela, Rourkela, Odisha, India
e-mail: sujitb@nitrkl.ac.in

Keywords

Mitochondria · Mitochondria biogenesis · Mitophagy · Mitophagic cell death · Cancer

Abbreviations

AMBRA1	Activating molecule in BECN1-regulated autophagy protein 1
AMPK	5'-AMP-activated protein kinase
BECN1	Beclin 1
Bnip3	BCL2/adenovirus E1B 19-kDa protein-interacting protein 3
Bnip3L	BCL2/adenovirus E1B 19-kDa protein-interacting protein 3-like
COX IV	Cytochrome c oxidase subunit IV
Drp1	Dynamin-related protein 1
ER	Endoplasmic reticulum
FIP200	FAK family kinase-interacting protein of 200 kDa
ILS	Insulin-like signaling
LC3	Microtubule-associated protein 1 light chain 3
LIR	LC3-interacting region
MDV	Mitochondria-derived vesicles
Mfn1/2	Mitofusin-1/2
mPTP	Mitochondrial permeability transition pore
mtDNA	Mitochondrial DNA
mTOR	Mechanistic target of rapamycin
nDNA	Nuclear DNA
NRF1	Nuclear respiratory factor 1
NRF2	Nuclear factor erythroid 2-related factor 2
Opa1	Optic atrophy protein 1
Parkin	E3 ubiquitin–protein ligase parkin
PGC-1α/1β	Peroxisome proliferator–activated receptor-γ coactivator 1α/1β
PINK1	Phosphatase and tensin homolog–induced putative kinase 1
PKA	Protein kinase A
PRC	PGC-1-related coactivator
ROS	Reactive oxygen species
SAM	Sorting and assembly machinery
SQSTM1	Sequestosome 1
TAX1BP1	Tax1-binding protein 1
TBK1	Serine/threonine-protein kinase TBK1
TFAM	Transcription factor A, mitochondrial
TIM	Translocase of the inner membrane
TOM	Translocase of the outer membrane
ULK1/2	Serine/threonine-protein kinase ULK1/2
UPS	Ubiquitin–proteasome system
VDAC	Voltage-dependent anion channel
VPS	Vacuolar protein sorting
VPS15	Phosphoinositide 3-kinase regulatory subunit 4
WIPI2	WD repeat domain phosphoinositide-interacting protein 2

7.1 Introduction

Mitochondria, well known as "energy factory," are double-layered membranous organelles with vital roles in modulating most of the essential cellular developments such as energy production, metabolite synthesis, lipid metabolism, calcium signaling, and cellular homeostasis needed for cell survival and apoptotic cell death (Bravo-Sagua et al. 2013; Pinton et al. 2008; Porporato et al. 2018; Yu and Pekkurnaz 2018). Similar to the prokaryotic organism, mitochondria have several copies of their own circular DNA. However, nuclear genes encode most of the mitochondrial proteins (Ploumi et al. 2017). The cellular mitochondrial contents, structure, and function are maintained through the two oppositely driving, evolutionary conserved cellular pathways viz. mitochondrial biogenesis and mitophagy, which coordinated regulation, is the prerequisite for sustainable energy metabolism, during various metabolic and cellular stress (Palikaras et al. 2015a; Ploumi et al. 2017; Vega et al. 2015). Any sort of instability or discrepancy between these two processes triggers the beginning and advanced unfolding of numerous pathophysiological disorders such as cancer, neurodegenerative diseases, and aging (Vafai and Mootha 2012). Therefore, mitochondrial quality control plays an imperative role in fine-tuning cellular and organismal homeostasis.

Mitochondrial biogenesis, is an intricate cellular pathway, which includes replication, transcription, and translation of mitochondrial DNA (mtDNA), with subsequent distribution of newly synthesized phospholipids, mitochondrial-, and nuclear-encoded proteins, which were then recruited, imported, and assembled in the mitochondrial sub-compartments (Ventura-Clapier et al. 2008; Zhu et al. 2013). This whole process is firmly regulated by different growth factor- or hormone-initiated intracellular signaling cascades, which leads to the stimulation of several nuclear transcription factors (Weitzel and Iwen 2011), for example, peroxisome proliferator-activated receptor gamma, coactivator-1α, -1β (PGC-1α, PGC-1β), transcription factor A, mitochondrial (TFAM), and the nuclear respiratory factors (NRF1, and NRF2) (Dominy and Puigserver 2013; Yun and Finkel 2014). Apart from these key factors, numerous other factors including hormones, kinase pathways (AMPK, MAPK, and PKA) (Herzig and Shaw 2018) and secondary messengers [e.g., cyclicAMP, calcium, and endothelial nitric oxide synthase (eNOS)], are also involved in the regulation of this process at different cellular levels (Nisoli and Carruba 2006; Ould Amer and Hebert-Chatelain 2018). These transcription factors selectively act to counter different environmental and/or intracellular stimuli such as hypoxia, nutrient deficiency, and unavailability of growth factors, hormones, and toxins. However, when mitochondria are formed in excess amounts, it will be detrimental for several cell types, causing various pathophysiological conditions such as cancer (Artal-Sanz and Tavernarakis 2009; Malpass 2013; Palikaras et al. 2015b; Wredenberg et al. 2002). Besides their indispensable roles in cellular metabolism and reprogramming, mitochondria also responsible for the formation of harmful reactive oxygen species (ROS) because of cellular respiration. Therefore, cancer cells have adopted several efficient mechanisms (e.g., autophagy the most efficient one) to remove the dysfunctional or damaged mitochondria following the prevention of cellular damage and eventual death and to preserve mitochondrial homeostasis (Lahiri et al. 2019).

Mitophagy, a selective form of autophagy, which involves targeted engulfment of dysfunctional or damaged mitochondria by double-layered membranous vesicles known as mitophagosome, with subsequent fusion with lysosome for degradation (Praharaj et al. 2019; Youle and Narendra 2011). Autophagic removal of damaged or dysfunctional mitochondria is accomplished either through general autophagy (non-selective) or through mitophagy (mitochondria targeting autophagy) (Chu 2019). In response to high-energy demands, cells trigger mitophagy, which eventually controls the size and quality of the functional mitochondrial pool. Moreover, it also stimulates resistance to several cellular or environmental stress, such as nutrient deprivation, oxidative, and genotoxic stress (Palikaras et al. 2015b). For the betterment of mitochondrial homeostasis, a coordinated regulation is required between the removal of dysfunctional and/or damaged mitochondria and genesis of functional and/or fresh mitochondria, which leads to the accurate functioning of mitochondria (Palikaras et al. 2015a; Ploumi et al. 2017). Eukaryotic cells control the level of mitophagy through upregulating many ATG genes at several points fine-tuning the proper timing of initiation and enormousness in the cell (Ding and Yin 2012). This coordinated regulation during cellular stress is important for optimum mitophagy efficacy and energy usage while preventing excessive mitophagy causing mitophagic cell death (Dorn and Kitsis 2015; Park et al. 2019). Here, we have discussed the updated research findings on the regulatory mechanisms involved in the two major evolutionary conserved mitochondrial pathways. Furthermore, we intended to establish the crosstalk among different key molecular pathways regulating smooth coordination amongst these processes, for defining where they meet to regulate both survival and death of the cancer cell in response to the functional abnormality of mitochondria.

7.2 Mitochondrial Biogenesis: A Sequential Event Toward a New Synthesis

Mitochondria are double-membrane bound organelles with the self-replicating genome, *i.e.* mtDNA, which is responsible for encoding 13 important components of the ETC along with rRNAs and tRNAs useful for translating the mtDNA-encoded proteins (Balaban 1990; Scarpulla et al. 2012). Apart from that, the rest of the mitochondrial proteins (>1000) (Lotz et al. 2014; Pagliarini et al. 2008) are translated from the nuclear DNA(nDNA). Hence, both mtDNA and nDNA are responsible for the successful execution of mitochondria biogenesis. Mitochondrial biogenesis is a firmly controlled process, where both nuclear and mitochondrial genes undergo synchronous transcription and translation and are necessary for the genesis of new mitochondria. Following transcription and translation, the mitochondrial proteins are then directed to different mitochondrial sub-compartments depending upon their sequence (Schulz and Rehling 2014). This whole process is well governed by the mechanistic target of rapamycin (mTOR), the

AMP-activated protein kinase (AMPK), insulin-like signaling (ILS), and calcium- and/or nitric oxide-associated signaling pathways (Cunningham et al. 2007; Lagouge et al. 2006; Nisoli et al. 2003; Wu et al. 2002; Zong et al. 2002).

7.2.1 Mitobiogenic Machinery: A Multifactorial Rewiring Network

When cells experience any sort of alterations in the abundance of fuel-substrates balance or energy status, a complex network of both mtDNA and nDNA activated to conduct vigorous and dynamic mitochondrial biogenesis. Amongst different key transcriptional coregulators such as peroxisome proliferator-activated receptor gamma, coactivator-1α, -1β (PGC-1α and PGC-1β), and PGC-1-related coactivator (PRC), PGC1α is considered as the chief regulating factor in mitochondrial biogenesis (Andersen et al. 2005; Andersson and Scarpulla 2001; Kressler et al. 2002; Lin et al. 2002; Puigserver and Spiegelman 2003; Puigserver et al. 1998). PGC1α is mainly localized in the cytoplasm, but, upon phosphorylation and/or deacetylation, it starts accumulating inside the nucleus to increase the mitochondrial biogenesis (Anderson et al. 2008; Chang et al. 2010). According to few reports, PGC1α is also found in the mitochondrial compartment, along with its interacting partner transcription factor A, mitochondrial (TFAM) at the mtDNA D-loop (Aquilano et al. 2010; Safdar et al. 2011). Both of these forms (viz. nuclear and mitochondrial pools) of PGC1α promote mitochondrial biogenesis with the response to exercise (Safdar et al. 2010). During physiological and/or pathological stress, as an adaptive response cell triggered PGC1α levels (Jones et al. 2012; Lopez-Lluch et al. 2008), which stimulates the transcription through direct interaction with the other transcription factors through its conserved LXXLL recognition domains. Then other factors are recruited mediating chromatin remodeling through histone acetylation, and helpful for the interaction with the TRAP/DRIP complex with subsequent recruitment of RNA polymerase II (Ge et al. 2002). These coactivators directly interact with a different transcription factor, allowing the synchronized control of the several signaling cascades involved in making new mitochondria (Vega et al. 2015). The critical involvement of the PGC-1 coactivators was again verified through both gain- and loss-of-function approach in the heart and other mitochondria-abundant organs, which are prone to higher mitochondrial biogenesis (Lehman et al. 2000). The activity of PGC1α is also controlled by its post-translational modification events (such as acetylation and phosphorylation), which modulate the mitochondrial function and biogenesis (Fig. 7.1).

Apart from PGC1α, another set of nuclear transcription factors viz. estrogen-related receptor (ERR), and the nuclear respiratory factors (NRF1 and NRF2) also modulate mitochondrial respiration, energy metabolism, and biogenesis via regulating the activity of genes responsible for mitochondrial proteins (Finck and Kelly 2006). NRF1 exists as a homodimer and positively regulate several mitochondrial target gene, for example, cytochrome c, COXIV subunits, TOMM34, TFB2M, TFB1M, TFAM, and metallothionein-1 and 2 (Biswas and Chan 2010; Gleyzer et al. 2005; Satoh et al. 2013; Scarpulla 2008a; Virbasius et al. 1993; Yang et al. 2014).

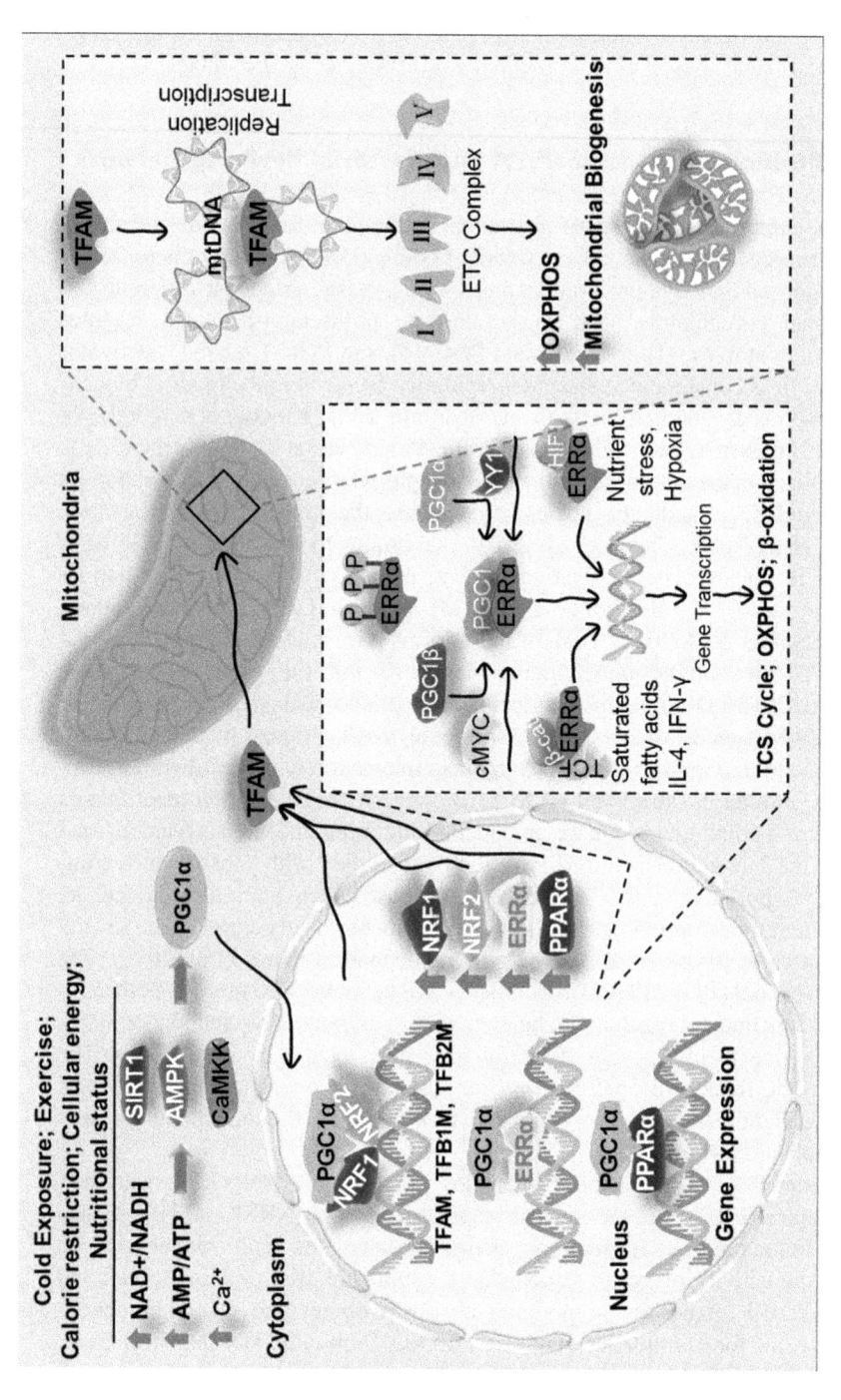

Fig. 7.1 Mitochondrial biogenesis in rescuing cellular stress. Upon nutrient stress, eukaryotic cells maintain mitochondrial mass through upregulating mitochondrial biogenesis. PGC-1α,-1β, ERRα, β, γ, PPARγ, and NRF1/2 are the key transcription factors responsible for the regulating mitochondrial biogenesis through upregulating transcription factor A, mitochondrial (TFAM) and mitochondrial polymerase along with several metabolic enzymes and imported protein for normal mitochondrial function. Moreover, YY1 and c-Myc also regulate mitochondrial biogenesis by regulating the activity of ERRα activity

For the first time, it was recognized through its capacity to binds to the promoter of cytochrome c (Scarpulla 2006). It undergoes phosphorylation at serine residues of its N-terminal domain, which increases its binding efficiency toward DNA (Gugneja and Scarpulla 1997) and trans-activation function (Herzig et al. 2000) under normal condition whereas, serum starvation facilitates dephosphorylation event (Gugneja and Scarpulla 1997). Moreover, ROS also triggered the phosphorylation of NRF1 via AKT1 in rat hepatoma cells, augmenting its nuclear translocation with a simultaneous increase in the expression of TFAM (Piantadosi and Suliman 2006). Unlikely to NRF1, NRF2 remains in the cytoplasm and interacts with its target genes of various mitochondria-associated biosynthetic pathways for the coordinated regulation of mitochondrial biogenesis (Jaramillo and Zhang 2013; Malhotra et al. 2010; Piantadosi et al. 2008; Taguchi et al. 2011). Additionally, skinhead-1 (SKN-1) (*Caenorhabditis elegans* homolog of NRF2) has also been found to be associated with MOM through its interaction with the phosphoglycerate mutase homolog-5 (PGAM-5), which gets translocated from mitochondria to the nucleus during cellular stress, modulating the activity of multiple target gene, different from those controlled by its cytoplasmic pool. It was mostly identified as a binding partner of cis-acting elements in the promoter region of cytochrome oxidase subunit IV (COX IV), (Scarpulla 2008b) with subsequent association with all subunits of cytochrome oxidase encoded by nDNA (Ongwijitwat and Wong-Riley 2005). Additionally, in association with NRF1, it also regulates mtDNA replication and transcription via controlling the expression of a key regulator of embryonic development and mitochondrial biogenesis viz. TFAM, TFB1M, and TFB2M (Bruni et al. 2010; Larsson et al. 1998; Scarpulla 2008b; Scarpulla et al. 2012). Apart from the nuclear-specific regulation, mitochondria-associated regulatory mechanisms, such as mtDNA replication, transcription, translation, and mitochondrial protein import, have a requisite role in mitochondrial biogenesis. The MOM, translocase of the outer membrane (TOM), MIM, translocase of the inner membrane (TIM), and sorting and assembly machinery (SAM) complexes, are the major contributor for mitochondrial biogenesis and protein import in normal as well as cancer cells.

7.3 Mitophagy: A Cellular Need for Mitochondrial Turnover and Clearance of Defective Mitochondria

When cancer cells experience any sort of stress such as hypoxia, nutrient unavailability, one of the important cellular function, that is, mitochondrial function is disturbed most because of damage accumulation. As a first step toward stress relief, cells initiate proteolysis of various misfolded and/or oxidized polypeptides formed in different sub-compartments of mitochondria through induction of intramitochondrial ATP-dependent proteases whereas OMM proteins are cleaved through the cytosolic ubiquitin-proteasome system (UPS) as a damage repaired mechanism (Tatsuta and Langer 2008). At the same time, mitochondrial unfolded protein response (UPRmt) starts the retrograde signaling pathway to the nucleus, to improve mitochondrial proteotoxic stress through upregulation of specific mitochondrial chaperones

(Qureshi et al. 2017; Zhang et al. 2018). In another mechanism, mitochondrial-derived vesicles (MDVs), a mitochondrial membrane enclosing damaged components of mitochondria, which are then chopped through a fission-independent mechanism followed by its cytosolic release for lysosomal based degradation (Sugiura et al. 2014). Moreover, mitochondrial dynamics, a continuous cycle of fission and fusion events contributing to the reduction of damaged mitochondria. When cells failed to activate any of these mitochondrial quality control mechanisms, it activates mitophagy for the removal of injured mitochondria (Tatsuta and Langer 2008).

7.3.1 Mitophagy: In Detail

Mitophagy, an evolutionary conserved catabolic process that selectively targets damaged or superfluous mitochondria for lysosomal-based degradation in eukaryotic cells (Abdrakhmanov et al. 2020; Youle and Narendra 2011). As mitochondria are relatively large organelles, their degradation process requires significantly higher lysosomal and autophagic capacity. Mitophagy is mostly regulated by two evolutionary conserved serine/threonine kinase phosphatase and tensin homolog–induced kinase 1 (PINK1) and Parkin having E3 ubiquitin ligase activity. The whole process starts with the accumulation of PINK1 on the mitochondrial outer membrane (MOM) upon misfolded protein accumulation and the collapse of mitochondrial membrane potential (Jin et al. 2010; Villa et al. 2018). Then there is subsequent recruitment of the E3 ubiquitin ligase Parkin to the MOM from the cytosol (Vives-Bauza et al. 2010). PINK1 then triggers and phosphorylates Parkin on Ser65 in its ubiquitin-like domain and phosphorylates its bound ubiquitin, thereby activating its ligase function (Kane et al. 2014; Koyano et al. 2014), in this manner marking the damaged and/or dysfunctional mitochondria for the association with different autophagy adaptors/receptors. Any mutation to the genes encoding these proteins triggers a hereditary neurodegenerative disorder known as autosomal-recessive juvenile Parkinsonism (AR-JP) (Kawajiri et al. 2011). Apart from the mutational status of these genes, posttranslational modification such as S-nitrosylation of PINK1 (SNO-PINK1) impairs mitophagy causing cell death of dopaminergic neurons upon nitrosative stress (Oh et al. 2017). Similarly, other E3 ubiquitin ligases confined to the MOM such as ARIH1, SIAH, MUL1, Gp78, MARCH 5, and SMURF1 also participate in driving mitophagy (Chen et al. 2017; Fu et al. 2013; Li et al. 2015; Orvedahl et al. 2011; Rojansky et al. 2016; Szargel et al. 2016; Villa et al. 2017). Unraveling the detailed mechanism linked with these ubiquitin ligases dependent mitophagic regulation would be a new target for the disease treatment (Fig. 7.2).

Along with this, several new specific mitophagy receptors bearing the LIR/GIM motif also regulate the mitophagy, stimulated not only during depolarization of mitochondria but also in response to hypoxia and chemotherapeutic drugs (Johansen and Lamark 2020; Shaid et al. 2013; Zaffagnini and Martens 2016). For example, during the hypoxic condition, BNIP3 and BNIP3L (also known as Nix) were

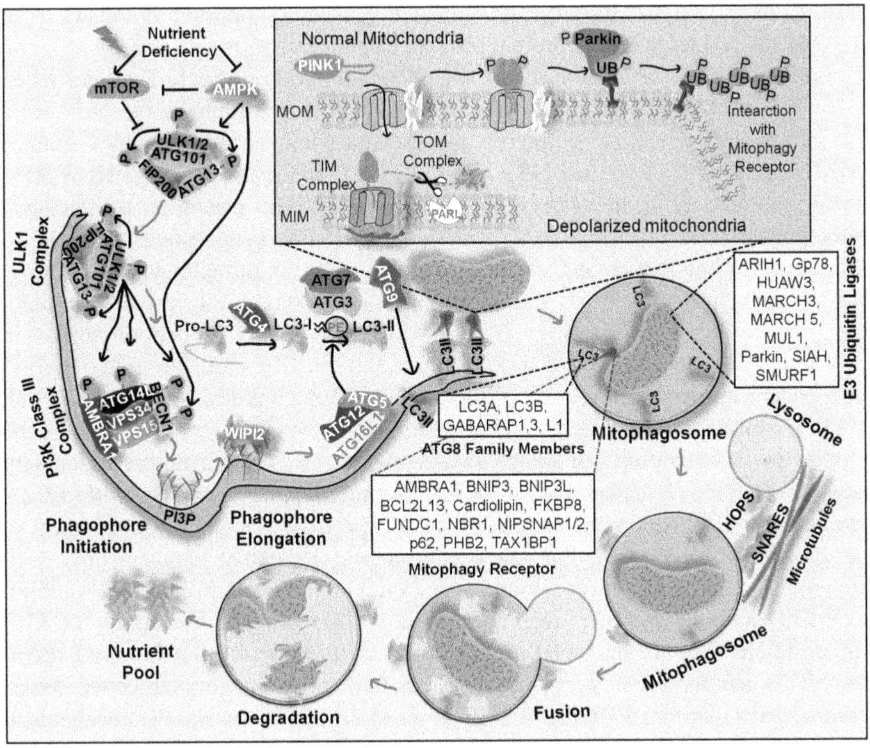

Fig. 7.2 Mitophagy, a cellular clearance act for damaged mitochondria. Cells maintained a healthy mitochondrial pool through eliminating damaged mitochondria through autophagic machinery, a process well known as mitophagy. In functional mitochondria, PTEN-induced putative kinase 1 (PINK1) is imported through translocase of the outer membrane (TOM) complex and translocate to translocase of the inner membrane (TIM) complex, where it undergoes quick degradation through presenilins-associated rhomboid-like protein (PARL) and Parkin remains in the cytosol in autoinhibition mode. During mitochondrial damage, PINK1 gets stabilized on the TOM complex, where it accesses its substrates such as ubiquitin chains and/or parkin and phosphorylates them. Parkin then accumulates ubiquitin chains on numerous MOM proteins, which can recruit ubiquitin-binding mitophagy receptors recruiting the LC3-positive phagophore, which enclosed the damaged mitochondria inside mitophagosome and allow it for lysosomal based degradation. The whole process involves the production of phosphatidylinositol-3-phosphate (PtdIns3P) on donor membranes, serine/threonine-protein kinase ULK1 complex which controls phagophore initiation and expansion and ATG8 conjugation pathway involving ATG7 (E1), ATG3 (E2), and the ATG5/ATG12–ATG16L1 (E3) complex

accumulated on the MOM, and after stabilization and dimerization, BNIP3 interacts with LC3B, and Nix interacts with GABARAP-L1 to mediate mitophagy (Hanna et al. 2012; Shi et al. 2014). In another mechanism, FUNDC1 also triggered hypoxia-mediated mitophagy with the combined kinase activity of CK2 and SRC kinases, disabling the phosphorylation of FUNDC1 at Tyr18 and Ser13, which has happened in normal condition promoting the interaction of FUNDC1 with LC3 (Kuang et al.

2016; Lv et al. 2017). Moreover, this interaction is also supported by the action of PGAM5 the serine/threonine protein phosphatase mediating the de-phosphorylation at Ser13 and ULK1 complex dependent phosphorylation at Ser17 (van der Bliek 2016; Wei et al. 2015). Recently, NIPSNAP1/2 another mitophagy receptors were get accumulated on MOM to interact with several well-reported adaptors proteins (p62, NDP52, TAX1BP1, and NBR1) and ATG8 proteins (LC3s and GABARAPs) during mitochondrial depolarization to accomplish the clearance of damaged mitochondria (Princely Abudu et al. 2019). A recent report suggesting the existence of a crossover between several signaling pathways during mitophagy. For instance, Bcl2-L-13 (mammalian Atg32 homolog) plays a key role in mitochondrial homeostasis, where its Bcl-2 homology (BH) 1–4 domains trigger mitochondrial fission separating the damaged part of mitochondria and the LIR motif promotes Parkin-independent mitophagy to eliminate the damaged mitochondrion (Murakawa et al. 2015). Furthermore, the association between Bcl2-L-13, ULK1 complex, and LC3B is essential in activating mitophagy (Murakawa et al. 2019). Another mitophagy receptor FKBP8 (also known as FKBP38) preferentially interacts with LC3A (Bhujabal et al. 2017) and can selectively escape from impaired mitochondria to the endoplasmic reticulum through a microtubule-dependent pathway linking to attenuation of apoptosis during mitophagy (Saita et al. 2013). Similarly, Prohibitin 2 (PHB2) a mitochondrial inner membrane (MIM) protein, interacts with LC3 triggering clearance of damaged mitochondria in response to mitochondrial stress. The MOM gets rupture in a proteasome-dependent manner facilitating better association of LIR motif of PHB2 and LC3 (Wei et al. 2017). In another mechanism, mitochondrial inner membrane lipid cardiolipin was also found to promote mitophagy through direct interaction with LC3, after its get translocated to MOM (Chu et al. 2013). Mitochondria fission is the prerequisite events that lead to mitophagy, interruption of which produces long elongated mitochondria whereas mitochondria fusion has just the opposite effect inhibition of which leads to the accumulation of small, fragmented mitochondria (Dorn et al. 2015).

7.4 Mitochondrial Biogenesis and Mitophagy; Where Do They Encounter?

In order to maintain the mitochondrial pool, cancer cells coordinately regulate the two evolutionary conserved opposing processes, that is, mitochondrial biogenesis and mitophagy, and maintain a dynamic equilibrium among these two through modulation of key transcriptional and post-translational factors. In recent times, several signaling pathways have been studied showing the existence of crosstalk between mitochondrial biogenesis and mitophagy (Gottlieb and Carreira 2010; Settembre et al. 2011). In the following sections, we have discussed the regulatory network involved in these dynamic processes for the regulation of tumorigenesis (Fig. 7.3).

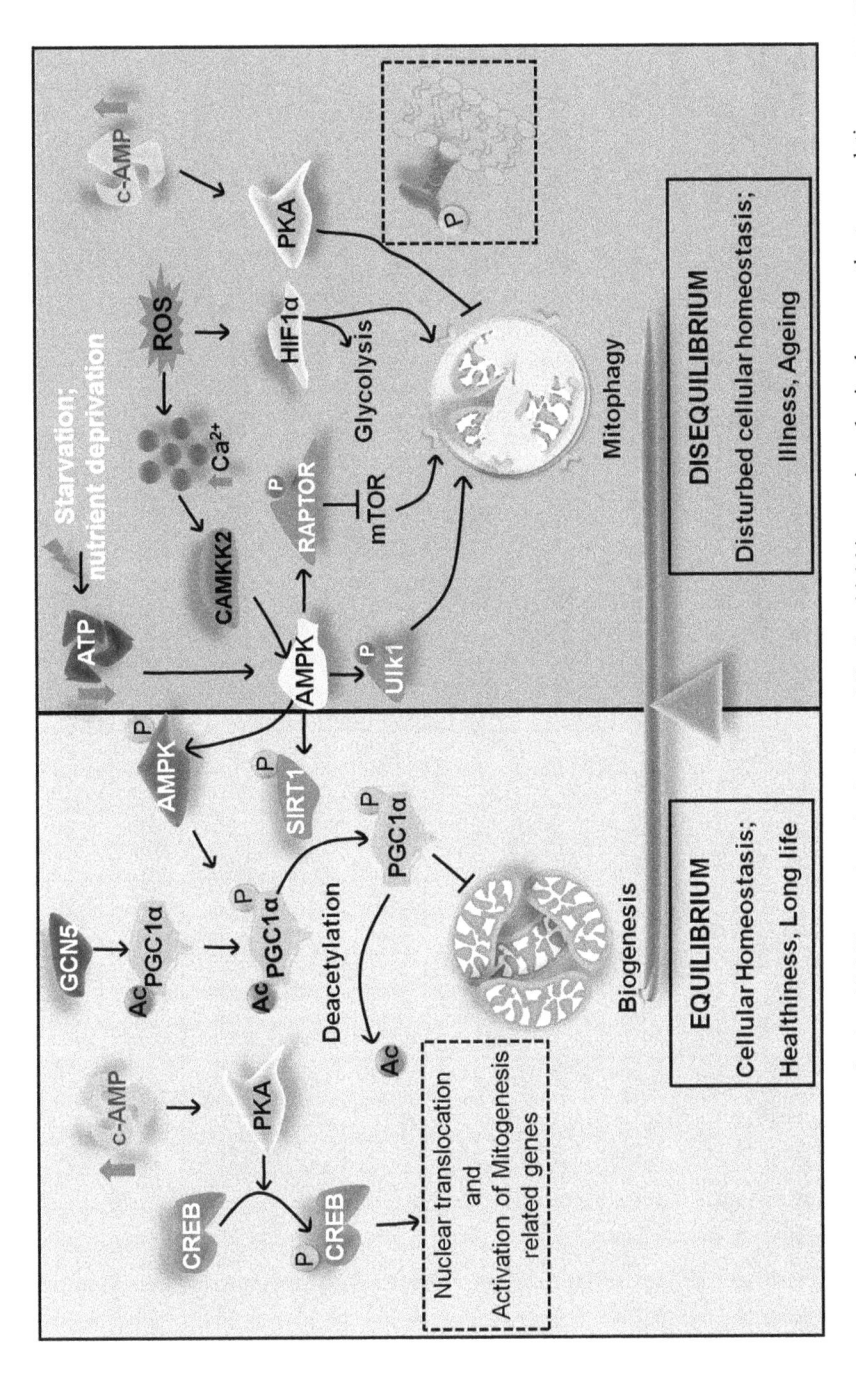

Fig. 7.3 The crossroad between mitochondrial biogenesis and mitophagy. Mitochondrial biogenesis and mitophagy are the two evolutionary conserved mitochondria-related pathways necessary for the maintenance of mitochondrial homeostasis and normal cellular function and existence. The balance amongst

Fig. 7.3 (continued) these processes is maintained through multiple kinase-associated signaling pathways such as PKA, MAPK1/3, AMPK, mTOR, which display counteracting effect in regulating both these processes result in either a net growth or reduction in total mitochondrial mass. AMPK signaling shows a dual role: triggering mitophagy via (a) phosphorylating RAPTOR result in mTOR inhibition (b) Ulk1 phosphorylation and in parallel promote mitochondrial biogenesis through SIRT1-PGC1α axis. CyclicAMP activates PKA signaling which favors mitochondrial biogenesis by upregulating PGC1α through phosphorylation and nuclear translocation of CREB and in parallel inhibits mitophagy via preventing LC3 lipidation. ROS also triggers intracellular Ca^{2+} level which activates CAMKK2 to accelerate AMPK activity. Proper equilibrium among these processes leads to cellular homeostasis resulting in healthiness and long life. However, any sort of disequilibrium leads to disturbed cellular homeostasis causing illness

7.4.1 PGC1α–TFEB Axis in Regulating the Crosstalk

In addition to its major involvement in mitochondrial biogenesis, PGC1α also acts as a positive regulator for mitophagy through direct induction of transcription factor EB (TFEB), which triggers lysosomal biogenesis (Settembre et al. 2011). Functional involvement of several key autophagy/mitophagy regulators such as ATG5, ATG9, and Parkin is also known to involve in the activation of TFEB and its translocation and transactivation of its associated target genes, explaining the interdependence between the phagophores initiation (Itakura et al. 2012; Kishi-Itakura et al. 2014) and the probable needs for mitophagy at the transcriptional level (Nezich et al. 2015). However, nutrient starvation leads to activation followed by translocation of TFEB to the nucleus, whose transcriptional activity is independent of key ATGs (such as ATG9, ATG5), and Parkin (Nezich et al. 2015). In turn, TFEB triggers PGC1α level, to establish a positive feedback loop to maintain the equilibrium between mitochondrial biogenesis and mitophagy. Likewise, the general control of amino acid synthesis 5-like 1 (GCN5L1) negatively regulates both mitophagy and mitochondrial biogenesis via modulating the expression of PGC1α and TFEB (Scott et al. 2014).

7.4.2 AMPK Regulated Signaling

During nutrient deprivation or CCCP treatment, AMP-activated protein kinase (AMPK), the metabolic sensor of the cells, is activated and stimulates the autophagy/mitophagy via attenuating mTOR pathway and activating ULK1 (Kim et al. 2011; Kwon et al. 2011). At the same time, AMPK also starts mitochondrial biogenesis via an increase in the intracellular NAD+ level and stimulates sirtuin1 (SIRT1) activity through phosphorylation, causing deacetylation (Canto et al. 2009; Rodgers et al. 2005) and subsequent activation and nuclear accumulation of PGC1α, possibly to compensate for higher mitochondrial turnover (Anderson et al. 2008; Reznick et al. 2007; Schulz et al. 2008). Moreover, when cells experience any fluctuation in the intracellular AMP/ATP ratio AMPK phosphorylates PGC1α at serine 538 and threonine 177 to initiate its regulatory function (Jager et al. 2007). On the other hand, SIRT1 can also trigger AMPK via its deacetylation and activate the STK11/LKB1 the kinase for AMPK, which promotes the PGC1α phosphorylation (Price et al. 2012). On the contrary, PGC1α activity is attenuated by its acetylation through the K (lysine) acetyltransferase 2 (KAT2/GCN5) attenuates (Kelly et al. 2009) and the steroid nuclear receptor coactivator 3 (NCOA3/SRC3) (Coste et al. 2008) in presence of excess calories. During acute exercise and/or high-energy requirement, PGC1α undergoes AMPK-mediated phosphorylation to regulate the genesis of new mitochondria (Jager et al. 2007). Moreover, these acetylations at different lysine residues make it a target for NAD+-dependent deacetylase SIRT1, possibly linking mitochondrial function, energy, and redox status of the cell (Canto et al. 2009; Coste et al. 2008; Gerhart-Hines et al. 2007). Similar to mitochondrial biogenesis, SIRT1 also fuels autophagy, whereas the SIRT1 deficiency leads to the higher accumulation of damaged mitochondria (Lee et al. 2008). It also triggered remodeling of the nucleosome and inhibit methylation at cytosine residues, which leads to the upregulation of TFAM, PGC1α, and uncoupling proteins 2 and 3 (UCP2 and UCP3) key players of mitochondrial function and biogenesis (Marin et al. 2017). Besides, activated AMPK also upregulates PGC1α expression during acute exercise in skeletal muscle (Little et al. 2010). Moreover, chronic MPP+ intoxication decreased mitochondrial PGC1α levels causing the impairment of mitochondrial biogenesis.

7.4.3 PINK-Parkin Signaling Pathway

PINK-Parkin signaling not only plays a vital role in mitophagy but also regulates mitochondrial biogenesis. PINK1 deficiency leads to impairment of mitochondrial biogenesis through multiple factors such as decreased mtDNA copy number, mitochondrial bioenergetics, mitochondrial ATP synthesis, cytochrome c oxidase activity, and fission/fusion proteins in both in vitro and in vivo conditions (Billia et al. 2011; Gegg et al. 2009). Similarly, apart from its central role in CCCP-triggered mitophagy, Parkin also stimulates mitochondrial biogenesis through its substrate, that is, Parkin-interacting substrate (PARIS) which undergoes ubiquitination and UPS-mediated elimination, an event negatively regulating PGC1α and its target NRF1 (Shin et al. 2011). Moreover, PARIS binds to insulin response sequence (IRS) present on the promoter region of the PGC1α suppressing PGC1α activity and higher accumulation of PARIS after the abolition of its usual role, that is, Parkin-mediated mitophagy it caused the degeneration of dopaminergic neurons and this effect could be rescued when PGC1α is overexpressed (Shin et al. 2011). Altogether, this study establishes the role of Parkin in PGC1α-mediated mitochondrial biogenesis through its substrate, PARIS. Additionally, Parkin directly associates with TFAM triggering the transcription of genes encoded by mtDNA with the subsequent promotion of mitochondrial biogenesis and mutations on PARK2 to eliminate this interaction with TFAM and mtDNA (Kuroda et al. 2006; Rothfuss et al. 2009).

7.4.4 cAMP-Responsive Element-Binding Protein (CREB) Associated Signaling

Cellular cyclicAMP is another key factor modulating both the processes through a different mechanism under different cellular stress. Upon different environmental stimuli exercise and/or upon exposure toward cold intracellular cAMP level increases which subsequently activate PKA, a kinase that acts upstream to both autophagy, and PGC1α and phosphorylate CREB. CREB then activates a series of downstream molecules (i.e., NRF1, NRF2, TFAM, and PGC1α) to modulate mitochondrial biogenesis (Baar et al. 2002; Chowanadisai et al. 2010; De Rasmo et al.

2010; Terada et al. 2002; Wu et al. 1999). In addition, CREB also transcriptionally triggers the expression of mtDNA-encoded genes (De Rasmo et al. 2009). On the other hand, the cAMP pathway negatively regulates mitophagy through its effector molecule, that is, PKA, which directly acts via phosphorylation of LC3 (Cherra et al. 2010; Dagda et al. 2011) and indirectly through phosphorylation of DNM1L/DRP1 inhibiting fission-promoting activity (Dickey and Strack 2011) and promotes biogenesis (De Rasmo et al. 2010).

7.4.5 MAPK Family Members

MAPK family members, for example, MAPK12 phosphorylate the PGC1α and promotes the strong binding between NRF1 promotor and cytochrome c promoter or with that of NRF2 promoter to the cytochrome oxidase subunit promoter (Barger et al. 2001; Wright et al. 2007b) or by disturbing the association between PGC1α and MYBBP1A [MYB binding protein (P160) 1a] (Fan et al. 2004). Particularly, the same pathway has also been involved in the regulating of mitochondrial degradation through MAPK14, and MAPK1, instead of p38 MAPK, triggering together hypoxia- and starvation-triggered mitophagy (Hirota et al. 2015). Moreover, calcium signaling also regulates mitochondrial biogenesis through PGC1α, MAPK, and calcium/calmodulin-dependent protein kinase II (CaMKII) (Wright et al. 2007a; Wu et al. 2002).

7.5 Mitophagic Cell Death: A Substantial Cause of Less Mitochondrial Biogenesis, Hyper Mitophagy

The "lifecycle" of mitochondria involves the synchronized genesis of both mtDNA-, and nDNA-encoded proteins, organized and sorting with all other essential components through mitogenesis and at the end, any malfunction leads to the removal of the whole and/or partial mitochondria from the cell through a well-orchestrated catabolic process known as mitophagy. It is very difficult to maintain the quantity and quality of total mitochondria without proper coordination between these two oppositely driving pathways (Palikaras et al. 2015a; Ploumi et al. 2017). More importantly, any disturbance in these anabolic-catabolic balances could detain stress release and contribute to cell death (Schapira 2012; Zhu et al. 2012). Autophagic cell death (ACD) symbolizes a type of cell death driven by autophagic machinery and mostly attenuated upon precise inhibition of the autophagic pathway (Galluzzi et al. 2015). Under minimal stress, mitophagy acts as a pro-survival mechanism providing longevity (Kubli and Gustafsson 2012; Praharaj et al. 2019). Whereas, once cells pass the threshold level of cellular stress the same autophagy turns out to be a "pro-death mechanism" leading to ACD (Denton and Kumar 2019; Knuppertz et al. 2017). Many research findings also suggest the existence of the pro-death role of autophagy as a backup cell death mechanism especially in cancer cells (i.e., highly resistant to apoptosis) upon exposure to several chemotherapeutic drugs and other pharmacological molecules (Fulda and Kogel 2015; Kogel et al. 2010).

The quick clearance of damaged or dysfunctional mitochondria attenuating the oxidized proteins, lipids, and DNA level inside the mitochondria, regulating the possibility of apoptosis (Hickson-Bick et al. 2008). On the other hand, excessive mitophagic degradation without proper compensatory biogenesis of new mitochondria may also give rise to the malfunction of mitochondria and leads to mitophagic cell death (Chu et al. 2007; Yan et al. 2012; Zhu et al. 2007). This essential mitochondrial protective signal strongly influences responses to therapy and the phenotypic evolution of cancer. AT-101 ([−]-gossypol), a natural compound from cottonseeds, stimulates the activation of heme oxygenase 1 (HMOX1) and the mitophagy receptors BNIP3L, and BNIP3 which induces an early mitochondrial depolarization with a subsequent reduction in mitochondrial mass/proteins (Meyer et al. 2018). Excessive mitophagy induction noticeably preceded the mitophagic type of cell death independent of the apoptosis in glioma cells (Meyer et al. 2018). Parkin-mediated mitophagy leads to cell death in insulin-deprived HCN cells. In the absence of insulin, Parkin triggers the mitochondrial accumulation of Ca^{2+}, initiating depolarization of mitochondria at the initial stages of mitophagy. This allowed the recruitment of Parkin and PINK1, which work in a mutually co-operative way to remove damaged mitochondria to initiate mitophagy (Park et al. 2019). Several stress signals can stimulate the intrinsic pathway of apoptosis, which involves the translocation of several inner mitochondrial proteins toward the cytosol or to the nucleus promoting cell death. Mitochondrial outer membrane permeabilization (MOMP) involves both Bax and Bak proteins, which undergoes oligomerization and assemble to form the Bax/Bak pore (Cosentino and Garcia-Saez 2017). Cells can prevent MOMP and initiation of apoptosis via blocking the assembly of Bax/Bak pores in a Parkin-dependent manner. Moreover, Parkin-dependent ubiquitination of the voltage-dependent anion channel (VDAC) on the MOM favors binding with Bax, which can prevent the association of Bax with mitochondria (Bernardini et al. 2019).

7.6 Modified Mitomass Affecting Cancer Development

According to several recent findings oncogenes and/or tumor suppressors can also predispose cellular mitochondrial mass by directly or indirectly regulating both mitochondrial biogenesis and mitophagy in different cancer types. However, whether this altered mitochondrial mass linked with tumor subtype, tumor grade, response toward therapy, and/or recurrence-free survival is yet to be established. Mitophagy receptors such as BNIP3L and BNIP3 and are mostly deregulated in human cancer with higher expression at pre-malignant stages and seem to be downregulated as cancer progressed to higher grade malignant cancer types (Okami et al. 2004; Tan et al. 2007). The most accepted mechanism explaining the downregulation with relation to tumor progression is epigenetic silencing of the promoter in several different cancer types such as pancreatic, hematologic, colorectal, lung, and liver cancers (Abe et al. 2005; Bacon et al. 2007; Calvisi et al. 2007; Castro et al. 2010; Murai et al. 2005). Moreover, loss of BNIP3 also promotes tumor growth and metastasis in the in-vivo mouse model system of breast cancer (Manka et al. 2005) confirming the tumor suppressor activity of mitophagy. Attenuation of mitophagy leads to an elevation in mtROS level, alter mitochondrial metabolism such as fatty acid oxidation and Krebs cycle might be responsible for promoting cell death in tumor cells. Noticeably, the more we know about how alteration in mitochondrial metabolism is deregulated in cancer, the more we expected to identify novel signaling pathways that are abnormally activated by the accumulation of specific metabolites. Mitochondrial biogenesis either promotes or suppresses cancer in a context-dependent manner, such as tissue type, tumor stage, or stress types in the tumor microenvironment (Bost and Kaminski 2019; Tan et al. 2016). Although the generation of new mitochondria causing higher metabolite and energy generation ideally promoting tumor progression. It also acts as a tumor-suppressive via stimulating oxidative metabolism, restraining cellular ROS level, and permitting stability to the HIF-1α. The c-Myc well known for its oncogenic activity also promotes mitochondrial biogenesis through stimulation of PGC-1β activity (Zhang et al. 2007) causing induction of different mitochondrial proteins, including TFAM, NRF1, and Polg (Kim et al. 2008; Li et al. 2005). In contrast, hypoxia inducible factor-1α (HIF-1α) attenuates biogenesis through triggering Mxi-1 (a repressor of c-Myc) that promotes the degradation of c-Myc, explaining by what means mitochondrial mass is concentrated during hypoxia (Zhang et al. 2007). In another mechanism, the stability of HIF-1α protein is also maintained by the loss of function of SIRT3, a downstream target of PGC1α (Finley et al. 2011), signifying the role of biogenesis promoting factors in preventing ROS-triggered HIF-1 stabilization. Higher PGC1α activity has been associated with the cause of melanomas subtype because of its stimulation by the melanocyte-specific transcription factor, MITF (Haq et al. 2013; Shoag et al. 2013; Vazquez et al. 2013). Melanomas expressing high PGC1α unveiled higher expression of mitochondrial proteins and dependency on oxidative phosphorylation whereas, melanomas expressing low PGC1α depend more on glycolysis (Haq et al. 2013; Vazquez et al. 2013). PGC1α is also needed for the growth and progression of melanomas via protection against ROS-induced apoptosis (Vazquez et al. 2013). Interestingly, oncogene-causing melanomas such as B-Raf suppress oxidative metabolism through attenuating MITF-triggered upregulation of PGC1α. However, when treated with vemurafenib (B-Raf inhibitors), melanomas critically reliant on oxidative metabolism for survival signifying the importance of combination therapy of the role of mitochondrial metabolism inhibitors and B-Raf inhibitors for effective treatment of melanoma (Haq et al. 2013). PGC1α is also promoting angiogenesis via co-activation of ERR-α driving HIF-independent expression of VEGF (Arany et al. 2008). The cellular activity of PGC1α also affects the cancer prognosis depending upon cancer subtypes such as high expression can predict a good outcome in patients suffering from prostate cancer (Torrano et al. 2016), however, it is poor when it comes to patients with breast cancer (Klimcakova et al. 2012). Furthermore, PGC1α could be an ideal biomarker in determining cancer's aggressiveness and response to treatment as it is a characteristic of cancer stem cells in pancreatic cancer (Sancho et al. 2015). According to another study, telomere dysfunction leads to p53-mediated suppression of PGC1α causing an elevated level of mitochondrial dysfunction and ROS production probably explaining less genesis of new mitochondria in ALTC tumors (Sahin et al. 2011). Telomerase inactivation causing slow growth in tumors but eventually, tumors getting more aggressive through activating the alternative lengthening of telomeres (ALT) pathway in atm null Tcell lymphomas (Hu et al. 2012). Interestingly, PGC-1β revealed constant copy-number alteration, and its elevated expression along with its target protein such as catalase, SOD2, TFAM, and NRF2 was also identified in ALTC tumors (Hu et al. 2012) signifying the selective benefit to emerging tumors through higher mitochondrial biogenesis. Several key transcription factors associated with mitochondrial biogenesis are found to be downregulated in various cancer subtypes (Bellance et al. 2009; Lee et al. 2011; Liu et al. 2019). PPAR agonists such as resveratrol, bezafibrate, enhances the cellular level of TFAM, and PGC1α, in several cancer cell types (e.g., osteosarcoma, breast cancer, and cervical carcinoma) inhibiting their proliferation and invasiveness (Wang and Moraes 2011). Similarly, in the case of human intestinal cancer cells, the gain of function of PGC1α drives mitochondrial biogenesis as well as trigger apoptosis mediated cell death (D'Errico et al. 2011). With these findings we can speculate the higher tendency of cancer cells to decrease the mitochondrial respiration in a predilection for aerobic energy production (glycolysis). Whereas the effect was just reversed in type I endometrial carcinoma, where the elevated genesis of new mitochondria has been associated with tumor growth (Cormio et al. 2012; Guerra et al. 2011). Hence, the impending role of operating mitochondrial homeostasis would be an ideal target for future cancer treatment options.

7.7 Conclusion and Future Prospective

Cancer cells develop multiple conserving strategies to control mitochondrial biogenesis, mitophagy, and mitophagic cell death, promoting longevity and resistance toward extreme stress, whereas any interruption in their balance accelerates several pathophysiological disorders. When cancer cells start removing slightly effective mitochondria in a rapid way than regenerating new efficient mitochondria, it is more harmful than helpful. In this perspective, attenuating the mitophagic activity while augmenting its biogenesis might be operational. It seems that mitophagy is evolving as a protective mechanism upon mild to moderate stress (Panigrahi et al. 2019). However, excessive activation also leads to mitophagic-cell death. Similarly, triggering the biogenesis of new mitochondria has been advantageous, although its hyperactivity may lead to higher oxygen consumption and oxidative stress. However, in the absence of mitophagy superfluous and/or damaged mitochondria are accumulated (Gusdon et al. 2012; Jennings et al. 2006; Suzuki et al. 2011), which are not efficient enough to buffer cytoplasmic calcium, and may go through the mitochondrial permeability transition releasing cytochrome c and other pro-apoptotic molecules to the cytosol for activation of downstream caspase for apoptosis. While substantial advancement has been made toward revealing the details of mitochondrial homeostasis, several key questions remain unanswered. What are the key transcription factors involved in the mitochondrial retrograde signaling pathways, triggering transcriptional reprogramming induced by mitochondrial damage in support of mitochondrial biogenesis? What are the fundamental mechanisms responsible for creating the imbalance in recycling responses and defining the methods to manipulate these cellular events for the development of new options for cancer treatment? How the key components of these separate pathways fit together in a whole network of mitochondrial homeostasis? The biggest encounter would be not only explaining the cellular commands necessary for synchronized regulations mitochondrial function, but also the feedback mechanism regulating both biogenesis and mitophagy in cancer prospective.

Acknowledgment Research support was partly provided by Council of Scientific & Industrial Research (CSIR) [Grant Number: 37(1715)/18/EMR-II], Government of India.
Conflict of interest: The authors disclose no conflict of interest.

References

Abdrakhmanov A, Gogvadze V, Zhivotovsky B (2020) To eat or to die: deciphering selective forms of autophagy. Trends Biochem Sci 45:347–364

Abe T, Toyota M, Suzuki H, Murai M, Akino K, Ueno M, Nojima M, Yawata A, Miyakawa H, Suga T, Ito H, Endo T, Tokino T, Hinoda Y, Imai K (2005) Upregulation of BNIP3 by 5-aza-2′-deoxycytidine sensitizes pancreatic cancer cells to hypoxia-mediated cell death. J Gastroenterol 40:504–510

Andersen G, Wegner L, Yanagisawa K, Rose CS, Lin J, Glümer C, Drivsholm T, Borch-Johnsen K, Jørgensen T, Hansen T, Spiegelman BM, Pedersen O (2005) Evidence of an association between genetic variation of the coactivator PGC-1beta and obesity. J Med Genet 42:402–407

Anderson RM, Barger JL, Edwards MG, Braun KH, O'Connor CE, Prolla TA, Weindruch R (2008) Dynamic regulation of PGC-1alpha localization and turnover implicates mitochondrial adaptation in calorie restriction and the stress response. Aging Cell 7:101–111

Andersson U, Scarpulla RC (2001) Pgc-1-related coactivator, a novel, serum-inducible coactivator of nuclear respiratory factor 1-dependent transcription in mammalian cells. Mol Cell Biol 21:3738–3749

Aquilano K, Vigilanza P, Baldelli S, Pagliei B, Rotilio G, Ciriolo MR (2010) Peroxisome proliferator-activated receptor gamma co-activator 1alpha (PGC-1alpha) and sirtuin 1 (SIRT1) reside in mitochondria: possible direct function in mitochondrial biogenesis. J Biol Chem 285:21590–21599

Arany Z, Foo SY, Ma Y, Ruas JL, Bommi-Reddy A, Girnun G, Cooper M, Laznik D, Chinsomboon J, Rangwala SM, Baek KH, Rosenzweig A, Spiegelman BM (2008) HIF-independent regulation of VEGF and angiogenesis by the transcriptional coactivator PGC-1alpha. Nature 451:1008–1012

Artal-Sanz M, Tavernarakis N (2009) Prohibitin couples diapause signalling to mitochondrial metabolism during ageing in C. elegans. Nature 461:793–797

Baar K, Wende AR, Jones TE, Marison M, Nolte LA, Chen M, Kelly DP, Holloszy JO (2002) Adaptations of skeletal muscle to exercise: rapid increase in the transcriptional coactivator PGC-1. FASEB J 16:1879–1886

Bacon AL, Fox S, Turley H, Harris AL (2007) Selective silencing of the hypoxia-inducible factor 1 target gene BNIP3 by histone deacetylation and methylation in colorectal cancer. Oncogene 26:132–141

Balaban RS (1990) Regulation of oxidative phosphorylation in the mammalian cell. Am J Phys 258: C377–C389

Barger PM, Browning AC, Garner AN, Kelly DP (2001) p38 mitogen-activated protein kinase activates peroxisome proliferator-activated receptor alpha: a potential role in the cardiac metabolic stress response. J Biol Chem 276:44495–44501

Bellance N, Benard G, Furt F, Begueret H, Smolkova K, Passerieux E, Delage JP, Baste JM, Moreau P, Rossignol R (2009) Bioenergetics of lung tumors: alteration of mitochondrial biogenesis and respiratory capacity. Int J Biochem Cell Biol 41:2566–2577

Bernardini JP, Brouwer JM, Tan IK, Sandow JJ, Huang S, Stafford CA, Bankovacki A, Riffkin CD, Wardak AZ, Czabotar PE, Lazarou M, Dewson G (2019) Parkin inhibits BAK and BAX apoptotic function by distinct mechanisms during mitophagy. EMBO J 38:e99916

Bhujabal Z, Birgisdottir AB, Sjottem E, Brenne HB, Overvatn A, Habisov S, Kirkin V, Lamark T, Johansen T (2017) FKBP8 recruits LC3A to mediate Parkin-independent mitophagy. EMBO Rep 18:947–961

Billia F, Hauck L, Konecny F, Rao V, Shen J, Mak TW (2011) PTEN-inducible kinase 1 (PINK1)/Park6 is indispensable for normal heart function. Proc Natl Acad Sci U S A 108:9572–9577

Biswas M, Chan JY (2010) Role of Nrf1 in antioxidant response element-mediated gene expression and beyond. Toxicol Appl Pharmacol 244:16–20

Bost F, Kaminski L (2019) The metabolic modulator PGC-1α in cancer. Am J Cancer Res 9:198–211

Bravo-Sagua R, Rodriguez AE, Kuzmicic J, Gutierrez T, Lopez-Crisosto C, Quiroga C, Diaz-Elizondo J, Chiong M, Gillette TG, Rothermel BA, Lavandero S (2013) Cell death and survival through the endoplasmic reticulum-mitochondrial axis. Curr Mol Med 13:317–329

Bruni F, Polosa PL, Gadaleta MN, Cantatore P, Roberti M (2010) Nuclear respiratory factor 2 induces the expression of many but not all human proteins acting in mitochondrial DNA transcription and replication. J Biol Chem 285:3939–3948

Calvisi DF, Ladu S, Gorden A, Farina M, Lee JS, Conner EA, Schroeder I, Factor VM, Thorgeirsson SS (2007) Mechanistic and prognostic significance of aberrant methylation in the molecular pathogenesis of human hepatocellular carcinoma. J Clin Invest 117:2713–2722

Canto C, Gerhart-Hines Z, Feige JN, Lagouge M, Noriega L, Milne JC, Elliott PJ, Puigserver P, Auwerx J (2009) AMPK regulates energy expenditure by modulating NAD+ metabolism and SIRT1 activity. Nature 458:1056–1060

Castro M, Grau L, Puerta P, Gimenez L, Venditti J, Quadrelli S, Sanchez-Carbayo M (2010) Multiplexed methylation profiles of tumor suppressor genes and clinical outcome in lung cancer. J Transl Med 8:86

Chang JS, Huypens P, Zhang Y, Black C, Kralli A, Gettys TW (2010) Regulation of NT-PGC-1alpha subcellular localization and function by protein kinase A-dependent modulation of nuclear export by CRM1. J Biol Chem 285:18039–18050

Chen Z, Liu L, Cheng Q, Li Y, Wu H, Zhang W, Wang Y, Sehgal SA, Siraj S, Wang X, Wang J, Zhu Y, Chen Q (2017) Mitochondrial E3 ligase MARCH5 regulates FUNDC1 to fine-tune hypoxic mitophagy. EMBO Rep 18:495–509

Cherra SJ 3rd, Kulich SM, Uechi G, Balasubramani M, Mountzouris J, Day BW, Chu CT (2010) Regulation of the autophagy protein LC3 by phosphorylation. J Cell Biol 190:533–539

Chowanadisai W, Bauerly KA, Tchaparian E, Wong A, Cortopassi GA, Rucker RB (2010) Pyrroloquinoline quinone stimulates mitochondrial biogenesis through cAMP response element-binding protein phosphorylation and increased PGC-1alpha expression. J Biol Chem 285:142–152

Chu CT (2019) Mechanisms of selective autophagy and mitophagy: Implications for neurodegenerative diseases. Neurobiol Dis 122:23–34

Chu CT, Zhu J, Dagda R (2007) Beclin 1-independent pathway of damage-induced mitophagy and autophagic stress: implications for neurodegeneration and cell death. Autophagy 3:663–666

Chu CT, Ji J, Dagda RK, Jiang JF, Tyurina YY, Kapralov AA, Tyurin VA, Yanamala N, Shrivastava IH, Mohammadyani D, Wang KZQ, Zhu J, Klein-Seetharaman J, Balasubramanian K, Amoscato AA, Borisenko G, Huang Z, Gusdon AM, Cheikhi A, Steer EK, Wang R, Baty C, Watkins S, Bahar I, Bayir H, Kagan VE (2013) Cardiolipin externalization to the outer mitochondrial membrane acts as an elimination signal for mitophagy in neuronal cells. Nat Cell Biol 15:1197–1205

Cormio A, Guerra F, Cormio G, Pesce V, Fracasso F, Loizzi V, Resta L, Putignano G, Cantatore P, Selvaggi LE, Gadaleta MN (2012) Mitochondrial DNA content and mass increase in progression from normal to hyperplastic to cancer endometrium. BMC Res Notes 5:279

Cosentino K, Garcia-Saez AJ (2017) Bax and bak pores: are we closing the circle? Trends Cell Biol 27:266–275

Coste A, Louet JF, Lagouge M, Lerin C, Antal MC, Meziane H, Schoonjans K, Puigserver P, O'Malley BW, Auwerx J (2008) The genetic ablation of SRC-3 protects against obesity and improves insulin sensitivity by reducing the acetylation of PGC-1{alpha}. Proc Natl Acad Sci U S A 105:17187–17192

Cunningham JT, Rodgers JT, Arlow DH, Vazquez F, Mootha VK, Puigserver P (2007) mTOR controls mitochondrial oxidative function through a YY1-PGC-1alpha transcriptional complex. Nature 450:736–740

D'Errico I, Salvatore L, Murzilli S, Lo Sasso G, Latorre D, Martelli N, Egorova AV, Polishuck R, Madeyski-Bengtson K, Lelliott C, Vidal-Puig AJ, Seibel P, Villani G, Moschetta A (2011) Peroxisome proliferator-activated receptor-gamma coactivator 1-alpha (PGC1alpha) is a metabolic regulator of intestinal epithelial cell fate. Proc Natl Acad Sci U S A 108:6603–6608

Dagda RK, Gusdon AM, Pien I, Strack S, Green S, Li C, Van Houten B, Cherra SJ 3rd, Chu CT (2011) Mitochondrially localized PKA reverses mitochondrial pathology and dysfunction in a cellular model of Parkinson's disease. Cell Death Differ 18:1914–1923

De Rasmo D, Signorile A, Roca E, Papa S (2009) cAMP response element-binding protein (CREB) is imported into mitochondria and promotes protein synthesis. FEBS J 276:4325–4333

De Rasmo D, Signorile A, Papa F, Roca E, Papa S (2010) cAMP/Ca2+ response element-binding protein plays a central role in the biogenesis of respiratory chain proteins in mammalian cells. IUBMB Life 62:447–452

Denton D, Kumar S (2019) Autophagy-dependent cell death. Cell Death Differ 26:605–616

Dickey AS, Strack S (2011) PKA/AKAP1 and PP2A/Bbeta2 regulate neuronal morphogenesis via Drp1 phosphorylation and mitochondrial bioenergetics. J Neurosci 31:15716–15726

Ding WX, Yin XM (2012) Mitophagy: mechanisms, pathophysiological roles, and analysis. Biol Chem 393:547–564

Dominy JE, Puigserver P (2013) Mitochondrial biogenesis through activation of nuclear signaling proteins. Cold Spring Harb Perspect Biol 5:a015008

Dorn GW 2nd, Kitsis RN (2015) The mitochondrial dynamism-mitophagy-cell death interactome: multiple roles performed by members of a mitochondrial molecular ensemble. Circ Res 116:167–182

Dorn GW 2nd, Vega RB, Kelly DP (2015) Mitochondrial biogenesis and dynamics in the developing and diseased heart. Genes Dev 29:1981–1991

Fan M, Rhee J, St-Pierre J, Handschin C, Puigserver P, Lin J, Jaeger S, Erdjument-Bromage H, Tempst P, Spiegelman BM (2004) Suppression of mitochondrial respiration through recruitment of p160 myb binding protein to PGC-1alpha: modulation by p38 MAPK. Genes Dev 18:278–289

Finck BN, Kelly DP (2006) PGC-1 coactivators: inducible regulators of energy metabolism in health and disease. J Clin Invest 116:615–622

Finley LW, Carracedo A, Lee J, Souza A, Egia A, Zhang J, Teruya-Feldstein J, Moreira PI, Cardoso SM, Clish CB, Pandolfi PP, Haigis MC (2011) SIRT3 opposes reprogramming of cancer cell metabolism through HIF1alpha destabilization. Cancer Cell 19:416–428

Fu M, St-Pierre P, Shankar J, Wang PT, Joshi B, Nabi IR (2013) Regulation of mitophagy by the Gp78 E3 ubiquitin ligase. Mol Biol Cell 24:1153–1162

Fulda S, Kogel D (2015) Cell death by autophagy: emerging molecular mechanisms and implications for cancer therapy. Oncogene 34:5105–5113

Galluzzi L, Bravo-San Pedro JM, Vitale I, Aaronson SA, Abrams JM, Adam D, Alnemri ES, Altucci L, Andrews D, Annicchiarico-Petruzzelli M, Baehrecke EH, Bazan NG, Bertrand MJ, Bianchi K, Blagosklonny MV, Blomgren K, Borner C, Bredesen DE, Brenner C, Campanella M, Candi E, Cecconi F, Chan FK, Chandel NS, Cheng EH, Chipuk JE, Cidlowski JA, Ciechanover A, Dawson TM, Dawson VL, De Laurenzi V, De Maria R, Debatin KM, Di Daniele N, Dixit VM, Dynlacht BD, El-Deiry WS, Fimia GM, Flavell RA, Fulda S, Garrido C, Gougeon ML, Green DR, Gronemeyer H, Hajnoczky G, Hardwick JM, Hengartner MO, Ichijo H, Joseph B, Jost PJ, Kaufmann T, Kepp O, Klionsky DJ, Knight RA, Kumar S, Lemasters JJ, Levine B, Linkermann A, Lipton SA, Lockshin RA, Lopez-Otin C, Lugli E, Madeo F, Malorni W, Marine JC, Martin SJ, Martinou JC, Medema JP, Meier P, Melino S, Mizushima N, Moll U, Munoz-Pinedo C, Nunez G, Oberst A, Panaretakis T, Penninger JM, Peter ME, Piacentini M, Pinton P, Prehn JH, Puthalakath H, Rabinovich GA, Ravichandran KS, Rizzuto R, Rodrigues CM, Rubinsztein DC, Rudel T, Shi Y, Simon HU, Stockwell BR, Szabadkai G, Tait SW, Tang HL, Tavernarakis N, Tsujimoto Y, Vanden Berghe T, Vandenabeele P, Villunger A, Wagner EF, Walczak H, White E, Wood WG, Yuan J, Zakeri Z, Zhivotovsky B, Melino G, Kroemer G (2015) Essential versus accessory aspects of cell death: recommendations of the NCCD 2015. Cell Death Differ 22:58–73

Ge K, Guermah M, Yuan CX, Ito M, Wallberg AE, Spiegelman BM, Roeder RG (2002) Transcription coactivator TRAP220 is required for PPAR gamma 2-stimulated adipogenesis. Nature 417:563–567

Gegg ME, Cooper JM, Schapira AH, Taanman JW (2009) Silencing of PINK1 expression affects mitochondrial DNA and oxidative phosphorylation in dopaminergic cells. PLoS One 4:e4756

Gerhart-Hines Z, Rodgers JT, Bare O, Lerin C, Kim SH, Mostoslavsky R, Alt FW, Wu Z, Puigserver P (2007) Metabolic control of muscle mitochondrial function and fatty acid oxidation through SIRT1/PGC-1alpha. EMBO J 26:1913–1923

Gleyzer N, Vercauteren K, Scarpulla RC (2005) Control of mitochondrial transcription specificity factors (TFB1M and TFB2M) by nuclear respiratory factors (NRF-1 and NRF-2) and PGC-1 family coactivators. Mol Cell Biol 25:1354–1366

Gottlieb RA, Carreira RS (2010) Autophagy in health and disease. 5. Mitophagy as a way of life. Am J Physiol Cell Physiol 299:C203–C210

Guerra F, Kurelac I, Cormio A, Zuntini R, Amato LB, Ceccarelli C, Santini D, Cormio G, Fracasso F, Selvaggi L, Resta L, Attimonelli M, Gadaleta MN, Gasparre G (2011) Placing mitochondrial DNA mutations within the progression model of type I endometrial carcinoma. Hum Mol Genet 20:2394–2405

Gugneja S, Scarpulla RC (1997) Serine phosphorylation within a concise amino-terminal domain in nuclear respiratory factor 1 enhances DNA binding. J Biol Chem 272:18732–18739

Gusdon AM, Zhu J, Van Houten B, Chu CT (2012) ATP13A2 regulates mitochondrial bioenergetics through macroautophagy. Neurobiol Dis 45:962–972

Hanna RA, Quinsay MN, Orogo AM, Giang K, Rikka S, Gustafsson AB (2012) Microtubule-associated protein 1 light chain 3 (LC3) interacts with Bnip3 protein to selectively remove endoplasmic reticulum and mitochondria via autophagy. J Biol Chem 287:19094–19104

Haq R, Shoag J, Andreu-Perez P, Yokoyama S, Edelman H, Rowe GC, Frederick DT, Hurley AD, Nellore A, Kung AL, Wargo JA, Song JS, Fisher DE, Arany Z, Widlund HR (2013) Oncogenic BRAF regulates oxidative metabolism via PGC1alpha and MITF. Cancer Cell 23:302–315

Herzig S, Shaw RJ (2018) AMPK: guardian of metabolism and mitochondrial homeostasis. Nat Rev Mol Cell Biol 19:121–135

Herzig RP, Scacco S, Scarpulla RC (2000) Sequential serum-dependent activation of CREB and NRF-1 leads to enhanced mitochondrial respiration through the induction of cytochrome c. J Biol Chem 275:13134–13141

Hickson-Bick DL, Jones C, Buja LM (2008) Stimulation of mitochondrial biogenesis and autophagy by lipopolysaccharide in the neonatal rat cardiomyocyte protects against programmed cell death. J Mol Cell Cardiol 44:411–418

Hirota Y, Yamashita S, Kurihara Y, Jin X, Aihara M, Saigusa T, Kang D, Kanki T (2015) Mitophagy is primarily due to alternative autophagy and requires the MAPK1 and MAPK14 signaling pathways. Autophagy 11:332–343

Hu J, Hwang SS, Liesa M, Gan B, Sahin E, Jaskelioff M, Ding Z, Ying H, Boutin AT, Zhang H, Johnson S, Ivanova E, Kost-Alimova M, Protopopov A, Wang YA, Shirihai OS, Chin L, DePinho RA (2012) Antitelomerase therapy provokes ALT and mitochondrial adaptive mechanisms in cancer. Cell 148:651–663

Itakura E, Kishi-Itakura C, Koyama-Honda I, Mizushima N (2012) Structures containing Atg9A and the ULK1 complex independently target depolarized mitochondria at initial stages of Parkin-mediated mitophagy. J Cell Sci 125:1488–1499

Jager S, Handschin C, St-Pierre J, Spiegelman BM (2007) AMP-activated protein kinase (AMPK) action in skeletal muscle via direct phosphorylation of PGC-1alpha. Proc Natl Acad Sci U S A 104:12017–12022

Jaramillo MC, Zhang DD (2013) The emerging role of the Nrf2-Keap1 signaling pathway in cancer. Genes Dev 27:2179–2191

Jennings JJ Jr, Zhu JH, Rbaibi Y, Luo X, Chu CT, Kiselyov K (2006) Mitochondrial aberrations in mucolipidosis type IV. J Biol Chem 281:39041–39050

Jin SM, Lazarou M, Wang C, Kane LA, Narendra DP, Youle RJ (2010) Mitochondrial membrane potential regulates PINK1 import and proteolytic destabilization by PARL. J Cell Biol 191:933–942

Johansen T, Lamark T (2020) Selective autophagy: ATG8 family proteins, LIR motifs and cargo receptors. J Mol Biol 432:80–103

Jones AW, Yao Z, Vicencio JM, Karkucinska-Wieckowska A, Szabadkai G (2012) PGC-1 family coactivators and cell fate: roles in cancer, neurodegeneration, cardiovascular disease and retrograde mitochondria-nucleus signalling. Mitochondrion 12:86–99

Kane LA, Lazarou M, Fogel AI, Li Y, Yamano K, Sarraf SA, Banerjee S, Youle RJ (2014) PINK1 phosphorylates ubiquitin to activate Parkin E3 ubiquitin ligase activity. J Cell Biol 205:143–153

Kawajiri S, Saiki S, Sato S, Hattori N (2011) Genetic mutations and functions of PINK1. Trends Pharmacol Sci 32:573–580

Kelly TJ, Lerin C, Haas W, Gygi SP, Puigserver P (2009) GCN5-mediated transcriptional control of the metabolic coactivator PGC-1beta through lysine acetylation. J Biol Chem 284:19945–19952

Kim J, Lee JH, Iyer VR (2008) Global identification of Myc target genes reveals its direct role in mitochondrial biogenesis and its E-box usage in vivo. PLoS One 3:e1798

Kim J, Kundu M, Viollet B, Guan KL (2011) AMPK and mTOR regulate autophagy through direct phosphorylation of Ulk1. Nat Cell Biol 13:132–141

Kishi-Itakura C, Koyama-Honda I, Itakura E, Mizushima N (2014) Ultrastructural analysis of autophagosome organization using mammalian autophagy-deficient cells. J Cell Sci 127:4089–4102

Klimcakova E, Chenard V, McGuirk S, Germain D, Avizonis D, Muller WJ, St-Pierre J (2012) PGC-1alpha promotes the growth of ErbB2/Neu-induced mammary tumors by regulating nutrient supply. Cancer Res 72:1538–1546

Knuppertz L, Warnsmann V, Hamann A, Grimm C, Osiewacz HD (2017) Stress-dependent opposing roles for mitophagy in aging of the ascomycete Podospora anserina. Autophagy 13:1037–1052

Kogel D, Fulda S, Mittelbronn M (2010) Therapeutic exploitation of apoptosis and autophagy for glioblastoma. Anti Cancer Agents Med Chem 10:438–449

Koyano F, Okatsu K, Kosako H, Tamura Y, Go E, Kimura M, Kimura Y, Tsuchiya H, Yoshihara H, Hirokawa T, Endo T, Fon EA, Trempe JF, Saeki Y, Tanaka K, Matsuda N (2014) Ubiquitin is phosphorylated by PINK1 to activate parkin. Nature 510:162–166

Kressler D, Schreiber SN, Knutti D, Kralli A (2002) The PGC-1-related protein PERC is a selective coactivator of estrogen receptor alpha. J Biol Chem 277:13918–13925

Kuang Y, Ma K, Zhou C, Ding P, Zhu Y, Chen Q, Xia B (2016) Structural basis for the phosphorylation of FUNDC1 LIR as a molecular switch of mitophagy. Autophagy 12:2363–2373

Kubli DA, Gustafsson ÅB (2012) Mitochondria and mitophagy: the yin and yang of cell death control. Circ Res 111:1208–1221

Kuroda Y, Mitsui T, Kunishige M, Shono M, Akaike M, Azuma H, Matsumoto T (2006) Parkin enhances mitochondrial biogenesis in proliferating cells. Hum Mol Genet 15:883–895

Kwon KY, Viollet B, Yoo OJ (2011) CCCP induces autophagy in an AMPK-independent manner. Biochem Biophys Res Commun 416:343–348

Lagouge M, Argmann C, Gerhart-Hines Z, Meziane H, Lerin C, Daussin F, Messadeq N, Milne J, Lambert P, Elliott P, Geny B, Laakso M, Puigserver P, Auwerx J (2006) Resveratrol improves mitochondrial function and protects against metabolic disease by activating SIRT1 and PGC-1alpha. Cell 127:1109–1122

Lahiri V, Hawkins WD, Klionsky DJ (2019) Watch what you (self-) eat: autophagic mechanisms that modulate metabolism. Cell Metab 29:803–826

Larsson NG, Wang J, Wilhelmsson H, Oldfors A, Rustin P, Lewandoski M, Barsh GS, Clayton DA (1998) Mitochondrial transcription factor A is necessary for mtDNA maintenance and embryogenesis in mice. Nat Genet 18:231–236

Lee IH, Cao L, Mostoslavsky R, Lombard DB, Liu J, Bruns NE, Tsokos M, Alt FW, Finkel T (2008) A role for the NAD-dependent deacetylase Sirt1 in the regulation of autophagy. Proc Natl Acad Sci U S A 105:3374–3379

Lee CH, Wu SB, Hong CH, Liao WT, Wu CY, Chen GS, Wei YH, Yu HS (2011) Aberrant cell proliferation by enhanced mitochondrial biogenesis via mtTFA in arsenical skin cancers. Am J Pathol 178:2066–2076

Lehman JJ, Barger PM, Kovacs A, Saffitz JE, Medeiros DM, Kelly DP (2000) Peroxisome proliferator-activated receptor gamma coactivator-1 promotes cardiac mitochondrial biogenesis. J Clin Invest 106:847–856

Li F, Wang Y, Zeller KI, Potter JJ, Wonsey DR, O'Donnell KA, Kim JW, Yustein JT, Lee LA, Dang CV (2005) Myc stimulates nuclearly encoded mitochondrial genes and mitochondrial biogenesis. Mol Cell Biol 25:6225–6234

Li J, Qi W, Chen G, Feng D, Liu J, Ma B, Zhou C, Mu C, Zhang W, Chen Q, Zhu Y (2015) Mitochondrial outer-membrane E3 ligase MUL1 ubiquitinates ULK1 and regulates selenite-induced mitophagy. Autophagy 11:1216–1229

Lin J, Puigserver P, Donovan J, Tarr P, Spiegelman BM (2002) Peroxisome proliferator-activated receptor gamma coactivator 1beta (PGC-1beta), a novel PGC-1-related transcription coactivator associated with host cell factor. J Biol Chem 277:1645–1648

Little JP, Safdar A, Cermak N, Tarnopolsky MA, Gibala MJ (2010) Acute endurance exercise increases the nuclear abundance of PGC-1alpha in trained human skeletal muscle. Am J Physiol Regul Integr Comp Physiol 298:R912–R917

Liu CL, Yang PS, Wang TY, Huang SY, Kuo YH, Cheng SP (2019) PGC1alpha downregulation and glycolytic phenotype in thyroid cancer. J Cancer 10:3819–3829

Lopez-Lluch G, Irusta PM, Navas P, de Cabo R (2008) Mitochondrial biogenesis and healthy aging. Exp Gerontol 43:813–819

Lotz C, Lin AJ, Black CM, Zhang J, Lau E, Deng N, Wang Y, Zong NC, Choi JH, Xu T, Liem DA, Korge P, Weiss JN, Hermjakob H, Yates JR 3rd, Apweiler R, Ping P (2014) Characterization, design, and function of the mitochondrial proteome: from organs to organisms. J Proteome Res 13:433–446

Lv M, Wang C, Li F, Peng J, Wen B, Gong Q, Shi Y, Tang Y (2017) Structural insights into the recognition of phosphorylated FUNDC1 by LC3B in mitophagy. Protein Cell 8:25–38

Malhotra D, Portales-Casamar E, Singh A, Srivastava S, Arenillas D, Happel C, Shyr C, Wakabayashi N, Kensler TW, Wasserman WW, Biswal S (2010) Global mapping of binding sites for Nrf2 identifies novel targets in cell survival response through ChIP-Seq profiling and network analysis. Nucleic Acids Res 38:5718–5734

Malpass K (2013) Neurodegenerative disease: defective mitochondrial dynamics in the hot seat-a therapeutic target common to many neurological disorders? Nat Rev Neurol 9:417

Manka D, Spicer Z, Millhorn DE (2005) Bcl-2/adenovirus E1B 19 kDa interacting protein-3 knockdown enables growth of breast cancer metastases in the lung, liver, and bone. Cancer Res 65:11689–11693

Marin TL, Gongol B, Zhang F, Martin M, Johnson DA, Xiao H, Wang Y, Subramaniam S, Chien S, Shyy JY (2017) AMPK promotes mitochondrial biogenesis and function by phosphorylating the epigenetic factors DNMT1, RBBP7, and HAT1. Sci Signal 10:eaaf7478

Meyer N, Zielke S, Michaelis JB, Linder B, Warnsmann V, Rakel S, Osiewacz HD, Fulda S, Mittelbronn M, Munch C, Behrends C, Kogel D (2018) AT 101 induces early mitochondrial dysfunction and HMOX1 (heme oxygenase 1) to trigger mitophagic cell death in glioma cells. Autophagy 14:1693–1709

Murai M, Toyota M, Suzuki H, Satoh A, Sasaki Y, Akino K, Ueno M, Takahashi F, Kusano M, Mita H, Yanagihara K, Endo T, Hinoda Y, Tokino T, Imai K (2005) Aberrant methylation and silencing of the BNIP3 gene in colorectal and gastric cancer. Clin Cancer Res 11:1021–1027

Murakawa T, Yamaguchi O, Hashimoto A, Hikoso S, Takeda T, Oka T, Yasui H, Ueda H, Akazawa Y, Nakayama H, Taneike M, Misaka T, Omiya S, Shah AM, Yamamoto A, Nishida K, Ohsumi Y, Okamoto K, Sakata Y, Otsu K (2015) Bcl-2-like protein 13 is a mammalian Atg32 homologue that mediates mitophagy and mitochondrial fragmentation. Nat Commun 6:7527

Murakawa T, Okamoto K, Omiya S, Taneike M, Yamaguchi O, Otsu K (2019) A mammalian mitophagy receptor, Bcl2-L-13, recruits the ULK1 complex to induce mitophagy. Cell Rep 26:338–345.e336

Nezich CL, Wang C, Fogel AI, Youle RJ (2015) MiT/TFE transcription factors are activated during mitophagy downstream of Parkin and Atg5. J Cell Biol 210:435–450

Nisoli E, Carruba MO (2006) Nitric oxide and mitochondrial biogenesis. J Cell Sci 119:2855–2862

Nisoli E, Clementi E, Paolucci C, Cozzi V, Tonello C, Sciorati C, Bracale R, Valerio A, Francolini M, Moncada S, Carruba MO (2003) Mitochondrial biogenesis in mammals: the role of endogenous nitric oxide. Science (New York, NY) 299:896–899

Oh CK, Sultan A, Platzer J, Dolatabadi N, Soldner F, McClatchy DB, Diedrich JK, Yates JR 3rd, Ambasudhan R, Nakamura T, Jaenisch R, Lipton SA (2017) S-Nitrosylation of PINK1 attenuates PINK1/Parkin-dependent mitophagy in hiPSC-based Parkinson's disease models. Cell Rep 21:2171–2182

Okami J, Simeone DM, Logsdon CD (2004) Silencing of the hypoxia-inducible cell death protein BNIP3 in pancreatic cancer. Cancer Res 64:5338–5346

Ongwijitwat S, Wong-Riley MT (2005) Is nuclear respiratory factor 2 a master transcriptional coordinator for all ten nuclear-encoded cytochrome c oxidase subunits in neurons? Gene 360:65–77

Orvedahl A, Sumpter R Jr, Xiao G, Ng A, Zou Z, Tang Y, Narimatsu M, Gilpin C, Sun Q, Roth M, Forst CV, Wrana JL, Zhang YE, Luby-Phelps K, Xavier RJ, Xie Y, Levine B (2011) Image-based genome-wide siRNA screen identifies selective autophagy factors. Nature 480:113–117

Ould Amer Y, Hebert-Chatelain E (2018) Mitochondrial cAMP-PKA signaling: what do we really know? Biochim Biophys Acta Bioenerg 1859:868–877

Pagliarini DJ, Calvo SE, Chang B, Sheth SA, Vafai SB, Ong SE, Walford GA, Sugiana C, Boneh A, Chen WK, Hill DE, Vidal M, Evans JG, Thorburn DR, Carr SA, Mootha VK (2008) A mitochondrial protein compendium elucidates complex I disease biology. Cell 134:112–123

Palikaras K, Lionaki E, Tavernarakis N (2015a) Balancing mitochondrial biogenesis and mitophagy to maintain energy metabolism homeostasis. Cell Death Differ 22:1399–1401

Palikaras K, Lionaki E, Tavernarakis N (2015b) Coordination of mitophagy and mitochondrial biogenesis during ageing in C. elegans. Nature 521:525–528

Panigrahi DP, Praharaj PP, Bhol CS, Mahapatra KK, Patra S, Behera BP, Mishra SR, Bhutia SK (2019) The emerging, multifaceted role of mitophagy in cancer and cancer therapeutics. Semin Cancer Biol. https://doi.org/10.1016/j.semcancer.2019.07.015

Park H, Chung KM, An HK, Gim JE, Hong J, Woo H, Cho B, Moon C, Yu SW (2019) Parkin promotes mitophagic cell death in adult hippocampal neural stem cells following insulin withdrawal. Front Mol Neurosci 12:46

Piantadosi CA, Suliman HB (2006) Mitochondrial transcription factor A induction by redox activation of nuclear respiratory factor 1. J Biol Chem 281:324–333

Piantadosi CA, Carraway MS, Babiker A, Suliman HB (2008) Heme oxygenase-1 regulates cardiac mitochondrial biogenesis via Nrf2-mediated transcriptional control of nuclear respiratory factor-1. Circ Res 103:1232–1240

Pinton P, Giorgi C, Siviero R, Zecchini E, Rizzuto R (2008) Calcium and apoptosis: ER-mitochondria Ca2+ transfer in the control of apoptosis. Oncogene 27:6407–6418

Ploumi C, Daskalaki I, Tavernarakis N (2017) Mitochondrial biogenesis and clearance: a balancing act. FEBS J 284:183–195

Porporato PE, Filigheddu N, Pedro JMB, Kroemer G, Galluzzi L (2018) Mitochondrial metabolism and cancer. Cell Res 28:265–280

Praharaj PP, Naik PP, Panigrahi DP, Bhol CS, Mahapatra KK, Patra S, Sethi G, Bhutia SK (2019) Intricate role of mitochondrial lipid in mitophagy and mitochondrial apoptosis: its implication in cancer therapeutics. Cell Mol Life Sci 76:1641–1652

Price NL, Gomes AP, Ling AJ, Duarte FV, Martin-Montalvo A, North BJ, Agarwal B, Ye L, Ramadori G, Teodoro JS, Hubbard BP, Varela AT, Davis JG, Varamini B, Hafner A, Moaddel R, Rolo AP, Coppari R, Palmeira CM, de Cabo R, Baur JA, Sinclair DA (2012) SIRT1 is required for AMPK activation and the beneficial effects of resveratrol on mitochondrial function. Cell Metab 15:675–690

Princely Abudu Y, Pankiv S, Mathai BJ, Hakon Lystad A, Bindesboll C, Brenne HB, Yoke Wui Ng M, Thiede B, Yamamoto A, Mutugi Nthiga T, Lamark T, Esguerra CV, Johansen T, Simonsen A (2019) NIPSNAP1 and NIPSNAP2 act as "Eat Me" signals for mitophagy. Dev Cell 49:509–525.e512

Puigserver P, Spiegelman BM (2003) Peroxisome proliferator-activated receptor-gamma coactivator 1 alpha (PGC-1 alpha): transcriptional coactivator and metabolic regulator. Endocr Rev 24:78–90

Puigserver P, Wu Z, Park CW, Graves R, Wright M, Spiegelman BM (1998) A cold-inducible coactivator of nuclear receptors linked to adaptive thermogenesis. Cell 92:829–839

Qureshi MA, Haynes CM, Pellegrino MW (2017) The mitochondrial unfolded protein response: Signaling from the powerhouse. J Biol Chem 292:13500–13506

Reznick RM, Zong H, Li J, Morino K, Moore IK, Yu HJ, Liu ZX, Dong J, Mustard KJ, Hawley SA, Befroy D, Pypaert M, Hardie DG, Young LH, Shulman GI (2007) Aging-associated reductions in AMP-activated protein kinase activity and mitochondrial biogenesis. Cell Metab 5:151–156

Rodgers JT, Lerin C, Haas W, Gygi SP, Spiegelman BM, Puigserver P (2005) Nutrient control of glucose homeostasis through a complex of PGC-1alpha and SIRT1. Nature 434:113–118

Rojansky R, Cha MY, Chan DC (2016) Elimination of paternal mitochondria in mouse embryos occurs through autophagic degradation dependent on PARKIN and MUL1. eLife 5:e17896

Rothfuss O, Fischer H, Hasegawa T, Maisel M, Leitner P, Miesel F, Sharma M, Bornemann A, Berg D, Gasser T, Patenge N (2009) Parkin protects mitochondrial genome integrity and supports mitochondrial DNA repair. Hum Mol Genet 18:3832–3850

Safdar A, Hamadeh MJ, Kaczor JJ, Raha S, Debeer J, Tarnopolsky MA (2010) Aberrant mitochondrial homeostasis in the skeletal muscle of sedentary older adults. PLoS One 5:e10778

Safdar A, Little JP, Stokl AJ, Hettinga BP, Akhtar M, Tarnopolsky MA (2011) Exercise increases mitochondrial PGC-1alpha content and promotes nuclear-mitochondrial cross-talk to coordinate mitochondrial biogenesis. J Biol Chem 286:10605–10617

Sahin E, Colla S, Liesa M, Moslehi J, Muller FL, Guo M, Cooper M, Kotton D, Fabian AJ, Walkey C, Maser RS, Tonon G, Foerster F, Xiong R, Wang YA, Shukla SA, Jaskelioff M,

Martin ES, Heffernan TP, Protopopov A, Ivanova E, Mahoney JE, Kost-Alimova M, Perry SR, Bronson R, Liao R, Mulligan R, Shirihai OS, Chin L, DePinho RA (2011) Telomere dysfunction induces metabolic and mitochondrial compromise. Nature 470:359–365

Saita S, Shirane M, Nakayama KI (2013) Selective escape of proteins from the mitochondria during mitophagy. Nat Commun 4:1410

Sancho P, Burgos-Ramos E, Tavera A, Bou Kheir T, Jagust P, Schoenhals M, Barneda D, Sellers K, Campos-Olivas R, Grana O, Viera CR, Yuneva M, Sainz B Jr, Heeschen C (2015) MYC/PGC-1alpha balance determines the metabolic phenotype and plasticity of pancreatic cancer stem cells. Cell Metab 22:590–605

Satoh J, Kawana N, Yamamoto Y (2013) Pathway analysis of ChIP-Seq-based NRF1 target genes suggests a logical hypothesis of their involvement in the pathogenesis of neurodegenerative diseases. Gene Regul Syst Biol 7:139–152

Scarpulla RC (2006) Nuclear control of respiratory gene expression in mammalian cells. J Cell Biochem 97:673–683

Scarpulla RC (2008a) Nuclear control of respiratory chain expression by nuclear respiratory factors and PGC-1-related coactivator. Ann N Y Acad Sci 1147:321–334

Scarpulla RC (2008b) Transcriptional paradigms in mammalian mitochondrial biogenesis and function. Physiol Rev 88:611–638

Scarpulla RC, Vega RB, Kelly DP (2012) Transcriptional integration of mitochondrial biogenesis. Trends Endocrinol Metab 23:459–466

Schapira AH (2012) Targeting mitochondria for neuroprotection in Parkinson's disease. Antioxid Redox Signal 16:965–973

Schulz C, Rehling P (2014) Remodelling of the active presequence translocase drives motor-dependent mitochondrial protein translocation. Nat Commun 5:4349

Schulz E, Dopheide J, Schuhmacher S, Thomas SR, Chen K, Daiber A, Wenzel P, Munzel T, Keaney JF Jr (2008) Suppression of the JNK pathway by induction of a metabolic stress response prevents vascular injury and dysfunction. Circulation 118:1347–1357

Scott I, Webster BR, Chan CK, Okonkwo JU, Han K, Sack MN (2014) GCN5-like protein 1 (GCN5L1) controls mitochondrial content through coordinated regulation of mitochondrial biogenesis and mitophagy. J Biol Chem 289:2864–2872

Settembre C, Di Malta C, Polito VA, Garcia Arencibia M, Vetrini F, Erdin S, Erdin SU, Huynh T, Medina D, Colella P, Sardiello M, Rubinsztein DC, Ballabio A (2011) TFEB links autophagy to lysosomal biogenesis. Science (New York, NY) 332:1429–1433

Shaid S, Brandts CH, Serve H, Dikic I (2013) Ubiquitination and selective autophagy. Cell Death Differ 20:21–30

Shi RY, Zhu SH, Li V, Gibson SB, Xu XS, Kong JM (2014) BNIP3 interacting with LC3 triggers excessive mitophagy in delayed neuronal death in stroke. CNS Neurosci Ther 20:1045–1055

Shin JH, Ko HS, Kang H, Lee Y, Lee YI, Pletinkova O, Troconso JC, Dawson VL, Dawson TM (2011) PARIS (ZNF746) repression of PGC-1alpha contributes to neurodegeneration in Parkinson's disease. Cell 144:689–702

Shoag J, Haq R, Zhang M, Liu L, Rowe GC, Jiang A, Koulisis N, Farrel C, Amos CI, Wei Q, Lee JE, Zhang J, Kupper TS, Qureshi AA, Cui R, Han J, Fisher DE, Arany Z (2013) PGC-1 coactivators regulate MITF and the tanning response. Mol Cell 49:145–157

Sugiura A, McLelland GL, Fon EA, McBride HM (2014) A new pathway for mitochondrial quality control: mitochondrial-derived vesicles. EMBO J 33:2142–2156

Suzuki SW, Onodera J, Ohsumi Y (2011) Starvation induced cell death in autophagy-defective yeast mutants is caused by mitochondria dysfunction. PLoS One 6:e17412

Szargel R, Shani V, Abd Elghani F, Mekies LN, Liani E, Rott R, Engelender S (2016) The PINK1, synphilin-1 and SIAH-1 complex constitutes a novel mitophagy pathway. Hum Mol Genet 25:3476–3490

Taguchi K, Motohashi H, Yamamoto M (2011) Molecular mechanisms of the Keap1-Nrf2 pathway in stress response and cancer evolution. Genes Cells 16:123–140

Tan EY, Campo L, Han C, Turley H, Pezzella F, Gatter KC, Harris AL, Fox SB (2007) BNIP3 as a progression marker in primary human breast cancer; opposing functions in in situ versus invasive cancer. Clin Cancer Res 13:467–474

Tan Z, Luo X, Xiao L, Tang M, Bode AM, Dong Z, Cao Y (2016) The role of PGC1alpha in cancer metabolism and its therapeutic implications. Mol Cancer Ther 15:774–782

Tatsuta T, Langer T (2008) Quality control of mitochondria: protection against neurodegeneration and ageing. EMBO J 27:306–314

Terada S, Goto M, Kato M, Kawanaka K, Shimokawa T, Tabata I (2002) Effects of low-intensity prolonged exercise on PGC-1 mRNA expression in rat epitrochlearis muscle. Biochem Biophys Res Commun 296:350–354

Torrano V, Valcarcel-Jimenez L, Cortazar AR, Liu X, Urosevic J, Castillo-Martin M, Fernandez-Ruiz S, Morciano G, Caro-Maldonado A, Guiu M, Zuniga-Garcia P, Graupera M, Bellmunt A, Pandya P, Lorente M, Martin-Martin N, Sutherland JD, Sanchez-Mosquera P, Bozal-Basterra L, Zabala-Letona A, Arruabarrena-Aristorena A, Berenguer A, Embade N, Ugalde-Olano A, Lacasa-Viscasillas I, Loizaga-Iriarte A, Unda-Urzaiz M, Schultz N, Aransay AM, Sanz-Moreno V, Barrio R, Velasco G, Pinton P, Cordon-Cardo C, Locasale JW, Gomis RR, Carracedo A (2016) The metabolic co-regulator PGC1alpha suppresses prostate cancer metastasis. Nat Cell Biol 18:645–656

Vafai SB, Mootha VK (2012) Mitochondrial disorders as windows into an ancient organelle. Nature 491:374–383

van der Bliek AM (2016) Mitochondria just wanna have FUN(DC1). EMBO J 35:1365–1367

Vazquez F, Lim JH, Chim H, Bhalla K, Girnun G, Pierce K, Clish CB, Granter SR, Widlund HR, Spiegelman BM, Puigserver P (2013) PGC1alpha expression defines a subset of human melanoma tumors with increased mitochondrial capacity and resistance to oxidative stress. Cancer Cell 23:287–301

Vega RB, Horton JL, Kelly DP (2015) Maintaining ancient organelles: mitochondrial biogenesis and maturation. Circ Res 116:1820–1834

Ventura-Clapier R, Garnier A, Veksler V (2008) Transcriptional control of mitochondrial biogenesis: the central role of PGC-1alpha. Cardiovasc Res 79:208–217

Villa E, Proics E, Rubio-Patino C, Obba S, Zunino B, Bossowski JP, Rozier RM, Chiche J, Mondragon L, Riley JS, Marchetti S, Verhoeyen E, Tait SWG, Ricci JE (2017) Parkin-independent mitophagy controls chemotherapeutic response in cancer cells. Cell Rep 20:2846–2859

Villa E, Marchetti S, Ricci JE (2018) No Parkin zone: mitophagy without Parkin. Trends Cell Biol 28:882–895

Virbasius CA, Virbasius JV, Scarpulla RC (1993) NRF-1, an activator involved in nuclear-mitochondrial interactions, utilizes a new DNA-binding domain conserved in a family of developmental regulators. Genes Dev 7:2431–2445

Vives-Bauza C, Zhou C, Huang Y, Cui M, de Vries RL, Kim J, May J, Tocilescu MA, Liu W, Ko HS, Magrané J, Moore DJ, Dawson VL, Grailhe R, Dawson TM, Li C, Tieu K, Przedborski S (2010) PINK1-dependent recruitment of Parkin to mitochondria in mitophagy. Proc Natl Acad Sci U S A 107:378–383

Wang X, Moraes CT (2011) Increases in mitochondrial biogenesis impair carcinogenesis at multiple levels. Mol Oncol 5:399–409

Wei H, Liu L, Chen Q (2015) Selective removal of mitochondria via mitophagy: distinct pathways for different mitochondrial stresses. Biochim Biophys Acta 1853:2784–2790

Wei Y, Chiang WC, Sumpter R Jr, Mishra P, Levine B (2017) Prohibitin 2 is an inner mitochondrial membrane mitophagy receptor. Cell 168:224–238.e210

Weitzel JM, Iwen KA (2011) Coordination of mitochondrial biogenesis by thyroid hormone. Mol Cell Endocrinol 342:1–7

Wredenberg A, Wibom R, Wilhelmsson H, Graff C, Wiener HH, Burden SJ, Oldfors A, Westerblad H, Larsson NG (2002) Increased mitochondrial mass in mitochondrial myopathy mice. Proc Natl Acad Sci U S A 99:15066–15071

Wright DC, Geiger PC, Han DH, Jones TE, Holloszy JO (2007a) Calcium induces increases in peroxisome proliferator-activated receptor gamma coactivator-1alpha and mitochondrial biogenesis by a pathway leading to p38 mitogen-activated protein kinase activation. J Biol Chem 282:18793–18799

Wright DC, Han DH, Garcia-Roves PM, Geiger PC, Jones TE, Holloszy JO (2007b) Exercise-induced mitochondrial biogenesis begins before the increase in muscle PGC-1alpha expression. J Biol Chem 282:194–199

Wu Z, Puigserver P, Andersson U, Zhang C, Adelmant G, Mootha V, Troy A, Cinti S, Lowell B, Scarpulla RC, Spiegelman BM (1999) Mechanisms controlling mitochondrial biogenesis and respiration through the thermogenic coactivator PGC-1. Cell 98:115–124

Wu H, Kanatous SB, Thurmond FA, Gallardo T, Isotani E, Bassel-Duby R, Williams RS (2002) Regulation of mitochondrial biogenesis in skeletal muscle by CaMK. Science (New York, NY) 296:349–352

Yan J, Feng Z, Liu J, Shen W, Wang Y, Wertz K, Weber P, Long J, Liu J (2012) Enhanced autophagy plays a cardinal role in mitochondrial dysfunction in type 2 diabetic Goto-Kakizaki (GK) rats: ameliorating effects of (−)-epigallocatechin-3-gallate. J Nutr Biochem 23:716–724

Yang ZF, Drumea K, Mott S, Wang J, Rosmarin AG (2014) GABP transcription factor (nuclear respiratory factor 2) is required for mitochondrial biogenesis. Mol Cell Biol 34:3194–3201

Youle RJ, Narendra DP (2011) Mechanisms of mitophagy. Nat Rev Mol Cell Biol 12:9–14

Yu SB, Pekkurnaz G (2018) Mechanisms orchestrating mitochondrial dynamics for energy homeostasis. J Mol Biol 430:3922–3941

Yun J, Finkel T (2014) Mitohormesis. Cell Metab 19:757–766

Zaffagnini G, Martens S (2016) Mechanisms of selective autophagy. J Mol Biol 428:1714–1724

Zhang H, Gao P, Fukuda R, Kumar G, Krishnamachary B, Zeller KI, Dang CV, Semenza GL (2007) HIF-1 inhibits mitochondrial biogenesis and cellular respiration in VHL-deficient renal cell carcinoma by repression of C-MYC activity. Cancer Cell 11:407–420

Zhang Q, Wu X, Chen P, Liu L, Xin N, Tian Y, Dillin A (2018) The mitochondrial unfolded protein response is mediated cell-non-autonomously by retromer-dependent Wnt signaling. Cell 174:870–883.e817

Zhu JH, Horbinski C, Guo F, Watkins S, Uchiyama Y, Chu CT (2007) Regulation of autophagy by extracellular signal-regulated protein kinases during 1-methyl-4-phenylpyridinium-induced cell death. Am J Pathol 170:75–86

Zhu JH, Gusdon AM, Cimen H, Van Houten B, Koc E, Chu CT (2012) Impaired mitochondrial biogenesis contributes to depletion of functional mitochondria in chronic MPP+ toxicity: dual roles for ERK1/2. Cell Death Dis 3:e312

Zhu J, Wang KZ, Chu CT (2013) After the banquet: mitochondrial biogenesis, mitophagy, and cell survival. Autophagy 9:1663–1676

Zong H, Ren JM, Young LH, Pypaert M, Mu J, Birnbaum MJ, Shulman GI (2002) AMP kinase is required for mitochondrial biogenesis in skeletal muscle in response to chronic energy deprivation. Proc Natl Acad Sci U S A 99:15983–15987

Mechanical Stress-Induced Autophagy: A Key Player in Cancer Metastasis

8

Joyjyoti Das and Tapas Kumar Maiti

Abstract

Metastasis is the leading cause of cancer-related mortality. The tumor microenvironment per se is a key player regulating the invasion and metastasis of cancer cells. Cancer cells residing in the tumor microenvironment as well as in transit during metastasis are exposed to various chemical and mechanical cues which contribute to their invasiveness. A plethora of studies since the last decade has shed light on the role of physical forces in tumor initiation and progression, iteratively underscoring the importance of cellular mechanobiology in the context of cancer. One of the emerging mechanobiological phenomena observed in cancer cells is autophagy. This chapter accounts for the various mechanical stimuli experienced by cancer cells in vivo and highlights the importance of mechanically-induced autophagy in the tumor milieu.

Keywords

Mechanical stress · Cancer · Autophagy · Invasion · Survival

J. Das (✉)
Nanoscience and Nanotechnology, School of Interdisciplinary Studies, University of Kalyani, Kalyani, West Bengal, India

T. K. Maiti
Department of Biotechnology, Indian Institute of Technology Kharagpur, Kharagpur, West Bengal, India
e-mail: tkmaiti@bt.iitkgp.ac.in

© Springer Nature Singapore Pte Ltd. 2020
S. K. Bhutia (ed.), *Autophagy in Tumor and Tumor Microenvironment*,
https://doi.org/10.1007/978-981-15-6930-2_8

8.1 Introduction

Solid tumors begin with the abnormal proliferation of cells in a confined area within the body. In due course of growth, cancerous cells from the periphery of the tumor start dislodging into the bloodstream and lymphatic ducts. What follows is the process called metastatic dissemination. Some of the dislodged cells survive through the bloodstream and migrate to the "foreign soil" of distant tissues to form a secondary tumor (Chaffer 2011). During this process, mechanical cues, alongside chemical signals, may activate myriad signaling pathways in metastatic cancer cells (Janmey and Miller 2011). While chemical signals include growth factors and soluble ligands, biophysical cues may arise in the form of matrix stiffness, confinement, topography, shear stress, compression, and mechanical stretching (Chaudhuri et al. 2018; Stylianopoulos et al. 2018). The subsequent section accounts for the genesis of mechanical cues in the tumor microenvironment.

8.2 Genesis of Mechanical Cues in the Tumor Microenvironment

As cancer progresses, solid tumors keep growing in confined spaces within normal tissue of a host and become increasingly rigid. An increase in the number of cells within a tumor causes stiffening of the tissue. This occurs due to the addition of cancer cells, stromal cells, and extracellular matrix constituents, resulting in higher elastic modulus than normal tissue, as high as one order of magnitude (Jain et al. 2014; Samani et al. 2007). Tumor growth generates compressive and stretching forces within the tumor and also between the tumor and the host tissue. Mathematical models have predicted that compressive stress at the tumor interior may exceed 40 kPa (Stylianopoulos et al. 2013; Voutouri et al. 2014; Roose et al. 2003). Confined growth of solid tumor leads to distortion of associated tumor vessels, which in turn induce both solid and fluid stresses, that foster tumor progression. While invading the dense matrix of the tumor stroma, cancer cells squeeze and deform through narrow pores. Naturally, when the cell passes through confinements of subnuclear dimensions, the nucleus is exposed to high deformation which may affect cellular behavior through a process called mechanotransduction. Biophysical properties thus influence the efficacy of cancer cell invasion and subsequent metastases (Talmadge and Fidler 2010). Inside the vasculature of the body, cancer cells experience shear stress generated by the interstitial flow at the tumor site and hemodynamic flow during metastasis. Fluid shear forces could affect both survival and invasiveness of circulating tumor cells (CTCs) (Mitchell and King 2013; Ma et al. 2017). Constricted lymphatic vessels may lead to enhanced interstitial fluid velocities (Chary and Jain 1989). Hemodynamic shear stresses, ranging from 0.5–30.0 dyn/cm^2, may arise due to the movement of blood along the surface of the cancer cell and is dependent on both fluid viscosity and flow rates (Mitchell and King 2013; Wirtz et al. 2011). Contact of tumor cells with endothelial cells, as tumor cells enter and leave the vasculature through the processes of intravasation and

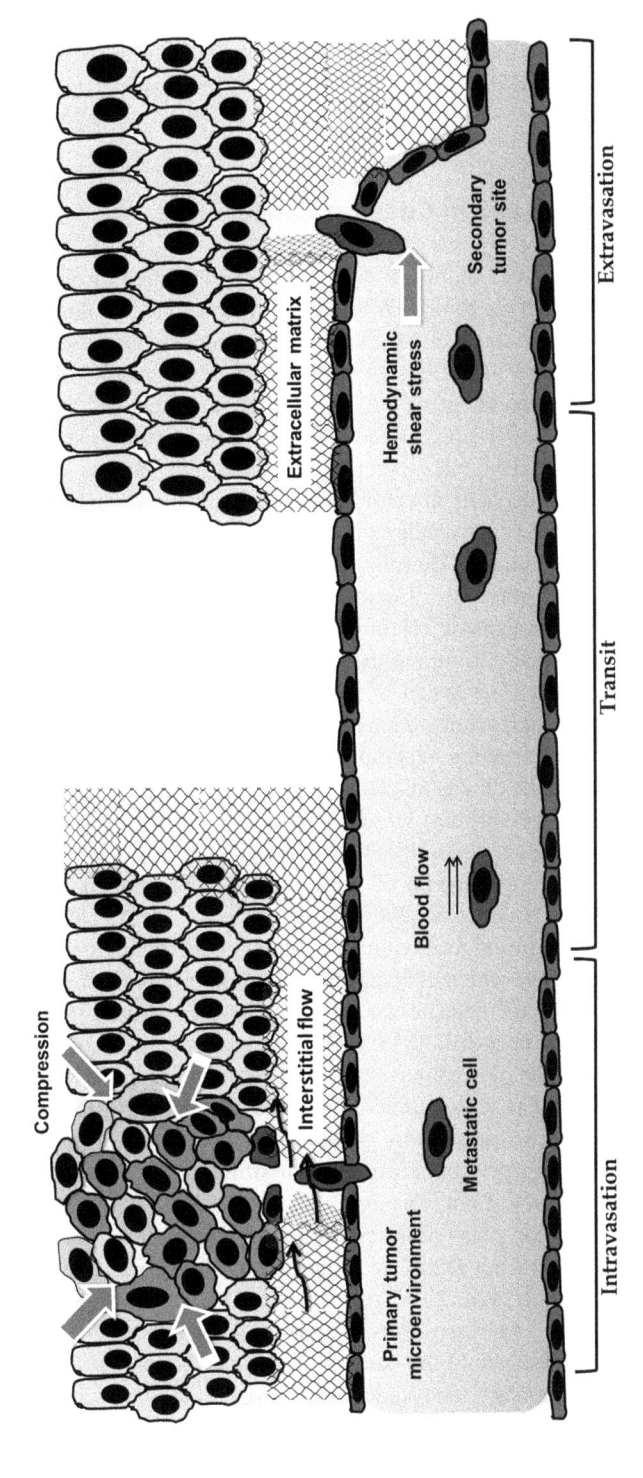

8 Mechanical Stress-Induced Autophagy: A Key Player in Cancer Metastasis 173

Fig. 8.1 Origin of various mechanical stresses in the tumor microenvironment. The growth of a primary tumor generates solid stress due to radially inward compression and circumferential tension or stretching of tumor cells by neighboring cells and the surrounding matrix. Also, the cancer cells are exposed to low shear stresses in the form of interstitial lymphatic flows within the stromal matrix. During hematogenous metastasis, intravasation of metastatic cells into the bloodstream exposes the cells to higher magnitudes of shear stress that the cells experience until they get arrested and extravasate at a secondary tumor site

extravasation respectively, may also lead to the generation of shear stresses (Northcott et al. 2018). A landscape of various mechanical forces in the tumor microenvironment is depicted in Fig. 8.1.

8.3 Impact of Mechanical Cues on Cancer Cells

The influence of physical forces on tumor growth has been studied through numerous in vitro models (Huang et al. 2018). This is intuitive given the fact that the tumor microenvironment is exposed to several factors like chemical signals, multiple cell–cell interactions, and mechanical stimuli. To study the exclusive effects of mechanical cues on tumor progression, scientists have resorted to a reductionist approach. A host of accumulating evidence has highlighted the role of biophysical cues in cancer development (Das et al. 2019a). This section discusses the various approaches employed to apply various mechanical stimuli to cancer cells in vitro and the effects observed thereof. One of the first challenges in this regard was to measure stress levels of developing solid tumors. Helmlinger et al. were the first to reveal such measurements using tumor spheroids of colon and breast cancer cells embedded in increasing concentrations of agarose (Helmlinger et al. 1997). Since then, several improvisations have been made to mimic the compressive environment of growing tumors in vitro, most of them being cancer spheroid models or their modifications (shown in Fig. 8.2a–c). Compression to cancer cells grown in monolayers have been applied using piston-based systems or simply by applying appropriate weights to bead encapsulated cancer cells (Tse et al. 2012; Kim et al. 2017). Compression alters gene expression in cancer cells in turn affecting invasion and metastasis (Tse et al. 2012; Koike et al. 2002; Kalli et al. 2018). However, excessive solid stress may play an inhibitory role by reducing the rate of proliferation while inducing apoptosis (Helmlinger et al. 1997; Cheng et al. 2009; Kaufman et al. 2005; Delarue et al. 2014). The compression of blood and lymphatic vessels may reduce perfusion and create a hypoxic microenvironment that promotes tumor progression (Jain et al. 2014; Jain 2014). In order to metastasize, tumor cells must survive through the circulation while migrating to a distant location. The time a cancer cell spends in circulation and the magnitude of shear stress it experiences on its way determines its survival (Fan et al. 2016). The shear-dominant microenvironment of metastasizing cancer cells has been mimicked in vitro by generating fluid flows though setups like parallel-plate flow chambers, peristaltic pumps, microfluidic platforms, and hypodermic needles (Ma et al. 2017; Lien et al. 2013; Das et al. 2011, 2018; Barnes et al. 2012) (see Fig. 8.2d–f). Shear stress of the physiological range (0.5–3 Pa) has been shown to inhibit proliferation but stimulate migration and adhesion of tumor cells (Mitchell and King 2013; Ma et al. 2017; Avvisato et al. 2007; Xiong et al. 2017). However, a high magnitude of stresses caused tumor cell death (Regmi et al. 2017). Tumor cells must show resistance to the various mechanical stresses if they have to travel and colonize at a secondary metastatic site. It is one of the major reasons why cancer cells must undergo various cellular adaptations (Northcott et al. 2018). The

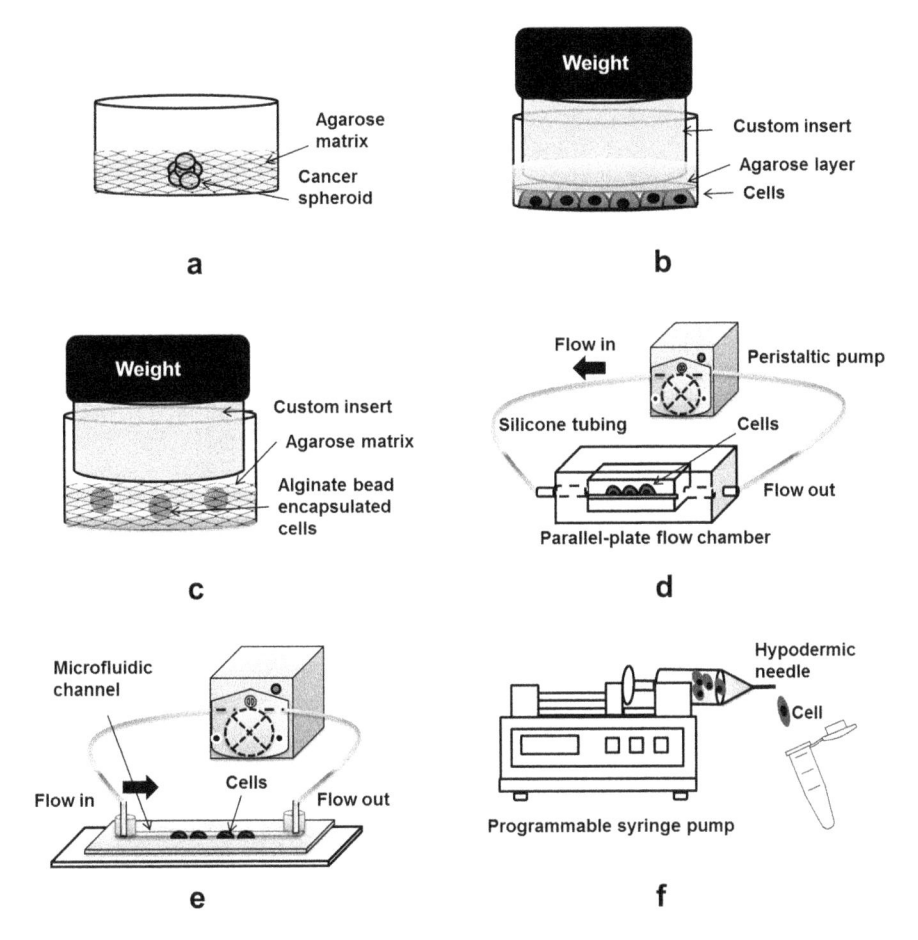

Fig. 8.2 (**a–f**) Schematic of the various experimental setups implemented for performing mechanobiological studies of cancer cells. (**a**) Cancer spheroids embedded in increasing concentrations of culture-media equilibrated agarose and allowed to grow in the compressive environment, (**b**) weight applied to exert compression on a monolayer of cancer cells grown in a well of a cell culture plate, (**c**) weight applied to exert compression on agarose scaffolded alginate bead-encapsulated cancer cells, (**d**) parallel-plate flow chamber connected via silicone tubings to a peristaltic pump for applying fluid shear stress to cancer cells, (**e**) peristaltic pump connected to a microfluidic channel for applying fluid shear stress to cancer cells adhered on the microchannel floor, (**f**) programmable syringe pump to flow cancer cell suspension through a hypodermic needle for several passes in order to apply shear stress to the cells

subsequent section discusses myriad cellular adaptive responses that may be elicited in cancer cells experiencing microenvironmental stresses.

8.4 Cellular Stress Response Mechanisms

A cell's response to any kind of stress is a reaction to perturbations of ambient conditions. Such unfavorable conditions may or may not damage cellular macromolecules. Cells may either perform homeostasis to attain the former state or adopt a changed state, depending upon the severity and duration of incumbent stress. For instance, mild or moderate stresses may result in enhanced defense and repair processes. It is thus essential to study stress-adaptive mechanisms in order to comprehend the processes that cancer cells may undergo during their metamorphosis into the malignant state, leading to the identification of critical therapeutic targets (Das et al. 2019a; Milisav et al. 2012). Adaptive stress responses may occur through several mechanisms like damage repair, synthesis of protective molecules, and control of apoptosis induction. Cellular repair is brought about by alterations in gene expression patterns, miRNA-transcription, growth arrest, and so on. Protective molecules may be antioxidant enzymes like catalase, peroxidase, superoxide dismutase, etc. In many forms of cancers, tumor initiation, progression, and resistance to current anticancer therapies may be attributed to the overexpression of the anti-apoptotic proteins. Another important strategy that a cell may adopt to react to stress is the clearance of damaged organelles. For soluble proteins, this may occur through the ubiquitin-proteasome pathway (Shang and Taylor 2011), whereas for other cellular material, the autophagic pathway for degradation may be activated. Autophagy is an evolutionarily conserved catabolic process whereby cytosolic components are enclosed in sealed bilayered vesicles and then digested through the action of lysosomes. Autophagy helps in increasing nutrient availability to the cells through the clearance of toxic cellular materials and unfolded proteins influencing numerous physiological processes including homeostasis during cellular stresses (Das et al. 2019a). However, in the mechanobiological aspect of cancer, the role of autophagy had been less explored until the seminal work by King et al. showed that autophagy is induced as an immediate response to compressive stress (King et al. 2011). Autophagy may either contribute to cellular adaptation and survival or cellular death (Maiuri et al. 2007). At lower pressures, autophagy may also instigate mechanical signaling (King 2012). Reportedly, tumor cells upregulate autophagy in response to increased metabolic demands and cellular stresses (Yang et al. 2011). Although autophagy is known to be involved in several processes like modulation of cancer stem cell viability and differentiation, epithelial-to-mesenchymal transition, tumor cell dormancy, motility and invasion, resistance to anoikis, escape from immune surveillance, and so on, the direct implications of mechanically induced autophagy in the cancer scenario was least investigated until recent times (Mowers et al. 2017).

8.5 Implications of Mechanical Stress-Induced Autophagy in Cancer Metastasis

The differential roles of flow-induced shear forces in modulating cancer progression have started coming in the limelight (Swartz and Lund 2012). Lien et al. demonstrated that laminar shear stress in the range of 0.5–12 dyn/cm^2, applied for more than 12 h, was able to induce autophagic and apoptotic death in cancer cells, but not in their normal counterparts (Lien et al. 2013). They found that laminar shear stress-induced autophagy acted not only as a parallel death-promoting mechanism but also as an independent death-inducing mechanism upstream of apoptosis. In contrast, Das et al. interestingly found that short pulses of laminar shear stress could elicit pro-survival autophagy in HeLa cells as an immediate response. Autophagy in this instance served as a protective mechanism that could delay apoptotic cell death (Das et al. 2018). In an independent study, Wang et al. found that inhibition of autophagy induced by fluid shear stress of 1.4 dyn/cm^2, suppressed cellular migration, and invasion in hepatocellular carcinoma cells (Wang et al. 2018; Yan et al. 2019). Transit through the vasculature followed by arrest and extravasation at a distant location are some of the key steps of metastasis. Since these phenomena are short timescale processes, the immediate pro-survival autophagic response due to fluid shear may prove to be a crucial escape route of metastasizing cancer cells (Follain et al. 2018). Very recently, Das et al. (2019b) recreated a mechanically-compressed tumor microenvironment, in vitro, by applying appropriate compression to agarose-scaffolded HeLa cell-encapsulated alginate beads. They demonstrated that compression upregulates autophagy, which promotes turnover of paxillin, a crucial protein involved in cell migration, and secretion of active-matrix metalloproteinase 2 (MMP 2), leading to enhanced migration of HeLa cells (Das et al. 2019b). These evidences hint at the fact that compressive and shear forces in the tumor milieu may foster cancer progression at least partially, by upregulating autophagy. At the molecular level, King et al. demonstrated that mechanical induction of autophagy is independent of classical TOR/Akt pathway and AMPK signaling, which is the conventional route of autophagy induction in cells (King et al. 2011). However, the molecular pathways governing mechanically-induced autophagy remained completely obscure until 2013 when Lien et al. showed that shear forces could elicit autophagy in cancer cells through the BMPRIB/Smad 1/5/p38 MAPK axis (Lien et al. 2013). Later, Das et al. (2018), demonstrated that shear stress causes membrane perturbation which triggers lipid rafts, that is, cholesterol-rich nanodomains of cell membranes, to mediate the phosphorylation of p38 MAPKs which in turn leads to LC3 II/I conversion and autophagy induction in HeLa cells (Das et al. 2018). Fluid shear stress also induced the expression of Rho GTPases and cytoskeleton remodeling via the integrin/FAK pathway, leading to the upregulation of autophagy in HepG2 cells (Yan et al. 2019). The various manifestations of mechanical stress-induced autophagy in cancer cells are depicted in Fig. 8.3.

Fig. 8.3 Manifestations of mechanical stress-induced autophagy in cancer cells. Shear stresses due to blood flows have been shown to trigger the phosphorylation of signaling molecules located on the cytoskeleton of cancer cells like integrins, focal adhesion kinases, and also of membrane lipid raft-associated cytoplasmic proteins like p38 MAPkinases, leading to activation of the autophagic cascade. Activation of focal adhesion kinases causes cytoskeletal rearrangement that also participates in the formation of autophagosomal components. Shear-induced autophagy imparts immediate resistance to shear-induced apoptosis while compression-induced autophagy upregulates secretion of MMP 2 and promote turnover of a crucial focal adhesion protein, Paxillin, thereby aiding in survival, invasion, and migration of cancer cells respectively

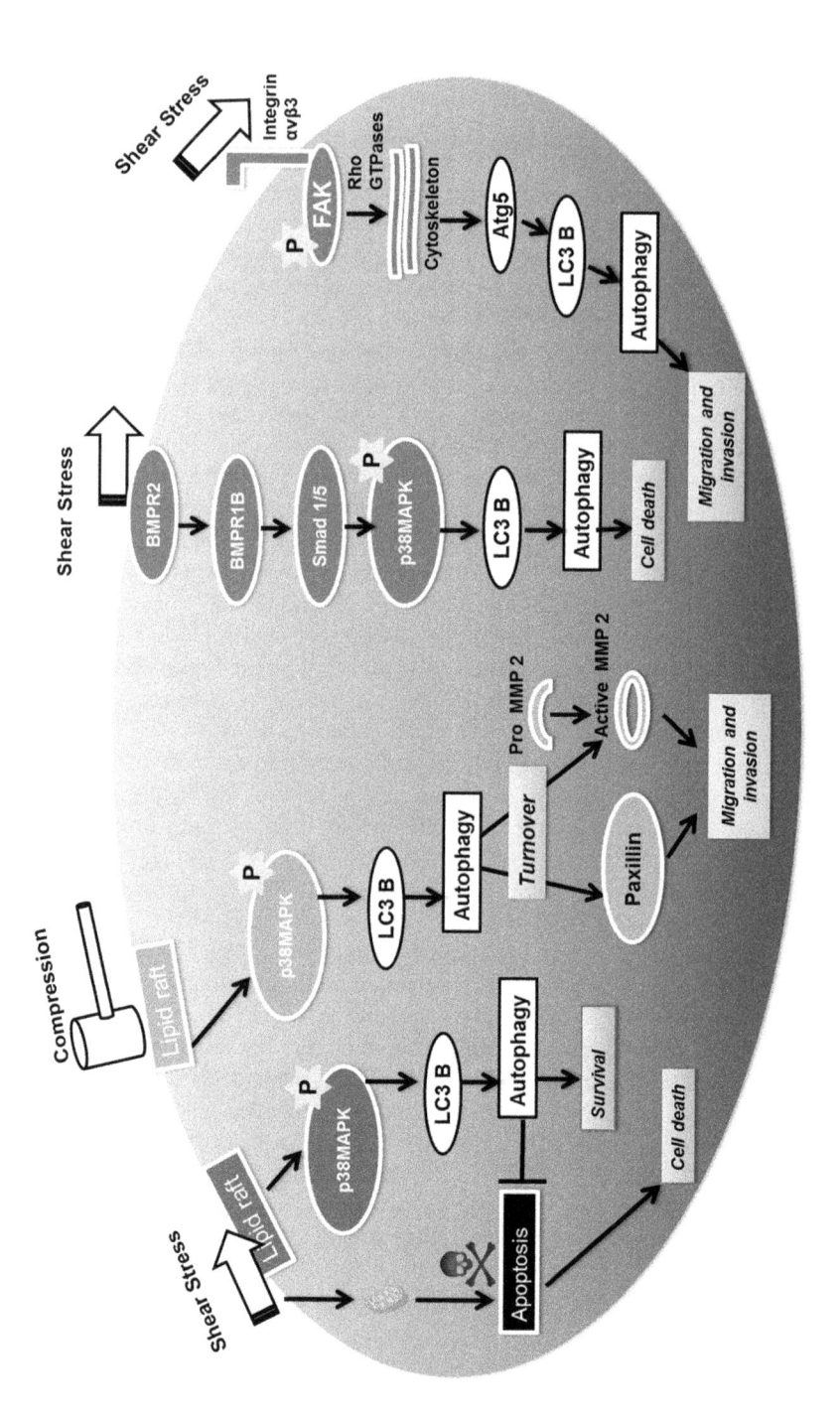

8.6 Conclusion and Perspective

Manipulation of mechanical forces in the tumor microenvironment to tame cancer is evolving as a new field termed "physical oncology". This may be done by alterations of the physical characteristics of the stroma or by inhibition of cellular responses to the stiffening of the stroma (Northcott et al. 2018). The underlying goal remains the alleviation of the solid and fluid stresses within the tumor microenvironment. To this effect, extracellular matrix-degrading enzymes like Hyaluronidases have been implemented to release immobilized fluid for improving tissue compliance (Whatcott et al. 2011). TGF-β blockers and MMP inhibitors are some of the drugs that target ECM synthesis (Chaudhuri et al. 2018). Also, Losartan, an angiotensin inhibitor, has been used for vessel dilation to reduce IFP (Chauhan et al. 2013). Based on the knowledge of signaling networks operating downstream of focal adhesions, drugs targeted toward reducing actomyosin contractility have shown successful repression of tumor progression. Reports on potent therapeutic targets associated with mechanically-induced autophagy have started cropping up. Depletion of cholesterol by Methyl-beta cyclodextrin (MBCD) could lead to impairment of lipid raft mediated-p38 MAPK phosphorylation under fluid shear stress, thereby impeding the induction of pro-survival autophagy (Das et al. 2018). Cliengitide, an integrin inhibitor, could inhibit the activation of downstream FAK, thereby attenuating fluid shear stress-induced autophagy in HepG2 cells (Yan et al. 2019). It is imperative to further investigate the arsenal of mechanosensory elements of cancer cells that participate in mechanical stress-induced autophagy. For example, reports have suggested that intracellular Ca^{2+}, a well-known regulator of autophagy, aids in the migration and proliferation of cancer cells (Filippi-chiela et al. 2016; Cui et al. 2017). Whether or not stretch-activated calcium channels participate in the mooted signal transduction pathway remains to be explored (Das et al. 2019a).

In summary, this chapter throws light on the origin of mechanical forces in the context of cancer and how these forces may govern cancer progression through the intervention of autophagy. Presently known implications of mechanical stress-induced autophagy in cancer include imparting immediate resistance to shear-induced apoptosis during metastasis and facilitating the migratory and invasive characteristics of cancer cells in the tumor microenvironment. It may thus be foreseen that the discovery of pathways related to mechanically-induced autophagy in cancer cells, may usher in an array of effective therapeutic molecules, in the near future.

References

Avvisato CL, Yang X, Shah S, Hoxter B, Li W, Gaynor R, Pestell R, Tozeren A, Byers SW (2007) Mechanical force modulates global gene expression and β-catenin signaling in colon cancer cells. J Cell Sci 120:2672–2682

Barnes JM, Nauseef JT, Henry MD (2012) Resistance to fluid shear stress is a conserved biophysical property of malignant cells. PLoS One 7:e50973

Chaffer CL (2011) A perspective on cancer cell metastasis. Science 331:1559–1564

Chary SR, Jain RK (1989) Direct measurement of interstitial convection and diffusion of albumin in normal and neoplastic tissues by fluorescence photobleaching. Proc Natl Acad Sci U S A 86:5385–5389

Chaudhuri PK, Low BC, Lim CT (2018) Mechanobiology of tumor growth. Chem Rev 118:6499–6515

Chauhan VP, Martin JD, Liu H, Lacorre DA, Jain SR, Kozin SV et al (2013) Angiotensin inhibition enhances drug delivery and potentiates chemotherapy by decompressing tumour blood vessels. Nat Commun 4:2516

Cheng G, Tse J, Jain RK, Munn LL (2009) Micro-environmental mechanical stress controls tumor spheroid size and morphology by suppressing proliferation and inducing apoptosis in cancer cells. PLoS One 4:e4632

Cui C, Merritt R, Fu L, Pan Z (2017) Targeting calcium signaling in cancer therapy. Acta Pharm Sin B 7:3–17

Das T, Maiti TK, Chakraborty S (2011) Augmented stress-responsive characteristics of cell lines in narrow confinements. Integr Biol (Camb) 3:684–695

Das J, Maji S, Agarwal T, Chakraborty S, Maiti TK (2018) Hemodynamic shear stress induces protective autophagy in HeLa cells through lipid raft-mediated mechanotransduction. Clin Exp Metastasis 35:135–148

Das J, Chakraborty S, Maiti TK (2019a) Seminars in cancer biology mechanical stress-induced autophagic response: a cancer-enabling characteristic? Semin Cancer Biol. https://doi.org/10.1016/j.semcancer.2019.05.017

Das J, Agarwal T, Chakraborty S, Maiti TK (2019b) Compressive stress-induced autophagy promotes invasion of HeLa cells by facilitating protein turnover in vitro. Exp Cell Res 381:201–207

Delarue M, Montel F, Vignjevic D, Prost J, Cappello G, Curie M, Curie I, De Recherche C (2014) Compressive stress inhibits proliferation in tumor spheroids through a volume limitation. Biophys J 107:1821–1828

Fan R, Emery T, Zhang Y, Xia Y, Sun J, Wan J (2016) Circulatory shear flow alters the viability and proliferation of circulating colon cancer cells. Sci Rep 6:27073

Filippi-chiela EC, Viegas MS, Thomé MP, Buffon A, Wink MR, Lenz G (2016) Modulation of autophagy by calcium signalosome in human disease. Mol Pharmacol 90:371–384

Follain G, Osmani N, Azevedo AS, Allio G, Mercier L, Karreman MA et al (2018) Hemodynamic forces tune the arrest, adhesion, and extravasation of circulating tumor cells. Dev Cell 45:33–52

Helmlinger G, Netti PA, Lichtenbeld HC, Melder RJ, Jain RK (1997) Solid stress inhibits the growth of multicellular tumor spheroids. Nat Biotechnol 15:778–783

Huang Q, Hu X, He W, Zhao Y, Hao S, Wu Q, Li S, Zhang S, Shi Fluid M (2018) Shear stress and tumor metastasis. Am J Cancer Res 8:763–777

Jain RK (2014) Perspective antiangiogenesis strategies revisited: from starving tumors to alleviating hypoxia. Cancer Cell 26:605–622

Jain RK, Martin JD, Stylianopoulos T (2014) The role of mechanical forces in tumor growth and therapy. Annu Rev Biomed Eng 16:321–346

Janmey PA, Miller RT (2011) Mechanisms of mechanical signaling in development and disease. J Cell Sci 124:9–18

Kalli M, Apageorgis PAP, Kretsi VAG, Tylianopoulos TRS (2018) Solid stress facilitates fibroblasts activation to promote pancreatic cancer cell migration. Ann Biomed Eng 46:657–669

Kaufman LJ, Brangwynne CP, Kasza KE, Filippidi E, Gordon VD, Deisboeck TS, Weitz DA (2005) Glioma expansion in collagen I matrices: analyzing collagen concentration-dependent growth and motility patterns. Biophys J 89:635–650

Kim BG, Gao M-Q, Kang S, Choi YP, Lee JH, Kim JE, Han HH, Mun SG, Cho NH (2017) Mechanical compression induces VEGFA overexpression in breast cancer via DNMT3A-dependent miR-9 downregulation. Cell Death Dis 8:e2646

King JS (2012) Mechanical stress meets autophagy: potential implications for physiology and pathology. Trends Mol Med 18:583–588

King JS, Veltman DM, Insall RH (2011) The induction of autophagy by mechanical stress. Autophagy 7:1490–1499

Koike C, Mckee TD, Pluen A, Ramanujan S, Burton K, Munn LL, Boucher Y, Jain RK (2002) Solid stress facilitates spheroid formation: potential involvement of hyaluronan. Br J Cancer 86:947–953

Lien S, Chang S, Lee P, Wei S, Chang MD, Chang J, Chiu J (2013) Mechanical regulation of cancer cell apoptosis and autophagy: roles of bone morphogenetic protein receptor, Smad1/5, and p38 MAPK. Biochim Biophys Acta Mol Cell Res 1833:3124–3133

Ma S, Fu A, Giap G, Chiew Y, Qian K (2017) Hemodynamic shear stress stimulates migration and extravasation of tumor cells by elevating cellular oxidative level. Cancer Lett 388:239–248

Maiuri MC, Zalckvar E, Kimchi A, Kroemer G (2007) Self-eating and self-killing: crosstalk between autophagy and apoptosis. Nat Rev Mol Cell Biol 8:741–752

Milisav I, Poljsak B, Šuput D (2012) Adaptive response, evidence of cross-resistance and its potential clinical use. Int J Mol Sci 13:10771–10806

Mitchell MJ, King MR (2013) Fluid shear stress sensitizes cancer cells to receptor-mediated apoptosis via trimeric death receptors. New J Phys 15:015008

Mowers EE, Sharifi MN, Macleod KF (2017) Autophagy in cancer metastasis. Oncogene 36:1619–1630

Northcott JM, Dean IS, Mouw JK, Weaver VM (2018) Feeling stress: the mechanics of cancer progression and aggression. Front Cell Dev Biol 6:1–12

Regmi S, Fu A, Luo KQ (2017) High shear stresses under exercise condition destroy circulating tumor cells in a microfluidic system. Sci Rep 7:1–12

Roose T, Netti PA, Munn LL, Boucher Y, Jain RK (2003) Solid stress generated by spheroid growth estimated using a linear poroelasticity model. Microvasc Res 66:204–212

Samani A, Plewes D, Samani A, Bishop J, Hagan JJO, Samani A (2007) Elastic moduli of normal and pathological human breast tissues: an inversion-technique-based investigation of 169 samples. Phys Med Biol 52:1565–1576

Shang F, Taylor A (2011) Ubiquitin-proteasome pathway and cellular responses to oxidative stress. Free Radic Biol Med 51:5–16

Stylianopoulos T, Martin JD, Snuderl M, Mpekris F, Jain SR, Jain RK (2013) Coevolution of solid stress and interstitial fluid pressure in tumors during progression: implications for vascular collapse. Cancer Res 73:3833–3841

Stylianopoulos T, Munn LL, Jain RK (2018) Reengineering the physical microenvironment of tumors to improve drug delivery and efficacy: from mathematical modeling to bench to bedside. Trends Cancer 4:292–319

Swartz MA, Lund AW (2012) Lymphatic and interstitial flow in the tumour microenvironment: linking mechanobiology with immunity. Nat Rev Cancer 12:210–219

Talmadge JE, Fidler IJ (2010) AACR centennial series: the biology of cancer metastasis: historical perspective. Cancer Res 70:5649–5669

Tse JM, Cheng G, Tyrrell JA, Wilcox-Adelman SA, Boucher Y, Jain RK, Munn LL (2012) Mechanical compression drives cancer cells toward invasive phenotype. Proc Natl Acad Sci U S A 109:911–916

Voutouri C, Mpekris F, Papageorgis P, Odysseos AD, Stylianopoulos T (2014) Role of constitutive behavior and tumor-host mechanical interactions in the state of stress and growth of solid tumors. PLoS One 9:e104717

Wang X, Zhang Y, Feng T, Su G, He J, Gao W, Shen Y, Liu X (2018) Fluid shear stress promotes autophagy in hepatocellular carcinoma cells. Int J Biol Sci 14:1277–1290

Whatcott CJ, Han H, Posner RG, Hostetter G, Von Hoff DD (2011) Targeting the tumor microenvironment in cancer: why hyaluronidase deserves a second look. Cancer Discov 1:291–296

Wirtz D, Konstantopoulos K, Searson PPC (2011) The physics of cancer: the role of physical interactions and mechanical forces in metastasis. Nat Rev Cancer 11:512–522

Xiong N, Li S, Tang K, Bai H, Peng Y, Yang H, Wu C, Liu Y (2017) Involvement of caveolin-1 in low shear stress-induced breast cancer cell motility and adhesion: roles of FAK/Src and ROCK/p-MLC pathways. Biochim Biophys Acta Mol Cell Res 1864:12–22

Yan Z, Su G, Gao W, He J, Shen Y, Zeng Y (2019) Fluid shear stress induces cell migration and invasion via activating autophagy in HepG2 cells. Cell Adhes Migr 13:1–12

Yang S, Wang X, Contino G, Liesa M, Sahin E, Mautner J, Tonon G, Haigis M, Shirihai OS, Doglioni C (2011) Pancreatic cancers require autophagy for tumor growth pancreatic cancers require autophagy for tumor growth. Genes Dev 25:717–729

The Interplay of Autophagy and the Immune System in the Tumor Microenvironment

9

Chandan Kanta Das, Bikash Chandra Jena, Ranabir Majumder, Himadri Tanaya Panda, and Mahitosh Mandal

Abstract

The tumor microenvironment (TME) is a very complicated ecosystem that consists of cancerous cells coexisting with various noncancerous and immune cells. TME shows exclusive cellular crosstalk between cancer and other cell types that have prominent consequences on tumor initiation, progression, and development. Of note, the immune system is an important determinant in the TME for tumor development, thus highlighting the importance of immunotherapy for better cancer treatment. Recently, the multifaceted role of autophagy in cancer immunity in the TME is extremely debated and exploited for the development of cutting-edge autophagy-based cancer immunotherapeutics. Interestingly, autophagy limits the immune responses by regulating the action of immune cells and the generation of cytokines. On the contrary, some immune cells and cytokines also manipulate the function of autophagy. A growing number of study spotlights the context-dependent role of autophagy in cancer immunity: it can activate the anti-tumor immunity by sustaining the integrity of the immune cells; however, it can also help the tumor cells to bypass the immune checkpoints by constraining the immune cell functions during hypoxia. In this chapter, we delineate the basic process of autophagy, the role autophagy in maintaining the crosstalk between cancer cells and stromal cells, and in particular, focusing more

Bikash Chandra Jena and Ranabir Majumder contributed equally to this work.

C. K. Das (✉) · B. C. Jena · R. Majumder · M. Mandal
School of Medical Science and Technology, Indian Institute of Technology Kharagpur, Kharagpur, West Bengal, India

H. T. Panda
Department of Life Science, National Institute of Technology Rourkela, Rourkela, Odisha, India

on the interaction of autophagy with cancer immunity. Finally, we highlight the role of autophagy as an ideal candidate for cancer immunotherapy.

Keywords

Cancer · Tumor microenvironment · Immune system · Autophagy · Tumor stroma · Immunotherapy

9.1 Introduction

The tumor mass is not a solitary structure, rather a diverse population of cancer cells, extracellular matrix proteins, secreted factors, resident, and infiltrating host cells, together known as the tumor microenvironment (TME). More specifically, the TME consists of tumor parenchymal cells, mesenchymal cells, fibroblasts, lymph vessels, blood, tumor-infiltrating immune cells, chemokines, and cytokines (Balkwill et al. 2012). TME has a vital role in tumor initiation, development, and regulation. The immune system is a critical player of the TME. In recent years, the significance of the interaction between the immune cells and cancer cells was constantly acknowledged and included in the rising hallmarks of cancer (Hanahan and Weinberg 2011). The cancer cells adopt several mechanisms that prompt them to evade immune surveillance and destruction. In the past decades, a large number of immunotherapy blueprints have been established based on the immune evasion mechanisms and their clinical significance has been validated. In contrast to the traditional approaches to cancer therapy, the immunotherapy acts through encountering the immune cells inside or outside the TME and attacks the cancer cells (Yost et al. 2019), thus making the immunotherapy strategy with higher specificity and with lower off-target effects. Recently, accumulating pieces of evidence support the function of autophagy in modulating the TME, including the immune system of the tumor cells (Deretic 2012), thus making it an ideal target for effective cancer therapy. Autophagy is essentially a eukaryotic homeostatic process that regulates cancer initiation and progression in many ways. It has a dual role in cancer as the fate of tumor cells regulated by autophagy is extremely ambience dependent which ranges among tumor types, stages, microenvironment, and genetic contents (White 2012). In the beginning, autophagy acts as a pro-death mechanism by removing the impaired organelles, proteins, lipids, and ROS, thus attenuating cancer by behaving as a cellular quality control process (Fulda and Kogel 2015). However, with the advancement of cancer and in therapy resistance, autophagy performs as a pro-survival mechanism to fulfill the substantial metabolic demands necessary for tumor survival (Das et al. 2018a, b, 2019a, b). So, more insight into the unraveling of the interplay of autophagy and the immune system will provide the major directions for future anticancer treatment strategies.

9.2 The Basic Process of Autophagy

Autophagy is an evolutionary conserved catabolic process that is often triggered by diverse cellular stress conditions. During this process, the damaged cellular components are degraded by the lysosomes and in turn, enrich the cellular nutrient pool (Das et al. 2018a). To date, autophagy is mainly classified into three types, like chaperone-mediated autophagy, microautophagy, and macroautophagy, depending on the mechanisms required for the selection and the delivery of cargos into the lysosomes (Yorimitsu and Klionsky 2005). In chaperone-mediated autophagy, the chaperone proteins are actively involved in the selection of the targeted substrates and their translocation into the lysosomes (Kaushik and Cuervo 2012). In microautophagy, the lysosomal membrane undergoes invagination to sequester the cytosolic components into intralysosomal vesicles (Mijaljica et al. 2011). Macroautophagy (herein referred to as autophagy) is the most commonly studied autophagy that sequesters the cytoplasmic damaged substances into the autophagosomes followed by their fusion into the lysosome to form endolysosome for hydrolytic degradation (Wong et al. 2011). The basic mechanism of autophagy is regulated by a group of ATG genes and it is preserved from yeast to mammals. The autophagy process is executed by the sequential development of the phagophore, autophagosome, and autolysosome regulated by the mammalian target of rapamycin (mTOR) (Bhol et al. 2019). Initially, upon stress various damaged cellular contents are entrapped by the initiation membrane or phagophore inside the cytosol; afterward, it gives rise to a complete double-membrane complex called the autophagosome. Further, the autophagosome merges with the lysosome to give rise to the autolysosome which helps in the hydrolytic deterioration of the damaged cellular entities, leading to enriching the nutrient pool (Mizushima et al. 2002) as shown in Fig. 9.1. However, the detailed molecular mechanisms of autophagy and its modulation in cancer cells were extensively depicted elsewhere in our recent review (Das et al. 2019a) and we refer the readers to this work for further details.

9.3 Autophagy in Maintaining the Crosstalk Between Cancer Cells and Tumor Stroma

Tumors occur from normal cells due to DNA mutations which lead to uninhibited cell growth. Initially, it has been thought that tumors are the collection of isolated diverse cell masses. But recent scientific results suggest that tumors are vastly heterogeneous and need to be considered as organs. It contains various types of special tumor cells and other tumor-related cell types such as immune cells, endothelial cells, and fibroblasts. These kinds of cellular components are called the tumor–stromal microenvironment (Maes et al. 2013; Denton et al. 2018). Tumor stroma comprises of various stress factors, like the absence of growth factors, intratumoral hypoxia, and tumor acidosis. Such kind of stress stimulates autophagy in tumor stroma to maintain energy homeostasis in cells by transporting intracellular damaged substances to lysosomes for degradation and recycling. Here, we are going

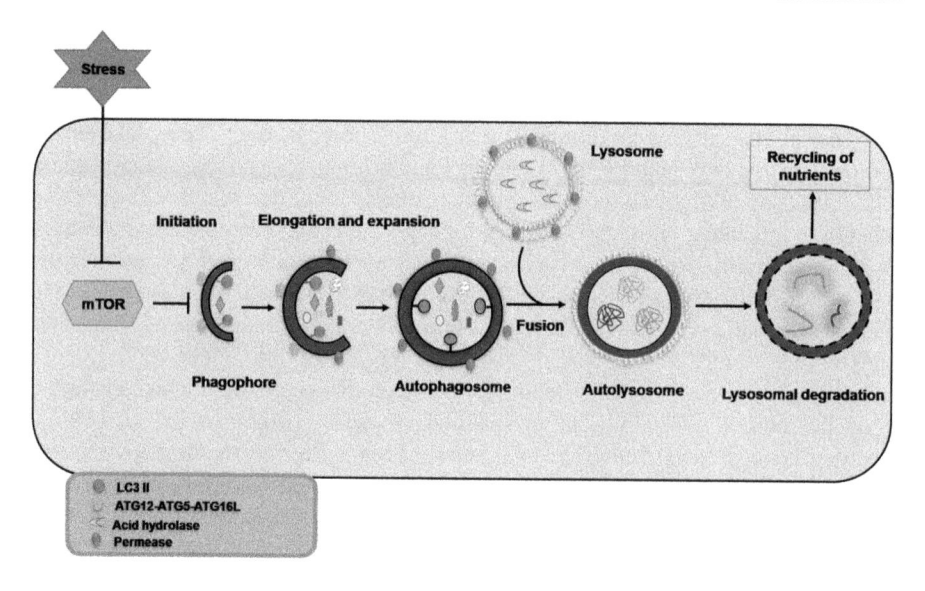

Fig. 9.1 The basic process of autophagy: Initially upon stress, the autophagy process begins with the inhibition of mTOR. Then the isolation membrane or phagophore is formed which further undergoes elongation and expansion by engulfing the damaged cargos and gives rise to a complete double-membrane structure called the autophagosome. Further, the autophagosome fuses with the lysosome to form the autolysosome. Finally, lysosomal degradation of damaged cargos takes place inside the autolysosome and the nutrients are recycled

to delineate the pivotal role of autophagy involved in the interactions between cancer cells and tumor stromal microenvironment.

9.3.1 Autophagy and Tumor-Associated Macrophages

Tumor-associated macrophages (TAMs) are classified as highly expressed innate immune cells in tumor stromal microenvironment (Lin et al. 2019). They are most important for cancer-related inflammation. Molecules secreted from tumor cells are accountable for the activation of macrophages. In the tumor stroma, TAMs perform a key role in the development of the tumor by generating cytokines, like IFN-γ, IL-6, IL-8, and IL-10 (Comito et al. 2014). Cytokines are important for the progression of chronic inflammation and the anti-tumor response, but they also initiate the advancement of cancer through inflammation. Chemokines, such as CCL2, engage monocytes from blood vessels in the tumor stromal microenvironment and later differentiate into TAMs (Chen and Bonaldo 2013). It has been illustrated that CCL2 plays a key role in apoptosis suppression in monocytes by stimulating antiapoptotic proteins (Roca et al. 2009). Besides that, CCL2 also hyper activates autophagy in monocytes to suppress apoptosis (Roca et al. 2009). This indicates that autophagy is very essential for the engagement of monocytes. CSF-1 is one of the important cytokines that differentiate macrophages from monocytes. As CSF-1 stimulates monocytes, autophagy is triggered by the phosphorylation of ULK1 (Jacquel et al. 2012a, b). Another cytokine CSF-2 contributes an important role in the differentiation of monocytes to macrophages through the MAPK/JNK pathway through inhibiting Atg5 cleavage and the synergy between BECN1 and BCL-2, both are essential for autophagy activation (Zhang et al. 2012). Such autophagy induction helps a critical conversion from monocyte apoptosis to differentiation. Overall, these findings suggest that autophagy operates a very critical function in every step of TAMs production and which leads the cancer progression through supporting tumor cells during the transition to malignancy (Noy and Pollard 2014) (Fig. 9.2).

9.3.2 Autophagy and Cancer-Associated Fibroblasts

Cancer-associated fibroblasts (CAFs) are the special kind of cells present in tumor stromal microenvironment, which induces tumorigenicity by instigating the remodeling of the extracellular matrix or by producing cytokines (Xing et al. 2010). The function of autophagy in CAF biology is very complicated. In the early stages of tumor development, normal fibroblasts cells nearby to the tumor go through a critical adaptation due to the vice versa interaction with tumor cells and evolve into a more myofibroblastic phenotype. Such evolved fibroblasts are usually recognized as CAFs. It has been observed that increased autophagy acts an important role in CAFs to support energy metabolism and the growth of neighboring epithelial cancer cells (Fig. 9.2). Such kind of paracrine crosstalk is led by the secretion of hydrogen peroxide from the cancer cells, which leads to the generation of oxidative stress and induces senescence in neighboring CAFs (Martinez-Outschoorn et al. 2011). Senescent CAFs lost mitochondrial function due to the increased autophagy, and mitophagy (induced by oxidative stress) and shifted toward aerobic glycolysis, called "Warburg Effect," a hallmark of the cancer phenotype (Hanahan and Weinberg 2011). This generates metabolic byproducts such as glutamine, lactate, free fatty acids, and ketone bodies that nourish oxidative phosphorylation in the tumor cells and contribute to anabolic growth (Jaboin et al. 2009; Tittarelli et al. 2015). Besides that, autophagic elimination of a tumor suppressor protein, the caveolin-1 negatively regulates Ras signaling in tumor stroma (Martinez-Outschoorn et al. 2010) which is connected with early development from DCIS to invasive cancer (Witkiewicz et al. 2009), metastatic disease in prostate cancer, and lymph node metastasis in breast cancer (Di Vizio et al. 2009). Additionally, autophagy may also stimulate the release of MMPs and pro-migratory cytokines from CAFs (Lock et al. 2014), which leads to further metastasis of the tumor cells. It has been illustrated that the growth and the metastasis of tumor cells are promoted when they are co-injected with senescent fibroblasts and overexpressed with pro-autophagic molecules that are genetically modified in the nude mice (Jaboin et al. 2009). But on the other side, autophagy is upregulated in tumor cells and suppresses growth when they are transplanted alone. Such results demonstrate that CAFs generated metabolic byproducts through autophagy play a critical role in tumor–stromal cell growth by fulfilling the high energy demands of the tumor cells.

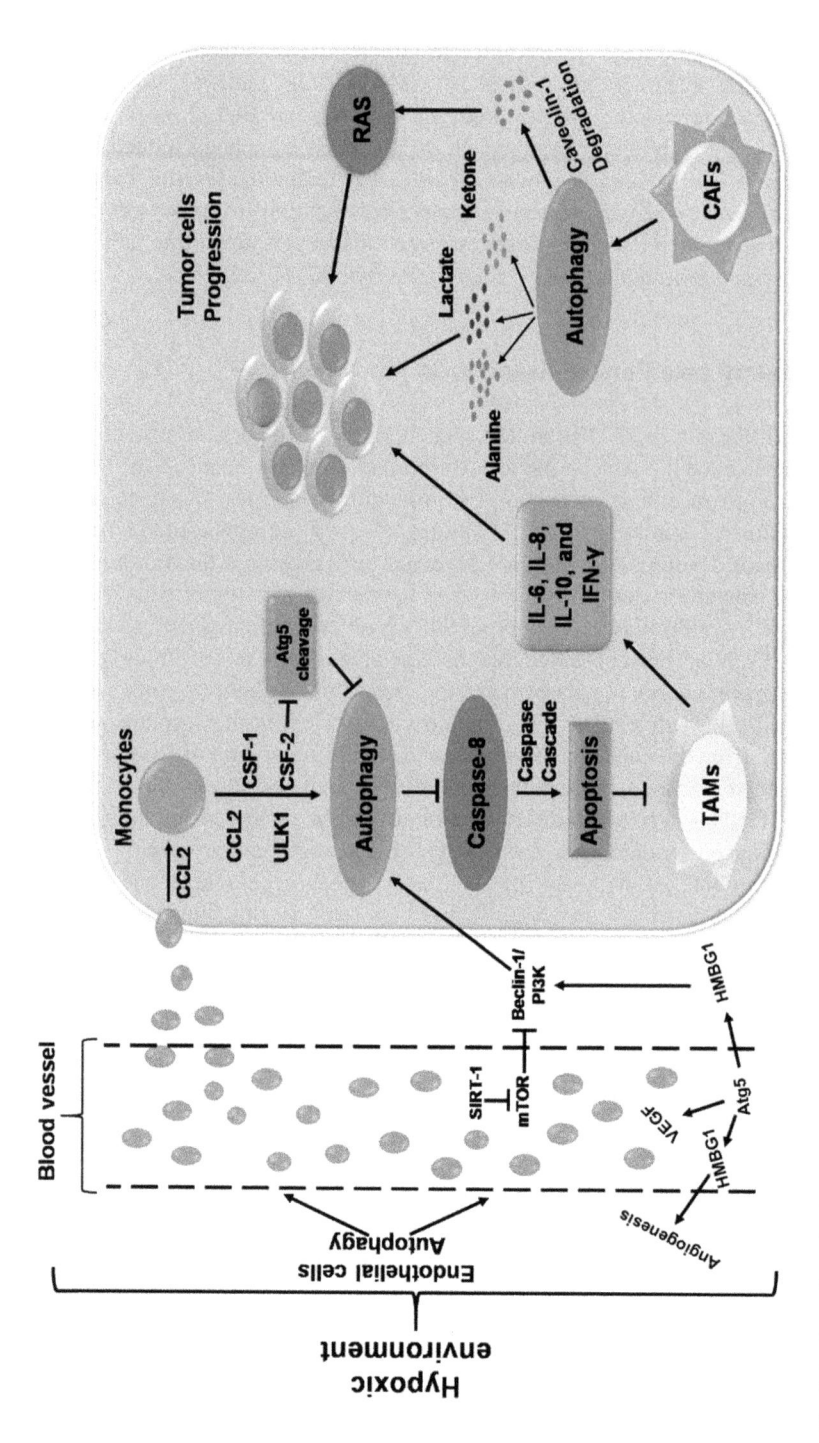

Fig. 9.2 Autophagy in the tumor stromal microenvironment. In the hypoxic and oxidative stress conditions in the tumor stroma, autophagy is stimulated by SIRT-1 and HMBG1 which helps endothelial and tumor cells to survive and grow. In the hypoxic tumor, monocytes from the endothelial cells are recruited into the cytosol by CCL2 chemokines. Monocytes are differentiated into TAMs via activated autophagy, which is stimulated by CCL2, CSF-1, CSF-2, and ULK1. Autophagy also suppresses apoptosis in monocytes. Later on, differentiated TAMs release cytokines, such as IL-6, IL-7, etc., which promote tumor progression. In oxidative stressed tumor stromal microenvironment, senescence CAFs go through autophagy and release metabolic byproducts, such as alanine, lactate, and ketone. Also, highly activated autophagy in such stressed condition degrades caveolin-1 which increases RAS activity and leads cancer development

9.3.3 Autophagy and Tumor-Associated Endothelial Cells

Emerging studies are indicating that autophagy is involved in the vital functions of the tumor-associated vasculature. In the high-stress tumor–stromal conditions, endothelial cells undergo nutrient deprivation and hypoxia due to low blood supply in solid tumors. Such conditions lead to vessel malfunction. For example, tumor vessels become more permeable and less stable as compared to normal vessels (Siemann 2011). Endothelial cells use autophagy as a survival mechanism to escape such stresses (Filippi et al. 2018). Study suggests that under the oxidative stress in the tumor stroma, autophagy is stimulated by SIRT-1, which helps endothelial cells to survive (Ou et al. 2014). SIRT-1, a NAD-dependent deacetylase, activates autophagy through PI3K/Beclin-1 and mTOR pathways. Under the metabolic stressors like hypoxia, such upregulation of autophagy in tumor-associated endothelial cells facilitates nutrients to them. Besides that, in the presence of high autophagic activity in tumor associate endothelial cells, Atg5 accelerates starvation–hypoxia evoked angiogenesis by alleviating α-subunit of hypoxia-inducible factor (HIF) complex and interfering with the VEGF signaling (Filippi et al. 2018; Du et al. 2012). Such interaction of Atg5 can also induce the secretion of a high mobility group box 1 (HMBG1) through an autophagy-modulated mechanism. HMBG1 is a leading chromatin-associated protein that is translocated into the cytoplasm and released to the outside of the cells due to increased metabolic stress in the endothelial cells (Sachdev et al. 2012). In the cytosol, HMBG1 binds to Beclin 1 and acts as a pro-autophagic factor (Kang et al. 2010). Besides that, it takes part in tissue remodeling and angiogenesis signaling (Sachdev et al. 2012). Thus it plays an important role in angiogenesis and also for the protection of tumor cells in the hypoxic microenvironment (Fig. 9.2). In another study, it has been found that Beclin $1^{+/-}$ knockout mice have shown elevated angiogenic activity in their endothelial cells only under hypoxic conditions (Lee et al. 2011). Such a result suggests that autophagy may also play an antiangiogenic role although the influence of Beclin 1 on HMGB1 secretion has not been evaluated in that study. These contradictory results could also indicate a distinct function of autophagy in tumor-associated endothelial cells.

9.3.4 Autophagy and the Immune System in the Tumor Microenvironment

The immune system of the human body, including innate immunity and adaptive immunity, operates a vital role in the immunosupervision of tumors. Autophagy works at the downstream of the pattern recognition receptors. In other words, these activated innate immune receptors upregulate autophagy. In the innate immune response, the crosstalk between the autophagy and the immune system begins with the innate immune receptors like toll-like receptors (TLRs) and nucleotide oligomerization domain (NOD)-like receptors (NLRs), thereby facilitating several effector responses, including NKT cell activation, cytokine production, and phagocytosis.

However, in innate immune response, autophagy provides a substantial amount of the antigens which in a later stage are loaded onto the MHC class II molecules for presentation to the dendritic cell-mediated cross-priming to CD8+ T cells (Jiang et al. 2019). In the coming sections, we will discuss both of the mechanisms of the immune system in the tumor immune microenvironment and their crosstalk with autophagy (Fig. 9.3) that will decide the fate of tumorigenesis.

9.3.4.1 Innate Immunity and Autophagy in Cancer

Innate immunity-mediated autophagy is largely dependent on the innate immune receptors, such as TLRs and NLRs. TLRs and NLRs are highly upregulated upon sensing the pathogenic or tumor antigens that ultimately activate the innate immune response (Zhong et al. 2016).

TLRs

TLRs are the most thoroughly characterized pattern recognition molecules of the innate immune surveillance. The immune system recognizes the tumor antigens by TLRs and infiltrates the tumor stroma resulting in tumor destruction through direct lysis or the involvement of cytokine (Shi et al. 2016). However, increasing pieces of evidence show quite opposing outcomes in cancer development. In the recent years, several developments regarding the use of potential TLRs toward therapeutic possibilities have been elucidated, of which the detailed mechanisms of these must be explored for better understanding. The study suggests that toll-like receptors are the group of innate immune receptors expressed in a wide range of cancer cells that activate several immune responses by regulating autophagy. TLRs are believed to be the autophagy inductors that activate autophagy. TLR7 with the help of a downstream signaling adapter MYD88 or TRIF recruits TRAF6 and Beclin-1 that further stimulate and develop the autophagosomes (Shi and Kehrl 2010). TLR2 induces phagocytosis and autophagy by enhancing the host innate immune system through the induction of c-JNK and ERK signaling cascade (Anand et al. 2011; Fang et al. 2014). TLR4 generally expressed in the innate immune cells, particularly in the dendritic cells and the macrophages, is one of the important targets for immune-modulating drugs. TLR4 stimulates autophagy by triggering TRIF (toll-IL-1 receptor (TIR) domain-containing adapter inducing IFN)/RIP1 (receptor-interacting protein), and p38-MAPK signaling pathway (Xu et al. 2007). Lipopolysaccharides and alpha-GalCer are reported to activate TLR4 signaling induced macrophage activation through mitogen-activated protein kinases and cytokines such as iNOS, IL-, and TNF-a (Xu et al. 2007; Hung et al. 2007). Zhan et al. suggested that TLR-3 and TLR-4 facilitate lung cancer invasion and migration by promoting TRAF6 (TNF receptor-associated factor 6, E3 ubiquitin-protein ligase)-regulated induction of autophagy and cytokine production (Zhan et al. 2014). In the breast cancer patients, it has been observed that higher TLR4 expression is associated with the upregulated LC3 II expression in CAFs which is correlated with the more aggressive relapse and poor prognosis of the tumors (Zhao et al. 2017). Lin et al. exhibited that TLR2 signaling plays a crucial role in the genotoxic carcinogen diethylnitrosamine (DEN) induced liver tissue damage. TLR2 activated intracellular senescence and autophagy

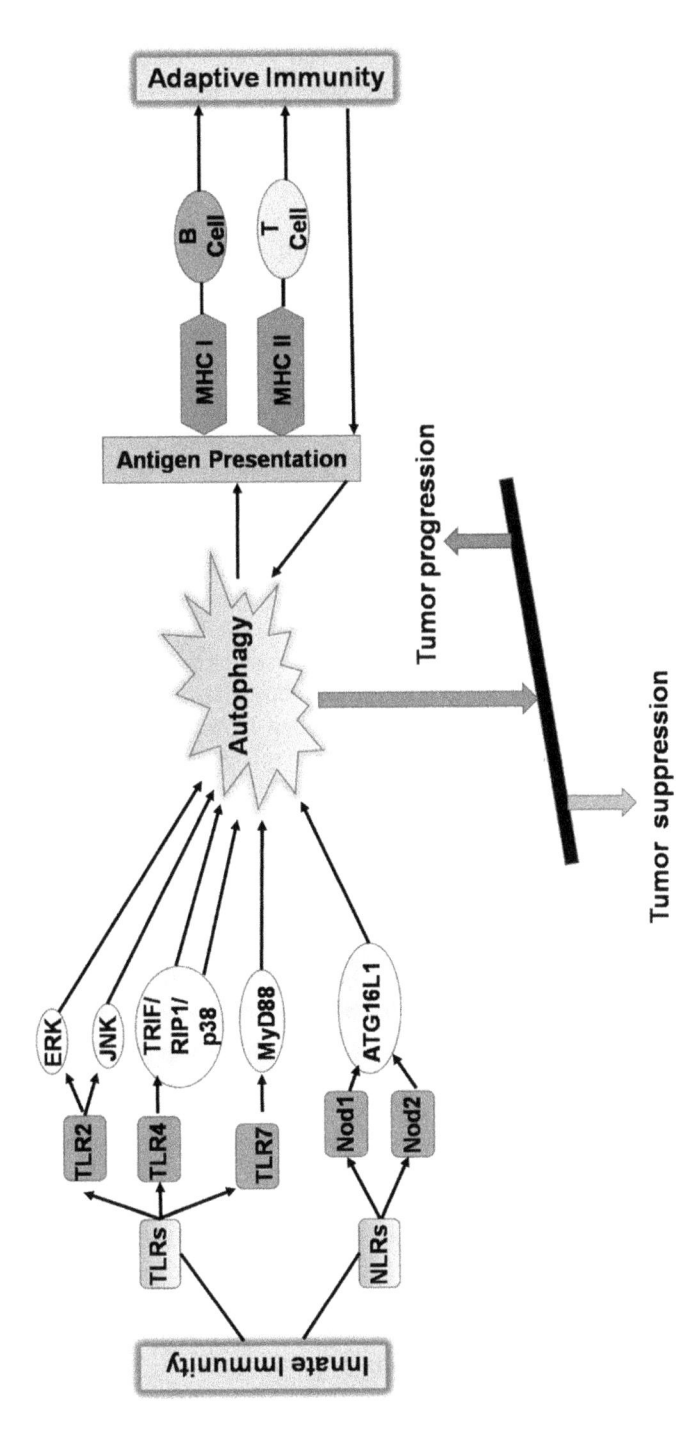

Fig. 9.3 Autophagy and the immune system in the tumor microenvironment. Activation of innate immune receptors like TLRs and NLRs induce autophagy. TLRs induce autophagy by JNK, ERK, MyD88, and TRIF/RIP1/P38 dependent pathway. NLRs up-regulate autophagy by engaging and interacting with ATG16L1. On the other hand, autophagy activates adaptive immunity by antigen presentation through MHCI and MHCII. Simultaneously, in adaptive immunity, antigen presentation also activates autophagy

eliminate the aggregation of ROS and DNA damage, thereby inhibiting the hepatocellular carcinoma development and progression (Lin et al. 2012, 2013). TLR3 in many instances acts as the possible therapeutic target for cancer immunotherapy. In human pharyngeal and oral squamous cell carcinoma cell lines, TLR3 induces apoptosis with the help of TLR3 ligand poly (I:C) (Estornes et al. 2012; Shatz et al. 2012). TLR3 ligand poly (I:C) not only destroys the tumors by apoptosis but rather they also destroys the TME by suppressing the angiogenesis in human hepatocellular carcinoma cells (Guo et al. 2012). Taking into account the dual effects of autophagy, the therapeutic options with TLRs agonists on autophagy cell death need an integrated consideration in clinical implications.

NLRs

NLRs are a group of cytoplasmic molecules that constitute a fundamental element of the innate immune response, generally recognize the bacterial cell wall components, and induce autophagy. Nod1 and Nod2 are the first and important NLRs recognized as the microbial associated molecular pattern (MAMP) detectors. ATG16L1, an important component of autophagosome formation, participates in the Nod1 and Nod2 directed autophagy by interacting with the plasma membrane (Travassos et al. 2010). Both of the Nod1 and Nod2 gene polymorphism is associated with an array of innate and adaptive immune response and autophagy in several cancer types (Kutikhin 2011). NOD1 is an intracellular receptor that induces autophagy and activates the NF–κB signaling in response to the Gram-negative bacterial peptidoglycan leading to the destruction of inflammation-based *Helicobacter pylori* which is believed as an imperative risk factor of gastric carcinogenesis (Suarez et al. 2015).

Others

Many stress-inducing factors activate interferon regulatory factor 8 (IRF8) which in turn activates the autophagy-related genes in dendritic cells. Autophagy-inducing stresses, such as IFNγ and TLR stimulation, macrophage colony-stimulating factor, and bacterial infection, activate IRF8 resulting in the activation of many genes involved in the formation of autophagosome and autophagy (Gupta et al. 2015). Furthermore, IFNγ contributes to the innate immune response and autophagy by the p38 MAPK signaling pathway (Matsuzawa et al. 2014). Inflammation-induced IFNγ attenuates gastric carcinogenesis by activating epithelial autophagy and T-cell apoptosis (Tu et al. 2011). Recent studies have shown that cytokines such as interleukins are also an important part of innate immunity that regulates autophagy. Furthermore, autophagy has a dominating role in the initiation and regulation of the inflammatory response by innate immune cells, mostly facilitated by IL-1 and its consequential effect on IL-23 secretion (Peral de Castro et al. 2012). Altogether, these studies have suggested the newer mechanisms that innate immune receptor-associated autophagy exhibits distinct regulation on carcinogenesis.

9.3.4.2 Adaptive Immunity and Autophagy in Cancer

In adaptive immunity, autophagy plays a pivotal role in anti-tumor effects through antigen presentation, cytokine release, thymus selection, lymphocyte development, and homeostasis. Major histocompatibility complex (MHCI and MHCII) molecules are essential in carrying the intracellular and extracellular peptide epitopes which are consequently recognized by the CD4+ and CD8+ T cells respectively for adaptive immune destruction (Zhong et al. 2016).

Autophagy and Antigen Presentation in Cancer

Cross presentation is one of the critical aspects of the adaptive immune response. Any foreign antigens when entered into the cells are taken up and fixed, which are finally presented to the specific T-cells by antigen-presenting cells (APCs) for immunogenic destruction (Baker et al. 2013). Currently, it has been identified that autophagy has a pivotal role in antigen sequestration in cross-presenting the antigens to the MHC I molecules (Li et al. 2008). Interestingly, TNF-α induces mitophagy (a form of autophagy in the mitochondria) that enabled the delivery of the mitochondrial antigens by the MHC-I molecules at the cell surface (Bell et al. 2013). There have been instances of the interrelationship between MHC-I-regulated autophagy and cancer immune response. Autophagy-induced lysosomal proteolysis and proteasomal degradation regulate the MHC-I molecules mediated cross-presentation of tumor antigens (Li et al. 2009). Apart from endogenous MHC-I peptides, the endogenous system also helps in the processing of MHC-II presentation. During the MHC-II presentation of the exogenous antigens, the lysosomal proteases degrade them and process them for the fusion with the MHC-II loading compartment for the immune surveillance (Gannage and Munz 2010). More specifically, autophagy also helps in the transport of the nuclear and cytosolic antigens for presentation to the CD4+ T cells by the MHC-II molecules (Crotzer and Blum 2009).

Role of Autophagy in the Regulation of T Cells Development and Function in Cancer

T cells depend on the basal autophagy to maintain their homeostasis and activation. Defects in autophagy by deletion of autophagic molecules such as Atg7, Atg3, Atg5, PI3K, and BECN1 can lead to improper T cell activation and differentiation (Pan et al. 2016). The survival of the naive T cells depends upon increased autophagy along with the stromal cells' interaction with the TCR and IL-7 signaling (Sena et al. 2013). After TCR stimulation, it has been noticed that autophagy is enhanced in the T cells along with increased calcium levels that further activate the AMPK via ULK1 complex phosphorylation (Kim et al. 2011; Botbol et al. 2015). There has been mounting evidence on the connection between the autophagy and the regulatory T cell manipulating antitumor immunity. Autophagy acts as a key regulatory mechanism for CD4+ T cell homeostasis. It is observed that autophagy is augmented in the murine CD+ T cells upon activation via JNK and Vps34 and importantly causes the growth factor-withdrawal cell death (Li et al. 2006). It has been also noticed that c-Met expressed on the tumors could act as a potential epitope against the helper T lymphocytes which is partly regulated by autophagy (Kumai et al. 2015). Many studies have delineated the effect of impaired autophagy on the development of T cells. Defective autophagy more frequently affects the CD8+ T cells than the CD4+ T cells. Inhibition of mTOR in the effector CD8+ T cells induces memory CD8+ T cell production in lymphoid rather than the mucosal tissue. Therefore, it can be predicted that CD8+ T cells are more reliant on autophagy (Kovacs et al. 2012; Sowell et al. 2014).

Role of Autophagy in the Regulation of B Cells Development and Function in Cancer

Autophagy acts a significant role in the development and the survival of the B cells. Autophagy is very much essential for the pro and pre-B cell transition and activation of the B cell in response to the stimulation of BCR. Autophagy is also very much necessary to sustain a normal number of peripheral B cells and their survival (Arnold et al. 2016). The development and the maturation of B cells require the pro-autophagy genes (Pan et al. 2016). Further, the knockdown of Atg5 prevents the transition between pro- and pre-B-cell stages in the bone marrow (Chanut et al. 2014). B cell activation is induced by the tumor-derived autophagosomes (termed "DRibbles") resulting in the production and secretion of cytokine. Moreover, DRibbles upregulates the CD40L expression on the macrophages with simultaneously enhances the expression of CD40 on the B cells. Macrophages play a significant role in the presentation of the antigens on the B cells for specific T cell activation (Zhou et al. 2015). Taken together, the current set of data indicates that autophagy serves a pivotal role in the advancement of the certain subgroups of B cells and memory B cells (Chen et al. 2014).

9.4 Autophagy as a Candidate for Cancer Immunotherapy

Among the diversified options available for cancer treatment, immunotherapy-based treatment options are gaining much attention nowadays. The basic mechanism involved in the cytotoxic effect of immunotherapy is based on the regulation of the response of the immune cells thereby preventing the binding with the immune suppressor or cancer cells (Chen and Mellman 2017; Galon and Bruni 2019). Several approaches have been implemented to enhance the immune system for better clinical outcomes. Nevertheless, partly due to tolerogenic effects, most of the strategies have been unsuccessful (Green et al. 2009). Recently, autophagy emerged as a potential mechanism connected to cancer immunotherapy. Targeting autophagy-mediated cross-presentation and immune responses may be considered as a potential therapeutic strategy for cancer treatment. In the subsequent sections, we will discuss the dual role of autophagy not only as a pro-death but also as a pro-survival inductor in the cancer immunotherapy.

9.4.1 Autophagy as a Pro-death Mechanism in Cancer Immunotherapy

In the TME, autophagy may act as a pro-death signal that retards the tumor progression and enhances the antitumor immunity in response to therapy. Several nanoparticle-based therapeutic options that trigger autophagy in cancer have been developed. These nanoparticles act as adjuvants that ultimately deliver the tumor antigens for the autophagosome formation and tumor destruction. In alpha-alumina (α-Al(2)O(3)) nanoparticle-based tumor regression, the former acts as the carrier of the tumor antigens that delivers the antigens to the autophagosomes in the tumor dendritic cells that further present the antigen to T cells by autophagy (Li et al. 2011). Similarly, monobenzone triggers melanosome autophagy by inhibiting the processing and the shedding of melanocyte differentiation antigens, leading to tyrosinase ubiquitination. The whole process activates the dendritic cells and cytotoxic T-cells which efficiently eliminate the melanoma in vivo (van den Boorn et al. 2011). Likewise, targeting folate receptors alpha (FRα) which is highly expressed on the human ovarian cancer cells suppressed cancer cell proliferation. MORAB-003 (farletuzumab) is a humanized mAB interferes with the folate metabolism by targeting the FRα, as a consequence of which, MORAB-003 induces autophagy and hinders cancer cell proliferation (Wen et al. 2015). Conventional chemotherapeutics administrations are more efficient when they elicit the immunogenic cell death (ICD) followed by a series of molecular events, like pre-apoptotic cell surface display of calreticulin, the release of high mobility group box 1 (HMGB1), and ATP secretion during the blebbing phase of apoptosis from the dying cells during the post-apoptotic stage (Pol et al. 2015). Autophagy is more effective in immunotherapy rather than conventional chemotherapy. Autophagy caused immunogenic cell death via T cell activation and mannose-6-phosphate receptor upregulation on the surface of the tumor cells (Ramakrishnan et al. 2012). Autophagy may act as the energy provider to the immune cells like DCs and T lymphocytes by the immunogenic release of ATP from the dying cells in the tumor bed whereas autophagy deficit hinders the ability of cancer cells to elicit an immune response (Michaud et al. 2011). BCG, a potential vaccine against TB, has shown its efficiency as an antitumor immunotherapy agent and it has been seen that a combinatorial approach of BCG and ionizing radiation effectively resulted in autophagic cell death in the colon cancer cells by the generation of ROS (Yuk et al. 2010). Moreover, cytokines also function as mediators of autophagy activation and cancer cell death. For instance, IFN-γ hinders gastric cancer progression by promoting epithelial cell autophagy (Tu et al. 2011). However, the role of IFN1 in anticancer treatment of chronic myeloid leukemia is not fully understood. Zhu et al. demonstrate that the active involvement of autophagy in IFN1-mediated cell death is through the upregulation of JAK1-STAT1 and the ReLA signaling pathway (Zhu et al. 2013).

9.4.2 Autophagy as a Pro-survival Mechanism in Cancer Immunotherapy

Besides the cancer regression, autophagy also plays an important role in tumor promotion, ignoring the immunotherapeutic administration. Autophagy impairs the anticancer immune response and avails tumor cells to evade immune surveillance, thereby promoting tumor growth and progression. In the inner tumoral region, hypoxic condition rises due to an inadequate supply of oxygen. Hypoxia in the tumor microenvironment has been proved as a mechanism of cancer cell survival by attenuating the therapeutic intervention by interfering with various signaling cascades. It has been reported that hypoxia-induced autophagy plays a major role in diminishing the effect of immunotherapy in cancer cells (Qiu et al. 2015). For example, hypoxia-stimulated autophagy can suppress T-cell-mediated cytotoxicity in lung cancer cells (Jaboin et al. 2009; Pan et al. 2016). Hypoxia-induced autophagy impairs CTLs-mediated tumor cell lysis that is associated with the hypoxia-dependent phosphorylation of STAT3 (pSTAT3), which in turn activates tumor cell survival, proliferation, and immune escape (Teng et al. 2014; Noman et al. 2012). Hypoxia-induced autophagy also interrupts the NK-mediated killing of the cancer cells by degrading the NK-derived Granzyme B (Baginska et al. 2013). During hypoxia, HIF-2α is localized to the nucleus and triggers the expression of autophagy sensor ITPR1 (inositol 1,4,5-trisphosphate receptor, type 1), ultimately deactivates the NK-mediated cell lysis and decreases immunotherapeutic effect (Hasmim et al. 2015; Messai et al. 2014). In the hypoxic melanoma cells, autophagy degrades the channel protein connexin 43, resulting in the destabilization of immune synapse that interferes with the NK-cell mediated lysis of the cell (Tittarelli et al. 2015). In some preclinical models, it has also been proved that inhibition of autophagy in combination with the other therapeutic approaches augments the cytotoxicity of cancer cells and inhibits the cancer progression. Recent studies have found that attenuating autophagy with chloroquine increases the efficacy of high-dose interleukin-2 (HDIL-2) in inhibiting cancer therapy by immunotherapeutic approach (Liang et al. 2012). Similarly, chloroquine enhances the HDIL-2-mediated antitumor immunity, triggering the NK cells, T-cells, and DCs in renal cell carcinoma. Administration of chloroquine blocks the autophagy and limits ATP production by inhibiting the Oxidative phosphorylation (Lotze et al. 2012). Moreover, chloroquine blocks the radiation-induced autophagy in breast cancer cells and promotes cell death via DCs mediated immunogenic cell death (Ratikan et al. 2013). Intrinsically, autophagy provides resistance to the immunological anticancer therapy by diminishing the immune effector mechanisms. Therefore, protective autophagy is activated against sepsis-induced T lymphocyte apoptosis and immunosuppression. Overall, the downregulation of autophagy in T lymphocytes may lead to an increased rate of apoptosis and decreased cell survival (Lin et al. 2014). Interleukin 24, an exclusive member of the IL-10 family, shows universal cancer-specific toxicity. A combination of autophagy inhibitors and IL-24 may be an encouraging strategy for tumor immunotherapy. In oral squamous cell carcinoma, 3-MA, a PI3K inhibitor enhances the IL-24 induced apoptosis by acting upon Vps34 and PI3Kγ (Li et al. 2015).

9.5 Conclusions

Despite significant advances in the detection techniques and the development of promising therapeutic approaches, like surgery, radiotherapy, and chemotherapy, cancer is still one of the major causes of death worldwide due to the adverse effects of these approaches and their inefficacy against all the tumors. However, prompt evolution of immunology, molecular biology, cell biology, and other relevant fields is believed to expedite the advancement of immunotherapy which successfully reduces tumor growth with minimal off-target effects on the host cells. In the complex tumor microenvironment, the fate of the tumor cells depends on the interactions among tumor cells with the immune cells. In line with this, autophagy has a significant role in the regulation of cancer development in the TME that makes it an ideal target for cancer therapy. Of note, autophagy has a multifaceted role in the TME. It can contribute to the survival as well as the destruction of the cancer cells depending upon the stages of cancer. Collectively, the study suggests that autophagy can stimulate antitumor immune responses through promoting differentiation, maturation, and also maintaining internal homeostasis in the immune cells. However, hypoxia-induced autophagy suppresses immune cell functions and facilitates tumor cell evasion from the immune surveillance. Thus, a better understanding of the in-depth molecular mechanisms associated with the crosstalk between the context-dependent roles of autophagy and the immune system in the TME will further magnify the therapeutic strategies against cancer.

Acknowledgment We would like to acknowledge Department of Science and Technology (DST-INSPIRE-IF130677), Council of Scientific and Industrial Research (CSIR), Ministry of Human Resource and Development (MHRD), Government of India; Indian Institute of Technology Kharagpur, India; and German Academic Exchange Service (DAAD), Germany for providing financial support.

Conflict of interest: The authors declare that they do not have any conflict of interest.

References

Anand PK, Tait SW, Lamkanfi M, Amer AO, Nunez G, Pages G, Pouyssegur J, McGargill MA, Green DR, Kanneganti TD (2011) TLR2 and RIP2 pathways mediate autophagy of Listeria monocytogenes via extracellular signal-regulated kinase (ERK) activation. J Biol Chem 286:42981–42991

Arnold J, Murera D, Arbogast F, Fauny JD, Muller S, Gros F (2016) Autophagy is dispensable for B-cell development but essential for humoral autoimmune responses. Cell Death Differ 23:853–864

Baginska J, Viry E, Berchem G, Poli A, Noman MZ, van Moer K, Medves S, Zimmer J, Oudin A, Niclou SP, Bleakley RC, Goping IS, Chouaib S, Janji B (2013) Granzyme B degradation by autophagy decreases tumor cell susceptibility to natural killer-mediated lysis under hypoxia. Proc Natl Acad Sci U S A 110:17450–17455

Baker K, Rath T, Lencer WI, Fiebiger E, Blumberg RS (2013) Cross-presentation of IgG-containing immune complexes. Cell Mol Life Sci 70:1319–1334

Balkwill FR, Capasso M, Hagemann T (2012) The tumor microenvironment at a glance. J Cell Sci 125:5591–5596

Bell C, English L, Boulais J, Chemali M, Caron-Lizotte O, Desjardins M, Thibault P (2013) Quantitative proteomics reveals the induction of mitophagy in tumor necrosis factor-alpha-activated (TNFalpha) macrophages. Mol Cell Proteomics 12:2394–2407

Bhol CS, Panigrahi DP, Praharaj PP, Mahapatra KK, Patra S, Mishra SR, Behera BP, Bhutia SK (2019) Epigenetic modifications of autophagy in cancer and cancer therapeutics. Semin Cancer Biol. https://doi.org/10.1016/j.semcancer.2019.05.020

Botbol Y, Patel B, Macian F (2015) Common gamma-chain cytokine signaling is required for macroautophagy induction during CD4+ T-cell activation. Autophagy 11:1864–1877

Chanut A, Duguet F, Marfak A, David A, Petit B, Parrens M, Durand-Panteix S, Boulin-Deveza M, Gachard N, Youlyouz-Marfak I, Bordesoulle D, Feuillard J, Faumont N (2014) RelA and RelB cross-talk and function in Epstein-Barr virus transformed B cells. Leukemia 28:871–879

Chen P, Bonaldo P (2013) Role of macrophage polarization in tumor angiogenesis and vessel normalization: implications for new anticancer therapies. Int Rev Cell Mol Biol 301:1–35

Chen DS, Mellman I (2017) Elements of cancer immunity and the cancer-immune set point. Nature 541:321–330

Chen M, Hong MJ, Sun H, Wang L, Shi X, Gilbert BE, Corry DB, Kheradmand F, Wang J (2014) Essential role for autophagy in the maintenance of immunological memory against influenza infection. Nat Med 20:503–510

Comito G, Giannoni E, Segura CP, Barcellos-de-Souza P, Raspollini MR, Baroni G, Lanciotti M, Serni S, Chiarugi P (2014) Cancer-associated fibroblasts and M2-polarized macrophages synergize during prostate carcinoma progression. Oncogene 33:2423–2431

Crotzer VL, Blum JS (2009) Autophagy and its role in MHC-mediated antigen presentation. J Immunol 182:3335–3341

Das CK, Mandal M, Kogel D (2018a) Pro-survival autophagy and cancer cell resistance to therapy. Cancer Metastasis Rev 37:749–766

Das CK, Linder B, Bonn F, Rothweiler F, Dikic I, Michaelis M, Cinatl J, Mandal M, Kogel D (2018b) BAG3 overexpression and cytoprotective autophagy mediate apoptosis resistance in chemoresistant breast cancer cells. Neoplasia 20:263–279

Das CK, Banerjee I, Mandal M (2019a) Pro-survival autophagy: an emerging candidate of tumor progression through maintaining hallmarks of cancer. Semin Cancer Biol. https://doi.org/10.1016/j.semcancer.2019.08.020

Das CK, Parekh A, Parida PK, Bhutia SK, Mandal M (2019b) Lactate dehydrogenase A regulates autophagy and tamoxifen resistance in breast cancer. Biochim Biophys Acta Mol Cell Res 1866:1004–1018

Denton AE, Roberts EW, Fearon DT (2018) Stromal cells in the tumor microenvironment. Adv Exp Med Biol 1060:99–114

Deretic V (2012) Autophagy: an emerging immunological paradigm. J Immunol 189:15–20

Di Vizio D, Morello M, Sotgia F, Pestell RG, Freeman MR, Lisanti MP (2009) An absence of stromal caveolin-1 is associated with advanced prostate cancer, metastatic disease and epithelial Akt activation. Cell Cycle 8:2420–2424

Du J, Teng RJ, Guan T, Eis A, Kaul S, Konduri GG, Shi Y (2012) Role of autophagy in angiogenesis in aortic endothelial cells. Am J Physiol Cell Physiol 302:C383–C391

Estornes Y, Toscano F, Virard F, Jacquemin G, Pierrot A, Vanbervliet B, Bonnin M, Lalaoui N, Mercier-Gouy P, Pacheco Y, Salaun B, Renno T, Micheau O, Lebecque S (2012) dsRNA induces apoptosis through an atypical death complex associating TLR3 to caspase-8. Cell Death Differ 19:1482–1494

Fang L, Wu HM, Ding PS, Liu RY (2014) TLR2 mediates phagocytosis and autophagy through JNK signaling pathway in Staphylococcus aureus-stimulated RAW264.7 cells. Cell Signal 26:806–814

Filippi I, Saltarella I, Aldinucci C, Carraro F, Ria R, Vacca A, Naldini A (2018) Different adaptive responses to hypoxia in normal and multiple myeloma endothelial cells. Cell Physiol Biochem 46:203–212

Fulda S, Kogel D (2015) Cell death by autophagy: emerging molecular mechanisms and implications for cancer therapy. Oncogene 34:5105–5113

Galon J, Bruni D (2019) Approaches to treat immune hot, altered and cold tumours with combination immunotherapies. Nat Rev Drug Discov 18:197–218

Gannage M, Munz C (2010) MHC presentation via autophagy and how viruses escape from it. Semin Immunopathol 32:373–381

Green DR, Ferguson T, Zitvogel L, Kroemer G (2009) Immunogenic and tolerogenic cell death. Nat Rev Immunol 9:353–363

Guo Z, Chen L, Zhu Y, Zhang Y, He S, Qin J, Tang X, Zhou J, Wei Y (2012) Double-stranded RNA-induced TLR3 activation inhibits angiogenesis and triggers apoptosis of human hepatocellular carcinoma cells. Oncol Rep 27:396–402

Gupta M, Shin DM, Ramakrishna L, Goussetis DJ, Platanias LC, Xiong H, Morse HC 3rd, Ozato K (2015) IRF8 directs stress-induced autophagy in macrophages and promotes clearance of Listeria monocytogenes. Nat Commun 6:6379

Hanahan D, Weinberg RA (2011) Hallmarks of cancer: the next generation. Cell 144:646–674

Hasmim M, Messai Y, Ziani L, Thiery J, Bouhris JH, Noman MZ, Chouaib S (2015) Critical role of tumor microenvironment in shaping NK cell functions: implication of hypoxic stress. Front Immunol 6:482

Hung LC, Lin CC, Hung SK, Wu BC, Jan MD, Liou SH, Fu SL (2007) A synthetic analog of alpha-galactosylceramide induces macrophage activation via the TLR4-signaling pathways. Biochem Pharmacol 73:1957–1970

Jaboin JJ, Hwang M, Lu B (2009) Autophagy in lung cancer. Methods Enzymol 453:287–304

Jacquel A, Obba S, Boyer L, Dufies M, Robert G, Gounon P, Lemichez E, Luciano F, Solary E, Auberger P (2012a) Autophagy is required for CSF-1-induced macrophagic differentiation and acquisition of phagocytic functions. Blood 119:4527–4531

Jacquel A, Obba S, Solary E, Auberger P (2012b) Proper macrophagic differentiation requires both autophagy and caspase activation. Autophagy 8:1141–1143

Jiang GM, Tan Y, Wang H, Peng L, Chen HT, Meng XJ, Li LL, Liu Y, Li WF, Shan H (2019) The relationship between autophagy and the immune system and its applications for tumor immunotherapy. Mol Cancer 18:17

Kang R, Livesey KM, Zeh HJ, Loze MT, Tang D (2010) HMGB1: a novel Beclin 1-binding protein active in autophagy. Autophagy 6:1209–1211

Kaushik S, Cuervo AM (2012) Chaperone-mediated autophagy: a unique way to enter the lysosome world. Trends Cell Biol 22:407–417

Kim J, Kundu M, Viollet B, Guan KL (2011) AMPK and mTOR regulate autophagy through direct phosphorylation of Ulk1. Nat Cell Biol 13:132–141

Kovacs JR, Li C, Yang Q, Li G, Garcia IG, Ju S, Roodman DG, Windle JJ, Zhang X, Lu B (2012) Autophagy promotes T-cell survival through degradation of proteins of the cell death machinery. Cell Death Differ 19:144–152

Kumai T, Matsuda Y, Ohkuri T, Oikawa K, Ishibashi K, Aoki N, Kimura S, Harabuchi Y, Celis E, Kobayashi H (2015) c-Met is a novel tumor associated antigen for T-cell based immunotherapy against NK/T cell lymphoma. Oncoimmunology 4:e976077

Kutikhin AG (2011) Role of NOD1/CARD4 and NOD2/CARD15 gene polymorphisms in cancer etiology. Hum Immunol 72:955–968

Lee SJ, Kim HP, Jin Y, Choi AM, Ryter SW (2011) Beclin 1 deficiency is associated with increased hypoxia-induced angiogenesis. Autophagy 7:829–839

Li C, Capan E, Zhao Y, Zhao J, Stolz D, Watkins SC, Jin S, Lu B (2006) Autophagy is induced in CD4+ T cells and important for the growth factor-withdrawal cell death. J Immunol 177:5163–5168

Li Y, Wang LX, Yang G, Hao F, Urba WJ, Hu HM (2008) Efficient cross-presentation depends on autophagy in tumor cells. Cancer Res 68:6889–6895

Li Y, Wang LX, Pang P, Twitty C, Fox BA, Aung S, Urba WJ, Hu HM (2009) Cross-presentation of tumor associated antigens through tumor-derived autophagosomes. Autophagy 5:576–577

Li H, Li Y, Jiao J, Hu HM (2011) Alpha-alumina nanoparticles induce efficient autophagy-dependent cross-presentation and potent antitumour response. Nat Nanotechnol 6:645–650

Li J, Yang D, Wang W, Piao S, Zhou J, Saiyin W, Zheng C, Sun H, Li Y (2015) Inhibition of autophagy by 3-MA enhances IL-24-induced apoptosis in human oral squamous cell carcinoma cells. J Exp Clin Cancer Res 34:97

Liang X, De Vera ME, Buchser WJ, de Vivar Chavez AR, Loughran P, Stolz DB, Basse P, Wang T, Van Houten B, Zeh HJ 3rd, Lotze MT (2012) Inhibiting systemic autophagy during interleukin 2 immunotherapy promotes long-term tumor regression. Cancer Res 72:2791–2801

Lin H, Hua F, Hu ZW (2012) Autophagic flux, supported by toll-like receptor 2 activity, defends against the carcinogenesis of hepatocellular carcinoma. Autophagy 8:1859–1861

Lin H, Yan J, Wang Z, Hua F, Yu J, Sun W, Li K, Liu H, Yang H, Lv Q, Xue J, Hu ZW (2013) Loss of immunity-supported senescence enhances susceptibility to hepatocellular carcinogenesis and progression in Toll-like receptor 2-deficient mice. Hepatology 57:171–182

Lin CW, Lo S, Hsu C, Hsieh CH, Chang YF, Hou BS, Kao YH, Lin CC, Yu ML, Yuan SS, Hsieh YC (2014) T-cell autophagy deficiency increases mortality and suppresses immune responses after sepsis. PLoS One 9:e102066

Lin Y, Xu J, Lan H (2019) Tumor-associated macrophages in tumor metastasis: biological roles and clinical therapeutic applications. J Hematol Oncol 12:76

Lock R, Kenific CM, Leidal AM, Salas E, Debnath J (2014) Autophagy-dependent production of secreted factors facilitates oncogenic RAS-driven invasion. Cancer Discov 4:466–479

Lotze MT, Buchser WJ, Liang X (2012) Blocking the interleukin 2 (IL2)-induced systemic autophagic syndrome promotes profound antitumor effects and limits toxicity. Autophagy 8:1264–1266

Maes H, Rubio N, Garg AD, Agostinis P (2013) Autophagy: shaping the tumor microenvironment and therapeutic response. Trends Mol Med 19:428–446

Martinez-Outschoorn UE, Pavlides S, Whitaker-Menezes D, Daumer KM, Milliman JN, Chiavarina B, Migneco G, Witkiewicz AK, Martinez-Cantarin MP, Flomenberg N, Howell A, Pestell RG, Lisanti MP, Sotgia F (2010) Tumor cells induce the cancer associated fibroblast phenotype via caveolin-1 degradation: implications for breast cancer and DCIS therapy with autophagy inhibitors. Cell Cycle 9:2423–2433

Martinez-Outschoorn UE, Lin Z, Trimmer C, Flomenberg N, Wang C, Pavlides S, Pestell RG, Howell A, Sotgia F, Lisanti MP (2011) Cancer cells metabolically "fertilize" the tumor microenvironment with hydrogen peroxide, driving the Warburg effect: implications for PET imaging of human tumors. Cell Cycle 10:2504–2520

Matsuzawa T, Fujiwara E, Washi Y (2014) Autophagy activation by interferon-gamma via the p38 mitogen-activated protein kinase signalling pathway is involved in macrophage bactericidal activity. Immunology 141:61–69

Messai Y, Noman MZ, Hasmim M, Janji B, Tittarelli A, Boutet M, Baud V, Viry E, Billot K, Nanbakhsh A, Ben Safta T, Richon C, Ferlicot S, Donnadieu E, Couve S, Gardie B, Orlanducci F, Albiges L, Thiery J, Olive D, Escudier B, Chouaib S (2014) ITPR1 protects renal cancer cells against natural killer cells by inducing autophagy. Cancer Res 74:6820–6832

Michaud M, Martins I, Sukkurwala AQ, Adjemian S, Ma Y, Pellegatti P, Shen S, Kepp O, Scoazec M, Mignot G, Rello-Varona S, Tailler M, Menger L, Vacchelli E, Galluzzi L, Ghiringhelli F, di Virgilio F, Zitvogel L, Kroemer G (2011) Autophagy-dependent anticancer immune responses induced by chemotherapeutic agents in mice. Science 334:1573–1577

Mijaljica D, Prescott M, Devenish RJ (2011) Microautophagy in mammalian cells: revisiting a 40-year-old conundrum. Autophagy 7:673–682

Mizushima N, Ohsumi Y, Yoshimori T (2002) Autophagosome formation in mammalian cells. Cell Struct Funct 27:421–429

Noman MZ, Janji B, Berchem G, Mami-Chouaib F, Chouaib S (2012) Hypoxia-induced autophagy: a new player in cancer immunotherapy? Autophagy 8:704–706

Noy R, Pollard JW (2014) Tumor-associated macrophages: from mechanisms to therapy. Immunity 41:49–61

Ou X, Lee MR, Huang X, Messina-Graham S, Broxmeyer HE (2014) SIRT1 positively regulates autophagy and mitochondria function in embryonic stem cells under oxidative stress. Stem Cells 32:1183–1194

Pan H, Chen L, Xu Y, Han W, Lou F, Fei W, Liu S, Jing Z, Sui X (2016) Autophagy-associated immune responses and cancer immunotherapy. Oncotarget 7:21235–21246

Peral de Castro C, Jones SA, Ni Cheallaigh C, Hearnden CA, Williams L, Winter J, Lavelle EC, Mills KH, Harris J (2012) Autophagy regulates IL-23 secretion and innate T cell responses through effects on IL-1 secretion. J Immunol 189:4144–4153

Pol J, Vacchelli E, Aranda F, Castoldi F, Eggermont A, Cremer I, Sautes-Fridman C, Fucikova J, Galon J, Spisek R, Tartour E, Zitvogel L, Kroemer G, Galluzzi L (2015) Trial watch: immunogenic cell death inducers for anticancer chemotherapy. Onco Targets Ther 4:e1008866

Qiu Y, Li P, Ji C (2015) Cell death conversion under hypoxic condition in tumor development and therapy. Int J Mol Sci 16:25536–25551

Ramakrishnan R, Huang C, Cho HI, Lloyd M, Johnson J, Ren X, Altiok S, Sullivan D, Weber J, Celis E, Gabrilovich DI (2012) Autophagy induced by conventional chemotherapy mediates tumor cell sensitivity to immunotherapy. Cancer Res 72:5483–5493

Ratikan JA, Sayre JW, Schaue D (2013) Chloroquine engages the immune system to eradicate irradiated breast tumors in mice. Int J Radiat Oncol Biol Phys 87:761–768

Roca H, Varsos ZS, Sud S, Craig MJ, Ying C, Pienta KJ (2009) CCL2 and interleukin-6 promote survival of human CD11b+ peripheral blood mononuclear cells and induce M2-type macrophage polarization. J Biol Chem 284:34342–34354

Sachdev U, Cui X, Hong G, Namkoong S, Karlsson JM, Baty CJ, Tzeng E (2012) High mobility group box 1 promotes endothelial cell angiogenic behavior in vitro and improves muscle perfusion in vivo in response to ischemic injury. J Vasc Surg 55:180–191

Sena LA, Li S, Jairaman A, Prakriya M, Ezponda T, Hildeman DA, Wang CR, Schumacker PT, Licht JD, Perlman H, Bryce PJ, Chandel NS (2013) Mitochondria are required for antigen-specific T cell activation through reactive oxygen species signaling. Immunity 38:225–236

Shatz M, Menendez D, Resnick MA (2012) The human TLR innate immune gene family is differentially influenced by DNA stress and p53 status in cancer cells. Cancer Res 72:3948–3957

Shi CS, Kehrl JH (2010) TRAF6 and A20 regulate lysine 63-linked ubiquitination of Beclin-1 to control TLR4-induced autophagy. Sci Signal 3:ra42

Shi M, Chen X, Ye K, Yao Y, Li Y (2016) Application potential of toll-like receptors in cancer immunotherapy: systematic review. Medicine 95:e3951

Siemann DW (2011) The unique characteristics of tumor vasculature and preclinical evidence for its selective disruption by tumor-vascular disrupting agents. Cancer Treat Rev 37:63–74

Sowell RT, Rogozinska M, Nelson CE, Vezys V, Marzo AL (2014) Cutting edge: generation of effector cells that localize to mucosal tissues and form resident memory CD8 T cells is controlled by mTOR. J Immunol 193:2067–2071

Suarez G, Romero-Gallo J, Piazuelo MB, Wang G, Maier RJ, Forsberg LS, Azadi P, Gomez MA, Correa P, Peek RM Jr (2015) Modification of Helicobacter pylori peptidoglycan enhances NOD1 activation and promotes cancer of the stomach. Cancer Res 75:1749–1759

Teng Y, Ross JL, Cowell JK (2014) The involvement of JAK-STAT3 in cell motility, invasion, and metastasis. Jak-Stat 3:e28086

Tittarelli A, Janji B, Van Moer K, Noman MZ, Chouaib S (2015) The selective degradation of synaptic Connexin 43 protein by hypoxia-induced autophagy impairs natural killer cell-mediated tumor cell killing. J Biol Chem 290:23670–23679

Travassos LH, Carneiro LA, Ramjeet M, Hussey S, Kim YG, Magalhaes JG, Yuan L, Soares F, Chea E, Le Bourhis L, Boneca IG, Allaoui A, Jones NL, Nunez G, Girardin SE, Philpott DJ (2010) Nod1 and Nod2 direct autophagy by recruiting ATG16L1 to the plasma membrane at the site of bacterial entry. Nat Immunol 11:55–62

Tu SP, Quante M, Bhagat G, Takaishi S, Cui G, Yang XD, Muthuplani S, Shibata W, Fox JG, Pritchard DM, Wang TC (2011) IFN-gamma inhibits gastric carcinogenesis by inducing epithelial cell autophagy and T-cell apoptosis. Cancer Res 71:4247–4259

van den Boorn JG, Picavet DI, van Swieten PF, van Veen HA, Konijnenberg D, van Veelen PA, van Capel T, Jong EC, Reits EA, Drijfhout JW, Bos JD, Melief CJ, Luiten RM (2011) Skin-depigmenting agent monobenzone induces potent T-cell autoimmunity toward pigmented cells by tyrosinase haptenation and melanosome autophagy. J Invest Dermatol 131:1240–1251

Wen Y, Graybill WS, Previs RA, Hu W, Ivan C, Mangala LS, Zand B, Nick AM, Jennings NB, Dalton HJ, Sehgal V, Ram P, Lee JS, Vivas-Mejia PE, Coleman RL, Sood AK (2015) Immunotherapy targeting folate receptor induces cell death associated with autophagy in ovarian cancer. Clin Cancer Res 21:448–459

White E (2012) Deconvoluting the context-dependent role for autophagy in cancer. Nat Rev Cancer 12:401–410

Witkiewicz AK, Dasgupta A, Nguyen KH, Liu C, Kovatich AJ, Schwartz GF, Pestell RG, Sotgia F, Rui H, Lisanti MP (2009) Stromal caveolin-1 levels predict early DCIS progression to invasive breast cancer. Cancer Biol Ther 8:1071–1079

Wong AS, Cheung ZH, Ip NY (2011) Molecular machinery of macroautophagy and its deregulation in diseases. Biochim Biophys Acta 1812:1490–1497

Xing F, Saidou J, Watabe K (2010) Cancer associated fibroblasts (CAFs) in tumor microenvironment. Front Biosci 15:166–179

Xu Y, Jagannath C, Liu XD, Sharafkhaneh A, Kolodziejska KE, Eissa NT (2007) Toll-like receptor 4 is a sensor for autophagy associated with innate immunity. Immunity 27:135–144

Yorimitsu T, Klionsky DJ (2005) Autophagy: molecular machinery for self-eating. Cell Death Differ 12(Suppl 2):1542–1552

Yost KE, Satpathy AT, Wells DK, Qi Y, Wang C, Kageyama R, McNamara KL, Granja JM, Sarin KY, Brown RA, Gupta RK, Curtis C, Bucktrout SL, Davis MM, Chang ALS, Chang HY (2019) Clonal replacement of tumor-specific T cells following PD-1 blockade. Nat Med 25:1251–1259

Yuk JM, Shin DM, Song KS, Lim K, Kim KH, Lee SH, Kim JM, Lee JS, Paik TH, Kim JS, Jo EK (2010) Bacillus Calmette-Guerin cell wall cytoskeleton enhances colon cancer radiosensitivity through autophagy. Autophagy 6:46–60

Zhan Z, Xie X, Cao H, Zhou X, Zhang XD, Fan H, Liu Z (2014) Autophagy facilitates TLR4- and TLR3-triggered migration and invasion of lung cancer cells through the promotion of TRAF6 ubiquitination. Autophagy 10:257–268

Zhang Y, Morgan MJ, Chen K, Choksi S, Liu ZG (2012) Induction of autophagy is essential for monocyte-macrophage differentiation. Blood 119:2895–2905

Zhao XL, Lin Y, Jiang J, Tang Z, Yang S, Lu L, Liang Y, Liu X, Tan J, Hu XG, Niu Q, Fu WJ, Yan ZX, Guo DY, Ping YF, Wang JM, Zhang X, Kung HF, Bian XW, Yao XH (2017) High-mobility group box 1 released by autophagic cancer-associated fibroblasts maintains the stemness of luminal breast cancer cells. J Pathol 243:376–389

Zhong Z, Sanchez-Lopez E, Karin M (2016) Autophagy, inflammation, and immunity: a Troika governing cancer and its treatment. Cell 166:288–298

Zhou M, Li W, Wen Z, Sheng Y, Ren H, Dong H, Cao M, Hu HM, Wang LX (2015) Macrophages enhance tumor-derived autophagosomes (DRibbles)-induced B cells activation by TLR4/MyD88 and CD40/CD40L. Exp Cell Res 331:320–330

Zhu S, Cao L, Yu Y, Yang L, Yang M, Liu K, Huang J, Kang R, Livesey KM, Tang D (2013) Inhibiting autophagy potentiates the anticancer activity of IFN1@/IFNalpha in chronic myeloid leukemia cells. Autophagy 9:317–327

Relevance of Autophagy in Cancer Stem Cell and Therapeutic

10

Niharika Sinha

Abstract

Autophagy in cancer acts as a double-edged sword whose functional discrepancies precisely depend on cancerization, progression, and type. During stress, they promote cancer cell survival, induce carcinogenesis due to their accumulated genetic mutations or abnormal cell signaling, initiating fast replication capacity, promoting more aggressiveness, and resistant to programmed cell death. Consequently, the study has drawn focus on autophagy in cancer. However, convincing preclinical and clinical evidence on the cytoprotective in addition to the lethal roles of autophagy for cancer stem cells (CSCs) are missing. There are quite a lot of clinical trials ongoing to manipulate autophagy and in this manner decide the result of disease therapy. The clinical relevance of this work encompasses autophagy modifiers, such as rapamycin and chloroquine that control autophagy in anticancer therapy, since autophagy plays roles in both tumor suppression and promotion. Further detailed examination of autophagy in cancer is required to understand how an increased function of autophagy in the tumor microenvironment, stemness, migration and invasion, dormancy, and drug resistance could be tweaked for enhanced therapeutic benefit by eradicating minimal residual disease and preventing metastasis. Here, we recapitulate how autophagy modulates the therapeutic potential to exterminate CSCs.

Keywords

Cancer stem cells · Self-renewal · Autophagy · Mitophagy · Anticancer therapy

N. Sinha (✉)
Reproductive and Developmental Sciences Program, Department of Animal Sciences, Michigan State University, East Lansing, MI, USA
e-mail: sinhani1@msu.edu

10.1 Introduction

Pluripotent cancer stem cells (CSCs) are subset of cancer cells that accentuate their ability to self-renew and (Aponte and Caicedo 2017) differentiate into all somatic cell lineages by indefinite cell division giving rise to the heterogeneous tumor populations and maintain their undifferentiated state (Liu et al. 2013). When a very small population of CSCs was introduced into an immunocompromised mice, it initiated the formation of the original tumor (Ghiaur et al. 2012). They are phenotypically slow cycling and their self-renewing capacity is accountable for tumor growth, resistance to therapy, and recurrence after treatment.

Autophagy is a double-edged sword in the progression of neoplasia and has further produced immense hurdles for researchers to explore its impression on carcinogenesis and tumor development. It has labeled tumor-suppressive and tumor-promoting functions (White and DiPaola 2009). Cytoprotective role of autophagy prevents malignant transformation through the ability to empower the premalignant cells by efficiently meeting up with the increased energy requirements by recycling cellular components that are important in maintaining the physiological tissue homeostasis. This attribute propagates their accommodation within the stress (metabolic, genotoxic, and inflammatory) occurring after the malignant transformation induced in response to anticancer (chemo/targeted/radiotherapy) treatment. Stresses including nutrient and energy stress, ER stress, danger-associated molecular patterns (DAMPs) and pathogen-associated molecular patterns (PAMPs), hypoxia, redox stress, and mitochondrial damage induce autophagy, alongside EMT and stemness. The cytoprotective role of autophagy can turn into a cell-suicidal weapon causing cell death in cancer cells. Defective autophagy has been linked with increased oncogenesis. For instance, low expression of Beclin-1 (Atg6) in some types of cancers of the prostate, breast, and ovary because of monoallelic mutations (Qu et al. 2003). However, the presence of heterozygosity in mice for the *beclin-1* gene makes it cancer prone (Qu et al. 2003; Yue et al. 2003) due to absence of functional of Beclin-1.

Cancer progression shows a degree of dependency on the existence of CSCs. The role of autophagy in cancer is multifaceted and has been studied extensively. High levels of autophagy contribute to pluripotency of CSCs in other cancer types, including colorectal cancer (Kantara et al. 2014), pancreatic cancer (Rausch et al. 2012; Viale et al. 2014), glioblastoma (Galavotti et al. 2013), chronic myeloid leukemia (Bellodi et al. 2009), and bladder cancer (Ojha et al. 2016). Despite recent advancement in research, the underlying molecular mechanism inducing autophagy in CSCs remains to be determined. It is difficult to explain how autophagy promotes stemness, have been preserved across different cancer. Mitophagy is a selective autophagy that unambiguously plays an important role in the quality control and homeostasis of mitochondria. Mitochondrial functional pathways play a crucial role in a vital interaction between cancer cells and stromal cells for cancer cell initiation, progression, and treatment response. They emanate a profound role in sustaining CSCs in adverse conditions and initiating their metabolic reprogramming to support the increased bioenergetic demand of the tumor. Transcription factors like SMAD

(Nazio et al. 2019), NF-Kb (Zhang et al. 2016), MITF (Moller et al. 2019), STAT3 (Marcucci et al. 2017; Zhang et al. 2016), FOXO (Naka et al. 2010), ATF4 (Pallmann et al. 2019), NANOG (Liu et al. 2017), regulate autophagy and mitophagy in the induction of EMT and maintenance of CSCs. Like autophagy, mitophagy acts in cancer as bimodal processes. Unfortunately, there are unanswered roles of canonical autophagy in cancer (Gewirtz 2014). Therefore, does mitophagy has a role in cancer? CSCs play an unbiased role in promoting therapy resistance leading to tumor recurrence (Shibue and Weinberg 2017), and autophagy deliberately endorses disseminated tumor cells (DTCs) which further lead to the metastatic expansion of tumors (Sosa et al. 2014). To understand how autophagy and mitophagy can inhibit to repress both the above phenotypes are challenging task for translational cancer. Recent studies have linked CSCs with chemoresistance and cancer relapse, autophagy, mitophagy, and CSCs showcase novel perspectives on potential therapeutic targets for enhancing anticancer drug sensitivity. The study of autophagy in cancer has been therapeutically manipulated by many investigators and various clinical trials that are already ongoing to regulate the result of disease therapy.

10.2 Autophagy/Mitophagy Drives Cancer Stem Cells Fate

CSCs are a heterogeneous population; they escalate tumor growth and progression by accelerating the proliferative potential and constitute a source for recurrence of cancer. Functional properties of cancer cells are influenced by epigenetic, genetic, and microenvironmental factors. To proliferate in its microenvironment, CSCs have a functional correlation with autophagy and mitophagy. Autophagy, a catabolic pathway enables CSCs to show autophagy dependence and may act as an oncosuppressive depending on tumor stage and type. They exploit the pro-survival attribute of autophagy at the later stage of oncogenesis to meet up with high-energy demands by a supply of metabolites. ATG-encoded gene products play a significant role in CSCs of numerous cancers. Beclin 1/Atg6 modulates CSC plasticity and tumorigenesis in vivo. However, in different cancers, Beclin 1 acts as a tumor suppressor, like human prostate, breast, and ovarian tumors (Liang et al. 1999; Qu et al. 2003; Shen et al. 2008). Improved survival in patients is observed having high Beclin 1 levels affected by large B-cell lymphoma, high-grade gliomas, or hepatocellular carcinoma (Ding et al. 2008; Huang et al. 2011; Pirtoli et al. 2009). The stemness was augmented by the transformation of $CD133^-$ to $CD133^+$ cells due to the inhibition of mTOR affecting the liver tumor cells by interrupting the differentiation and stimulating the tumor development in vivo (Yang et al. 2011). Suppression of autophagy by knockdown of autophagic proteins Atg5 and Atg7, curtails stemness markers, such as Sox2, Nanog, and Oct4, resulting colorectal CSCs to undergo suppressed cell proliferation and improved cell senescence (Sharif et al. 2017). In colorectal cancers, mutations in Atg5, Atg12 have been described (Kang et al. 2009) while deletion of Atg5 or Atg7 is supporting the advancement of liver hepatomas (Takamura et al. 2011). Autophagy induction by overexpressing Atg4A

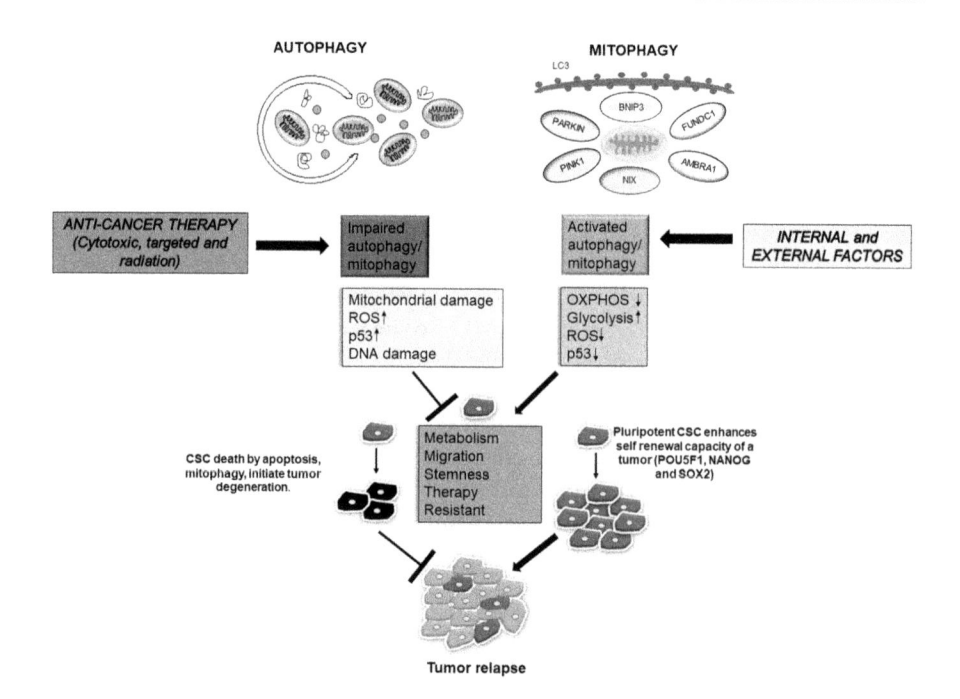

Fig. 10.1 The basal level autophagy and mitophagy are important for cell metabolism. When there is stress due to anticancer therapy, autophagy, and mitophagy get impaired, while they are activated due to internal and external factors leading to either suppression or progression of cancer

protein promotes mammosphere formation and hence increases CSC numbers and in vivo tumorigenesis (Wolf et al. 2013). The conditional knockout of Atg3 affected the continued existence of CML cells and leukemogenesis (Altman et al. 2011). Inhibition of Atg4B resulted in its increased phosphorylation followed by arresting the tumor growth in animal models in a subset of glioblastoma cancer (Huang et al. 2017). The depletion of ATG4B impaired the survivability of CML stem/progenitor cells (Rothe et al. 2014). Knockout of Atg4C in mice increased the propensity to develop fibrosarcomas induced by methylcholanthrene, hence play a tumor-suppressor role. Contrastingly, its tumorigenic role in breast cancer was delineated (Antonelli et al. 2017). Tumor suppressive role of Atg4D expression was observed in colorectal carcinogenesis (Gil et al. 2018). Moreover, its tumor-promoting role was highlighted when cancer cells were sensitized to chemotherapeutic drugs on ATG4D silencing (Betin and Lane 2009) (Fig. 10.1).

EMT (epithelial to mesenchymal transition) signaling is an important characteristic of CSCs (Shibue and Weinberg 2017). Autophagy signaling is strongly correlated to EMT in enhancing the metastatic potential of CSCs to migrate by maintaining their mesenchymal signature in the later stages of metastasis. Interestingly, during early metastasis autophagy decreases the invasion and migration of tumor cells in situ. In glioblastoma cells, blocked cell migration and invasion were caused by nutrient deprivations and mTOR inhibition (Catalano et al. 2015). Using specific siRNAs directed against the autophagy-related factors DRAM1 and p62 proteins, autophagy-controlled bioenergetic metabolism, migration/invasion of glioblastoma CSCs was thwarted while the mesenchymal phenotype was restored on autophagy upregulation (Galavotti et al. 2013). Furthermore, in glioblastoma cells, enhanced migration and invasion with EMT regulators continued with knockdown of Beclin 1, Atg5, and Atg7 (Catalano et al. 2015). EMT promotes stemness and can give rise to CSCs through the core stemness factors POU5F1, Sox2, and Nanog, including Slug and Twist that maintains the pluripotency of CSCs and tumor-propagating properties (Mani et al. 2008). Hypoxia and TGF-β through MITF (Caramel et al. 2013), Sox2, and Nanog (Sharif et al. 2017) promote EMT via activating autophagy. Autophagy may promote tumor cell dormancy, lipid metabolism, mitochondrial function, and CSCs existence in muscle stem cells and HSCs (Ho et al. 2017; Warr et al. 2013). It ensures a reversible dormant pool of CSCs potentially making a contribution to tumor repopulation and preventing irreversible senescence (Ho et al. 2017). Autophagy plays a decisive role in the survival of disseminated tumor cells (DTCs) at secondary location to establish drug resistance, minimum residual disease, and metastatic dormancy (Sosa et al. 2014). Interestingly, these DTCs are CSCs that are relatively quiescent and motile state expressing upregulated CSC markers in the bone marrow of breast cancer patients (Balic et al. 2006). Furthermore, a selective form of autophagy known as mitophagy promotes stemness. It abrogates senescence by disrupting the ROS-induced DNA damage and has a principal role in maintaining the stem cell population renewal and homeostasis. It has been reported to maintain hepatic CSCs by regulating p53 localization. Therefore, inhibition of mitophagy phosphorylates p53 by PINK1 leading to its translocation to the nucleus where Oct4 and Sox2 induction of Nanog get alienated. Mitophagy evokes CSCs dependence more on glycolysis for energy needs and hence contributes to its quiescent state. Recent evidence suggests that mitochondrial dysfunction also encourages oncogenesis (Boya et al. 2018). Mitochondrial ROS due to BNIP3 loss subsequently resulted from defects in mitophagy followed by mammary neoplastic progression to metastasis (Chourasia et al. 2015) (Table 10.1).

10.3 Targeting Autophagy/Mitophagy: New Therapeutic Strategies

CSC generation, differentiation, plasticity, migration/invasion, and immune resistance are very much dependent on the variation of autophagy/mitophagy. During anticancer therapy, CSCs remain at the dormant stage to cope with intracellular and environmental stress, involving oxidative stress triggered by overproduction of reactive oxygen species (ROS). These dormant cells arise from EMT tumor cells and become non-cycling autophagic CSC which are later maneuvered on the release of paracrine factors (like MET, TGF-β receptor, IL-6 receptor, PDGFR, EGFR, FGFR, Hedgehog/Smoothened, WNT/Frizzled, Gas6/AXL, and Notch ligands) to

Table 10.1 Role of autophagy in different types of cancer and genes targeted for anticancer therapy

Types of cancer	Animal model/ cells/CSCs	Autophagy as protective or lethal or both	Targeted genes involved in the induction of autophagy	References
Neuroblastoma, multiple myeloma cells	SH-SY5Y cells	Protective	NAMPT	Billington et al. (2008), Cea et al. (2012), Ghosh and Matsui (2009), Schneider et al. (2011), Sharif et al. (2017)
Colorectal cancer	HCT116, HT29, CaCO2, and DLD1CSCs; DCLK1-positive colon CSCs	Protective	Endolysosomal RAB5/7 regulating mitophagic pathway; LC3, Beclin1, Atg6	Kantara et al. (2014), Takeda et al. (2019)
Malignant pluripotent embryonal carcinoma	NT2/D1 CSCs	Protective	NAMPT	Sharif et al. (2017)
Breast cancer	MCF-7 CSCs; SUM149 CSCs	Both protective and lethal in MCF-7 and protective in SUM149	Protective: Beclin1, c-Jun NH2 terminal kinase (JNK/SAPK) in MCF-7, and Atg4A in SUM149 Lethal: Beclin1, Akt/mammalian target of rapamycin (mTOR) pathway in MCF-7	Protective: MCF-7 (Chaterjee and van Golen 2011; Sanchez et al. 2011) and SUM149 (Wolf et al. 2013) Lethal: MCF-7 (Liang et al. 1999; Lu et al. 2014)
Prostate and breast cancer	PC-3 and DU145 cells; MDA-MB-231 cells	Lethal	AMP-associated protein kinase (AMPK)/Unc-51 like autophagy activating kinase 1 (ULK1) pathway and inhibition of mTOR/Raptor complex 1 expression	Aryal et al. (2014)

(continued)

Table 10.1 (continued)

Types of cancer	Animal model/ cells/CSCs	Autophagy as protective or lethal or both	Targeted genes involved in the induction of autophagy	References
Pancreatic cancer	$CD133^+$ pancreatic CSCs, BxPc-3 (CSC^{low}) and MIA-PaCa2 (CSC^{high}), inducible mouse model of mutated *Kras*	Protective	HIF-1a; Beclin1, Atg4B, LC3, p62; AMPK, LC3	Rausch et al. (2012), Viale et al. (2014), Zhu et al. (2013)
Urinary bladder cancer	T24 and UM-UC-3 CSCs; T24 CSCs	Protective	Beclin1, Atg7, and p62; IFN-γ-mediated JAK2 and STAT3 pathway	Ojha et al. (2014), Ojha et al. (2016)
Brain tumor	CSCs: MDNSC11, MDNSC13, MDNSC23, MDNSC16; GBM stem cells–GSCs	Lethal and protective	p16INK4/Rb pathway, Atg5; DRAM1, SQSTM1, p62	Galavotti et al. (2013), Jiang et al. (2007)
Chronic myeloid leukemia	$p210^{BCR/ABL}$-expressing CML cells, CML lymphoid BC cell line BV173, K562 cells	Protective	LC3, Atg5, Atg7	Bellodi et al. (2009)

cycling CSC with low autophagy. Thus, autophagy and mitophagy enable CSCs to colonize, migrate and metastasize, defy apoptosis and antitumor drugs and hence become therapy-resistant by its self-renewal property and replace the pool of differentiated tumor cells (Marcucci et al. 2017) (Table 10.2).

Autophagy/mitophagy has an inevitable role in cancer cell survival, metastasis, and therapy resistance. The potentially new targeted therapeutic strategy is to use double or triple combinatorial doses of drugs or antibodies and/or radiation to modulate autophagic machinery to efficiently eradicate CSCs. Chemotherapy is a widespread treatment strategy for cancer therapy that engulfs dividing cells and disrupts cancer–cell division. However, several studies have revealed that the overall success rate of chemotherapy is often restricted via the upregulation of cytoprotective activation of autophagy in CSCs which protects cancer cells subjected to anticancer therapy. Cancer chemotherapeutic drugs 5-Fluorouracil (5FU) and cisplatin used in various solid cancers, like, gallbladder and colorectal cancers show autophagy-regulated chemoresistance (Ferreira et al. 2016; Liang et al. 2014;

Table 10.2 Anticancer therapy, targeted autophagy induced molecular pathway, and their current clinical status

Anticancer therapy	Type of cancer	Molecular pathway targeted	Current status	References
Liensinine, an extracted from the seed embryo of *Nelumbo nucifera* Gaertn	Breast cancer cells (MDA-MB-231 and MCF-7)	Inhibition of autophagosome–lysosome fusion	Preclinical trials	Zhou et al. (2015)
Resveratrol, a natural phenol	Breast CSCs (MCF-7 and SUM159)	Suppression of the Wnt/β-catenin signaling pathway	Preclinical trials	Fu et al. (2014)
Mefloquine, an anti-parasite used to treat malaria	Colorectal cancer cell lines (HCT116, HT29, CaCO$_2$, and DLD1)	Inhibited lysosomal activity by targeting RAB5 and RAB7 resulted in suppression of mitophagic PINK1/PARKIN	Preclinical trials	Takeda et al. (2019)
Metformin, used for the treatment of type 2 diabetes	Pancreatic intraepithelial neoplasia (PanIN)	Modulating the mTOR signaling pathway	Preclinical trials	Mohammed et al. (2013)
Rottlerin, a polyphenolic compound	Metastatic colorectal cancer cell lines (Tu12, Tu21, and Tu22 cells) breast CSCs, prostate CSCs (human prostate tumor samples)	mTOR inhibition; AMPK activation and proteasome inhibition; AMPK activation and inhibition of PI3K/Akt/mTOR pathway	Preclinical trials; preclinical trials; clinical + preclinical trials	Francipane and Lagasse (2013), Kumar et al. (2013, 2014)
Combinatorial treatment of Sorafenib, a quizartinib (AC220) and crenolamib, a FLT3-ITD inhibitors	AML	Lethal mitophagy	Preclinical trials	Stein and Tallman (2016)
Combinatorial treatment of FH535, a synthetic inhibitor of the Wnt/β-catenin pathway; FH535-N, a derivative of FH535; and sorafenib	HCC cell line (Huh7, Hep3B, and PLC)	Wnt/β-catenin pathway	Preclinical trials	Turcios et al. (2019)
LCL-461, a mitochondria-targeted ceramide analog drug	FLT3-ITD + AML	Lethal mitophagy	Preclinical trials	Dany et al. (2016)

CerS1/C18-ceramide, a central molecule of sphingolipid metabolism	Squamous head and neck cancer cell lines (UM-SCC-22A and UM-SCC-22B)	Lethal mitophagy	Preclinical trials	Sentelle et al. (2012)
Doxorubicin, a DNA damaging agent	Colorectal cancer stem cells (HCT8)	Inhibition of mitophagy by *BNIP3L* silencing	Preclinical trials	Yan et al. (2017)
Salinomycin, an antibacterial and coccidiostat ionophore drug	Prostate cancer cell line (PC3), breast cancer cell lines (SKBR3 and MDA-MB468)	Mitochondrial hyperpolarization	Preclinical trials	Jangamreddy et al. (2013)
UNBS1450, a sodium channel antagonist	Stromal neuroblastoma SH-N-AS cell line	Inhibition of autophagy by small inhibitory RNAs targeting Atg5, autophagy related 7 (Atg7), and Beclin-1	Preclinical trials	Radogna et al. (2016)
Combinatorial treatment of temozolomide, a DNA alkylating agent and ganciclovir, a synthetic guanine derivative	Glioblastoma multiforme	Arrested tumor growth	Preclinical trials	Chen et al. (2012)
Combinatorial treatment of photosan-II (PS-II)-mediated photodynamic therapy (PS-PDT) and autophagy inhibitors	Colorectal cancer cell lines (SW620 and HCT116)	Downregulated AKT-mTOR pathway	Preclinical trials	Xiong et al. (2017)
Rapamycin, an inhibitor of the mammalian target of rapamycin (mTOR)	Pulmonary diseases	Inhibition of mTOR	Clinical trials: NCT01462006	Galluzzi et al. (2017)
Temsirolimus, a rapamycin analog	Renal cell carcinoma	Inhibition of mTOR	Clinical trials: NCT01404104	Kwitkowski et al. (2010)
Everolimus, a rapamycin analog	Pancreatic NET	Inhibition of mTOR	Clinical trials: NCT02305810	Yao et al. (2013)

(continued)

Table 10.2 (continued)

Anticancer therapy	Type of cancer	Molecular pathway targeted	Current status	References
Combinatorial treatment of SAR405, a kinase inhibitor of Vps18 and Vps34, and everolimus	Renal cancer cell lines (ACHN and 786-O)	Impairs lysosomal function inhibition of cancer proliferation	Preclinical trials	Ronan et al. (2014)
Chloroquine or hydroxychloroquine, an aminoquinoline-a late-stage autophagy inhibitor	Bladder cancer cell lines (5637 and T24) and pancreatic adenocarcinoma cell lines MiaPaCa2 (nonmetastatic) and S2VP10 (metastatic)	Targeting basal autophagy	Preclinical trials	Frieboes et al. (2014), Lin et al. (2017)
Combinatorial treatment of temsirolimus and hydroxychloroquine	Advanced solid tumors and melanoma	Temsirolimus-associated induction of autophagy and hydroxychloroquine-associated block in the clearance of autophagic vacuoles	Clinical: NCT00909831 + preclinical trials	Rangwala et al. (2014)
Combinatorial treatment of Lys05, a water-soluble analog of hydroxychloroquine and PLX4720, a BRaf kinase inhibitor	Melanoma (MEL624)	Lysosomal autophagy inhibition	Preclinical trials	Ma et al. (2014)
Quinacrine (DQ661), derived from Lys05	Melanoma, colon cancer, and breast cancer	Inhibition of PPT1 causing mTOR inhibition	Preclinical trials	Rebecca et al. (2019)
Combinatorial treatment of vemurafenib, an inhibitor of the BRaf enzyme, and hydroxychloroquine	Metastatic BRaf V600E + melanoma	Inhibition of PERK arm of the ER stress response	Clinical trials: NCT01897116	Ma et al. (2014)
Combinatorial treatment of chloroquine or hydroxychloroquine and gemcitabine, an antimetabolite	Pancreatic cancer stem cells (primary PDAC tumors)	Inhibition of CXCL12/CXCR4 signaling	Clinical: NCT01777477, NCT01128296 + preclinical trials	Balic et al. (2014)

10 Relevance of Autophagy in Cancer Stem Cell and Therapeutic

Combinatorial treatment of chloroquine or bafilomycin A1, another late-stage autophagy inhibitor with IM and nilotinib, or dasatinib, a tyrosine kinase inhibitor	Chronic myeloid leukemia (K562, BV173, IM-treated 32D-p210$^{BCR/ABL}$)	Inhibition of autophagy	Preclinical trials	Bellodi et al. (2009)
Combinatorial treatment of Bevacizumab, a blocker of EGFR, or Temozolomide, with chloroquine	Glioblastoma stem cells (U87-MG)	Suppressing Akt/mTOR signaling	Preclinical trials	Huang et al. (2018)
Triple combinatorial treatment of 5-fluorouracil, chloroquine, and Notch inhibitor	Gastric CSCs (MGC-803 and MKN-45)	Inhibition of Notch signaling pathway	Preclinical trials	Li et al. (2018)
Triple combinatorial treatment of HCQ, radiation therapy (RT), and temozolomide (TMZ)	Glioblastoma (GB)	Autophagy inhibition	Clinical trials: NCT00486603	Rosenfeld et al. (2014)
Triple combinatorial treatment of chloroquine, PI3K/Akt pathway inhibitor along with gamma-irradiation	Primary stem-like glioma cells.	Inhibition of late autophagy	Clinical + preclinical trials	Firat et al. (2012)
Spautin-1, a novel autophagy inhibitor	Imatinib mesylate (IM)-resistant chronic myeloid leukemia (CML)	Inhibition of PI3K/AKT	Preclinical trials	Shao et al. (2014)
Combinatorial treatment of Dofequidar Fumarate, an inhibitor of ABC transporters and Docetaxel	Chronic myeloid leukemia (K562), breast cancer (BSY-1, HBC-4, and HBC-5), glioma human (U251), pancreatic cancer (Capan-1), colon cancer (KM12), and stomach cancer (MKN74)	Inhibit the efflux of chemotherapeutic drugs and increase the sensitivity to anticancer drugs in CSCs	Clinical: NCT00004886 + preclinical trials	Katayama et al. (2009)

(continued)

Table 10.2 (continued)

Anticancer therapy	Type of cancer	Molecular pathway targeted	Current status	References
Combinatorial treatment of Demcizumab, humanized monoclonal antibody with Pembrolizumab, a humanized antibody used in cancer immunotherapy	Metastatic solid tumors	Inhibits Delta-like ligand 4 (DLL4) in the Notch signaling pathway, target CSCs, bulk tumor, and angiogenesis	Clinical trials: NCT02722954	Previs et al. (2015)

Park et al. 2013). Additionally, the CXCL12/CXCR4 axis is prompted in colorectal cancer and is linked with potential progression of cancer, such as invasion, metastasis, and chemoresistance. Subsequently, grants 5-fluorouracil (5-FU) resistance by increasing autophagy *both* in vitro and in vivo (Yu et al. 2017).

Suppression of autophagy preferentially stimulated in multiple molecular pathways that govern CSCs growth and differentiation, includes Notch (Li et al. 2018), Sonic Hedgehog (Fan et al. 2019), Wnt/β-catenin (Pai et al. 2017), NF-kβ (Trocoli and Djavaheri-Mergny 2011), transforming growth factor-β (Kiyono et al. 2009), and fibroblast growth factor (Chen et al. 2018) signaling cascades lead to sensitization of cancer cells to anticancer therapy. The appreciating effect of the Wnt/β-catenin pathway is inhibited by FH535 and its derivative (FH535-N) alone and in combination with sorafenib through nullification of the autophagic flux in hepatocellular carcinoma (Turcios et al. 2019). Hyperactivation of PI3K/Akt/mTOR pathway in GBM and its inhibition exerts antineoplastic activity by targeting CSCs, supporting differentiation, and inhibiting cell migration and invasion prospective of GSCs (Li et al. 2016). Balance is the key between Beclin1 and Bcl2/Bcl-xL that supports the concept of the presence of a complex relationship between autophagy and apoptosis, which seems important in the context of cancer and cancer therapy (Kim et al. 2014). JNK-mediated protective autophagy increased Bcl2 expression followed by an increased autophagic flux and conferred chemoresistance in colon cancer (Sui et al. 2014).

Evolving clinical and experimental evidence indicates that CSCs have clinical significance as they are bestowed with intrinsic resistance to radio- and chemotherapy owing to the indulgence of autophagy (Chen et al. 2012; Vitale et al. 2015). Targeting components of the autophagic machinery can be recruited as the hopeful target to selectively eliminate CSCs facilitating cancer cell growth/progression/metastasis and enhancing the effectiveness of radio- and chemotherapy (Nazio et al. 2019; Ojha et al. 2015; Perez-Hernandez et al. 2019). Henceforth, these findings completely indicate that autophagy suppression and its activation, both, can be deemed to be promising approaches for sensitizing CSCs to anticancer therapy, evaluated by the reduction of the number of CSCs. So, the development of new anticancer drugs focuses on CSCs which is key to the problem required to be resolved in drug clinical trials (Fig. 10.2).

10.4 Conclusion

Development of autophagy inhibitors, specific mitophagy inhibitors have been proven beneficial, given the fears about global autophagy suppression for tissue homeostasis and that mitophagy has a crucial functional role earlier credited to general autophagy. Focusing on selective inhibitors will pave an unexplored path of how autophagy is responsible for determining stemness, dormancy-whether DTCs are autophagy-dependent CSCs, and which autophagy functions will be significant in promoting drug resistance and cancer recurrence. Further research is requisite before CSCs can be treated by regulating autophagy and mitophagy.

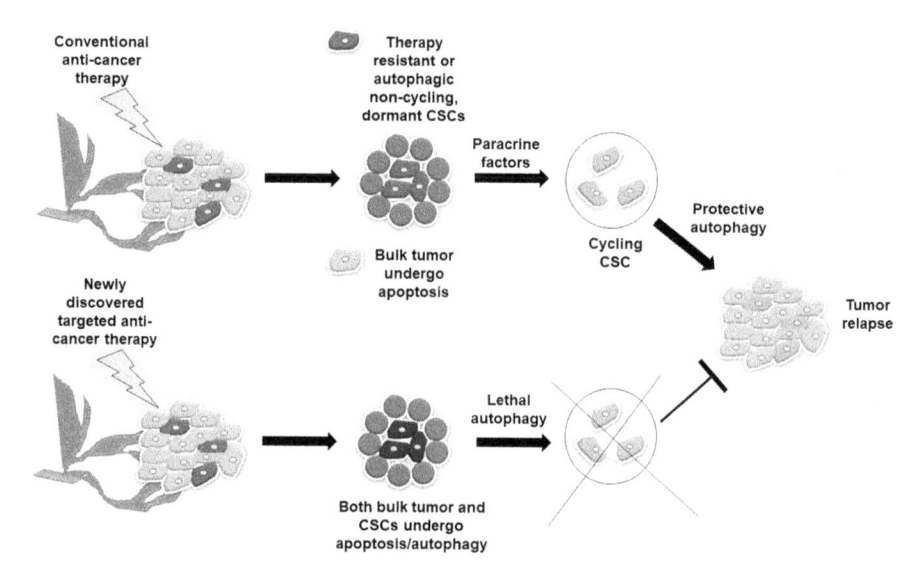

Fig. 10.2 The conventional anticancer therapy is incapable to target the CSCs that shoot to cancer relapse. Autophagy plays a Janus role in cancer cell modulation, acts protective during tumor relapse, and lethal via programmed cell death. Newly discovered combinatorial treatments target both the bulk tumor and cancer stem cells leading to elimination of persistent CSCs and tumor regression

Desirable therapeutic impacts of anticancer reagents have not been achieved by only targeting autophagy using autophagy modulators; to the contrary, it has enacted as a pro-survival response by supplying nutrients to cancer cells. Consequently, clinical trials that aim autophagy by a combination of autophagy alterations and anticancer components are appropriate to consider autophagy as a possible effectual therapeutic approach in anticancer therapy. The conjunction of these techniques hopefully deciphers the vital mechanisms necessary for maintaining cancer stemness and will play an important role in designing more efficient and effective personalized therapeutic strategies.

References

Altman BJ, Jacobs SR, Mason EF, Michalek RD, MacIntyre AN, Coloff JL, Ilkayeva O, Jia W, He YW, Rathmell JC (2011) Autophagy is essential to suppress cell stress and to allow BCR-Abl-mediated leukemogenesis. Oncogene 30:1855–1867

Antonelli M, Strappazzon F, Arisi I, Brandi R, D'Onofrio M, Sambucci M, Manic G, Vitale I, Barila D, Stagni V (2017) ATM kinase sustains breast cancer stem-like cells by promoting ATG4C expression and autophagy. Oncotarget 8:21692–21709

Aponte PM, Caicedo A (2017) Stemness in cancer: stem cells, cancer stem cells, and their microenvironment. Stem Cells Int 2017:5619472

Aryal P, Kim K, Park PH, Ham S, Cho J, Song K (2014) Baicalein induces autophagic cell death through AMPK/ULK1 activation and downregulation of mTORC1 complex components in human cancer cells. FEBS J 281:4644–4658

Balic M, Lin H, Young L, Hawes D, Giuliano A, McNamara G, Datar RH, Cote RJ (2006) Most early disseminated cancer cells detected in bone marrow of breast cancer patients have a putative breast cancer stem cell phenotype. Clin Cancer Res 12:5615–5621

Balic A, Sorensen MD, Trabulo SM, Sainz B Jr, Cioffi M, Vieira CR, Miranda-Lorenzo I, Hidalgo M, Kleeff J, Erkan M, Heeschen C (2014) Chloroquine targets pancreatic cancer stem cells via inhibition of CXCR4 and hedgehog signaling. Mol Cancer Ther 13:1758–1771

Bellodi C, Lidonnici MR, Hamilton A, Helgason GV, Soliera AR, Ronchetti M, Galavotti S, Young KW, Selmi T, Yacobi R, Van Etten RA, Donato N, Hunter A, Dinsdale D, Tirro E, Vigneri P, Nicotera P, Dyer MJ, Holyoake T, Salomoni P, Calabretta B (2009) Targeting autophagy potentiates tyrosine kinase inhibitor-induced cell death in Philadelphia chromosome-positive cells, including primary CML stem cells. J Clin Invest 119:1109–1123

Betin VM, Lane JD (2009) Caspase cleavage of Atg4D stimulates GABARAP-L1 processing and triggers mitochondrial targeting and apoptosis. J Cell Sci 122:2554–2566

Billington RA, Genazzani AA, Travelli C, Condorelli F (2008) NAD depletion by FK866 induces autophagy. Autophagy 4:385–387

Boya P, Codogno P, Rodriguez-Muela N (2018) Autophagy in stem cells: repair, remodelling and metabolic reprogramming. Development 145:dev146506

Caramel J, Papadogeorgakis E, Hill L, Browne GJ, Richard G, Wierinckx A, Saldanha G, Osborne J, Hutchinson P, Tse G, Lachuer J, Puisieux A, Pringle JH, Ansieau S, Tulchinsky E (2013) A switch in the expression of embryonic EMT-inducers drives the development of malignant melanoma. Cancer Cell 24:466–480

Catalano M, D'Alessandro G, Lepore F, Corazzari M, Caldarola S, Valacca C, Faienza F, Esposito V, Limatola C, Cecconi F, Di Bartolomeo S (2015) Autophagy induction impairs migration and invasion by reversing EMT in glioblastoma cells. Mol Oncol 9:1612–1625

Cea M, Cagnetta A, Fulciniti M, Tai YT, Hideshima T, Chauhan D, Roccaro A, Sacco A, Calimeri T, Cottini F, Jakubikova J, Kong SY, Patrone F, Nencioni A, Gobbi M, Richardson P, Munshi N, Anderson KC (2012) Targeting NAD+ salvage pathway induces autophagy in multiple myeloma cells via mTORC1 and extracellular signal-regulated kinase (ERK1/2) inhibition. Blood 120:3519–3529

Chaterjee M, van Golen KL (2011) Breast cancer stem cells survive periods of farnesyl-transferase inhibitor-induced dormancy by undergoing autophagy. Bone Marrow Res 2011:362938

Chen J, Li Y, Yu TS, McKay RM, Burns DK, Kernie SG, Parada LF (2012) A restricted cell population propagates glioblastoma growth after chemotherapy. Nature 488:522–526

Chen CH, Changou CA, Hsieh TH, Lee YC, Chu CY, Hsu KC, Wang HC, Lin YC, Lo YN, Liu YR, Liou JP, Yen Y (2018) Dual inhibition of PIK3C3 and FGFR as a new therapeutic approach to treat bladder cancer. Clin Cancer Res 24:1176–1189

Chourasia AH, Tracy K, Frankenberger C, Boland ML, Sharifi MN, Drake LE, Sachleben JR, Asara JM, Locasale JW, Karczmar GS, Macleod KF (2015) Mitophagy defects arising from BNip3 loss promote mammary tumor progression to metastasis. EMBO Rep 16:1145–1163

Dany M, Gencer S, Nganga R, Thomas RJ, Oleinik N, Baron KD, Szulc ZM, Ruvolo P, Kornblau S, Andreeff M, Ogretmen B (2016) Targeting FLT3-ITD signaling mediates ceramide-dependent mitophagy and attenuates drug resistance in AML. Blood 128:1944–1958

Ding ZB, Shi YH, Zhou J, Qiu SJ, Xu Y, Dai Z, Shi GM, Wang XY, Ke AW, Wu B, Fan J (2008) Association of autophagy defect with a malignant phenotype and poor prognosis of hepatocellular carcinoma. Cancer Res 68:9167–9175

Fan J, Zhang X, Wang S, Chen W, Li Y, Zeng X, Wang Y, Luan J, Li L, Wang Z, Sun X, Shen B, Ju D (2019) Regulating autophagy facilitated therapeutic efficacy of the sonic Hedgehog pathway inhibition on lung adenocarcinoma through GLI2 suppression and ROS production. Cell Death Dis 10:626

Ferreira JA, Peixoto A, Neves M, Gaiteiro C, Reis CA, Assaraf YG, Santos LL (2016) Mechanisms of cisplatin resistance and targeting of cancer stem cells: adding glycosylation to the equation. Drug Resist Updat 24:34–54

Firat E, Weyerbrock A, Gaedicke S, Grosu AL, Niedermann G (2012) Chloroquine or chloroquine-PI3K/Akt pathway inhibitor combinations strongly promote gamma-irradiation-induced cell death in primary stem-like glioma cells. PLoS One 7:e47357

Francipane MG, Lagasse E (2013) Selective targeting of human colon cancer stem-like cells by the mTOR inhibitor Torin-1. Oncotarget 4:1948–1962

Frieboes HB, Huang JS, Yin WC, McNally LR (2014) Chloroquine-mediated cell death in metastatic pancreatic adenocarcinoma through inhibition of autophagy. JOP 15:189–197

Fu Y, Chang H, Peng X, Bai Q, Yi L, Zhou Y, Zhu J, Mi M (2014) Resveratrol inhibits breast cancer stem-like cells and induces autophagy via suppressing Wnt/beta-catenin signaling pathway. PLoS One 9:e102535

Galavotti S, Bartesaghi S, Faccenda D, Shaked-Rabi M, Sanzone S, McEvoy A, Dinsdale D, Condorelli F, Brandner S, Campanella M, Grose R, Jones C, Salomoni P (2013) The autophagy-associated factors DRAM1 and p62 regulate cell migration and invasion in glioblastoma stem cells. Oncogene 32:699–712

Galluzzi L, Bravo-San Pedro JM, Levine B, Green DR, Kroemer G (2017) Pharmacological modulation of autophagy: therapeutic potential and persisting obstacles. Nat Rev Drug Discov 16:487–511

Gewirtz DA (2014) The four faces of autophagy: implications for cancer therapy. Cancer Res 74:647–651

Ghiaur G, Gerber JM, Matsui W, Jones RJ (2012) Cancer stem cells: relevance to clinical transplantation. Curr Opin Oncol 24:170–175

Ghosh N, Matsui W (2009) Cancer stem cells in multiple myeloma. Cancer Lett 277:1–7

Gil J, Ramsey D, Pawlowski P, Szmida E, Leszczynski P, Bebenek M, Sasiadek MM (2018) The influence of tumor microenvironment on ATG4D gene expression in colorectal cancer patients. Med Oncol 35:159

Ho TT, Warr MR, Adelman ER, Lansinger OM, Flach J, Verovskaya EV, Figueroa ME, Passegue E (2017) Autophagy maintains the metabolism and function of young and old stem cells. Nature 543:205–210

Huang JJ, Zhu YJ, Lin TY, Jiang WQ, Huang HQ, Li ZM (2011) Beclin 1 expression predicts favorable clinical outcome in patients with diffuse large B-cell lymphoma treated with R-CHOP. Hum Pathol 42:1459–1466

Huang T, Kim CK, Alvarez AA, Pangeni RP, Wan X, Song X, Shi T, Yang Y, Sastry N, Horbinski CM, Lu S, Stupp R, Kessler JA, Nishikawa R, Nakano I, Sulman EP, Lu X, James CD, Yin XM, Hu B, Cheng SY (2017) MST4 phosphorylation of ATG4B regulates autophagic activity, tumorigenicity, and radioresistance in glioblastoma. Cancer Cell 32(840–855):e848

Huang H, Song J, Liu Z, Pan L, Xu G (2018) Autophagy activation promotes bevacizumab resistance in glioblastoma by suppressing Akt/mTOR signaling pathway. Oncol Lett 15:1487–1494

Jangamreddy JR, Ghavami S, Grabarek J, Kratz G, Wiechec E, Fredriksson BA, Rao Pariti RK, Cieslar-Pobuda A, Panigrahi S, Los MJ (2013) Salinomycin induces activation of autophagy, mitophagy and affects mitochondrial polarity: differences between primary and cancer cells. Biochim Biophys Acta 1833:2057–2069

Jiang H, Gomez-Manzano C, Aoki H, Alonso MM, Kondo S, McCormick F, Xu J, Kondo Y, Bekele BN, Colman H, Lang FF, Fueyo J (2007) Examination of the therapeutic potential of Delta-24-RGD in brain tumor stem cells: role of autophagic cell death. J Natl Cancer Inst 99:1410–1414

Kang MR, Kim MS, Oh JE, Kim YR, Song SY, Kim SS, Ahn CH, Yoo NJ, Lee SH (2009) Frameshift mutations of autophagy-related genes ATG2B, ATG5, ATG9B and ATG12 in gastric and colorectal cancers with microsatellite instability. J Pathol 217:702–706

Kantara C, O'Connell M, Sarkar S, Moya S, Ullrich R, Singh P (2014) Curcumin promotes autophagic survival of a subset of colon cancer stem cells, which are ablated by DCLK1-siRNA. Cancer Res 74:2487–2498

Katayama R, Koike S, Sato S, Sugimoto Y, Tsuruo T, Fujita N (2009) Dofequidar fumarate sensitizes cancer stem-like side population cells to chemotherapeutic drugs by inhibiting ABCG2/BCRP-mediated drug export. Cancer Sci 100:2060–2068

Kim SY, Song X, Zhang L, Bartlett DL, Lee YJ (2014) Role of Bcl-xL/Beclin-1 in interplay between apoptosis and autophagy in oxaliplatin and bortezomib-induced cell death. Biochem Pharmacol 88:178–188

Kiyono K, Suzuki HI, Matsuyama H, Morishita Y, Komuro A, Kano MR, Sugimoto K, Miyazono K (2009) Autophagy is activated by TGF-beta and potentiates TGF-beta-mediated growth inhibition in human hepatocellular carcinoma cells. Cancer Res 69:8844–8852

Kumar D, Shankar S, Srivastava RK (2013) Rottlerin-induced autophagy leads to the apoptosis in breast cancer stem cells: molecular mechanisms. Mol Cancer 12:171

Kumar D, Shankar S, Srivastava RK (2014) Rottlerin induces autophagy and apoptosis in prostate cancer stem cells via PI3K/Akt/mTOR signaling pathway. Cancer Lett 343:179–189

Kwitkowski VE, Prowell TM, Ibrahim A, Farrell AT, Justice R, Mitchell SS, Sridhara R, Pazdur R (2010) FDA approval summary: temsirolimus as treatment for advanced renal cell carcinoma. Oncologist 15:428–435

Li X, Wu C, Chen N, Gu H, Yen A, Cao L, Wang E, Wang L (2016) PI3K/Akt/mTOR signaling pathway and targeted therapy for glioblastoma. Oncotarget 7:33440–33450

Li LQ, Pan D, Zhang SW, Xie D-Y, Zheng XL, Chen H (2018) Autophagy regulates chemoresistance of gastric cancer stem cells via the Notch signaling pathway. Eur Rev Med Pharmacol Sci 22:3402–3407

Liang XH, Jackson S, Seaman M, Brown K, Kempkes B, Hibshoosh H, Levine B (1999) Induction of autophagy and inhibition of tumorigenesis by beclin 1. Nature 402:672–676

Liang X, Tang J, Liang Y, Jin R, Cai X (2014) Suppression of autophagy by chloroquine sensitizes 5-fluorouracil-mediated cell death in gallbladder carcinoma cells. Cell Biosci 4:10

Lin YC, Lin JF, Wen SI, Yang SC, Tsai TF, Chen HE, Chou KY, Hwang TI (2017) Chloroquine and hydroxychloroquine inhibit bladder cancer cell growth by targeting basal autophagy and enhancing apoptosis. Kaohsiung J Med Sci 33:215–223

Liu A, Yu X, Liu S (2013) Pluripotency transcription factors and cancer stem cells: small genes make a big difference. Chin J Cancer 32:483–487

Liu K, Lee J, Kim JY, Wang L, Tian Y, Chan ST, Cho C, Machida K, Chen D, Ou JJ (2017) Mitophagy controls the activities of tumor suppressor p53 to regulate hepatic cancer stem cells. Mol Cell 68(281–292):e285

Lu J, Sun D, Gao S, Gao Y, Ye J, Liu P (2014) Cyclovirobuxine D induces autophagy-associated cell death via the Akt/mTOR pathway in MCF-7 human breast cancer cells. J Pharmacol Sci 125:74–82

Ma XH, Piao SF, Dey S, McAfee Q, Karakousis G, Villanueva J, Hart LS, Levi S, Hu J, Zhang G, Lazova R, Klump V, Pawelek JM, Xu X, Xu W, Schuchter LM, Davies MA, Herlyn M, Winkler J, Koumenis C, Amaravadi RK (2014) Targeting ER stress-induced autophagy overcomes BRAF inhibitor resistance in melanoma. J Clin Invest 124:1406–1417

Mani SA, Guo W, Liao MJ, Eaton EN, Ayyanan A, Zhou AY, Brooks M, Reinhard F, Zhang CC, Shipitsin M, Campbell LL, Polyak K, Brisken C, Yang J, Weinberg RA (2008) The epithelial-mesenchymal transition generates cells with properties of stem cells. Cell 133:704–715

Marcucci F, Ghezzi P, Rumio C (2017) The role of autophagy in the cross-talk between epithelial-mesenchymal transitioned tumor cells and cancer stem-like cells. Mol Cancer 16:3

Mohammed A, Janakiram NB, Brewer M, Ritchie RL, Marya A, Lightfoot S, Steele VE, Rao CV (2013) Antidiabetic drug metformin prevents progression of pancreatic cancer by targeting in part cancer stem cells and mTOR signaling. Transl Oncol 6:649–659

Moller K, Sigurbjornsdottir S, Arnthorsson AO, Pogenberg V, Dilshat R, Fock V, Brynjolfsdottir SH, Bindesboll C, Bessadottir M, Ogmundsdottir HM, Simonsen A, Larue L, Wilmanns M,

Thorsson V, Steingrimsson E, Ogmundsdottir MH (2019) MITF has a central role in regulating starvation-induced autophagy in melanoma. Sci Rep 9:1055

Naka K, Hoshii T, Muraguchi T, Tadokoro Y, Ooshio T, Kondo Y, Nakao S, Motoyama N, Hirao A (2010) TGF-beta-FOXO signalling maintains leukaemia-initiating cells in chronic myeloid leukaemia. Nature 463:676–680

Nazio F, Bordi M, Cianfanelli V, Locatelli F, Cecconi F (2019) Autophagy and cancer stem cells: molecular mechanisms and therapeutic applications. Cell Death Differ 26:690–702

Ojha R, Jha V, Singh SK, Bhattacharyya S (2014) Autophagy inhibition suppresses the tumorigenic potential of cancer stem cell enriched side population in bladder cancer. Biochim Biophys Acta 1842:2073–2086

Ojha R, Bhattacharyya S, Singh SK (2015) Autophagy in cancer stem cells: a potential link between chemoresistance, recurrence, and metastasis. Biores Open Access 4:97–108

Ojha R, Singh SK, Bhattacharyya S (2016) JAK-mediated autophagy regulates stemness and cell survival in cisplatin resistant bladder cancer cells. Biochim Biophys Acta 1860:2484–2497

Pai SG, Carneiro BA, Mota JM, Costa R, Leite CA, Barroso-Sousa R, Kaplan JB, Chae YK, Giles FJ (2017) Wnt/beta-catenin pathway: modulating anticancer immune response. J Hematol Oncol 10:101

Pallmann N, Livgard M, Tesikova M, Zeynep Nenseth H, Akkus E, Sikkeland J, Jin Y, Koc D, Kuzu OF, Pradhan M, Danielsen HE, Kahraman N, Mokhlis HM, Ozpolat B, Banerjee PP, Uren A, Fazli L, Rennie PS, Jin Y, Saatcioglu F (2019) Regulation of the unfolded protein response through ATF4 and FAM129A in prostate cancer. Oncogene 38:6301–6318

Park JM, Huang S, Wu TT, Foster NR, Sinicrope FA (2013) Prognostic impact of Beclin 1, p62/sequestosome 1 and LC3 protein expression in colon carcinomas from patients receiving 5-fluorouracil as adjuvant chemotherapy. Cancer Biol Ther 14:100–107

Perez-Hernandez M, Arias A, Martinez-Garcia D, Perez-Tomas R, Quesada R, Soto-Cerrato V (2019) Targeting autophagy for cancer treatment and tumor chemosensitization. Cancers (Basel) 11:1599

Pirtoli L, Cevenini G, Tini P, Vannini M, Oliveri G, Marsili S, Mourmouras V, Rubino G, Miracco C (2009) The prognostic role of Beclin 1 protein expression in high-grade gliomas. Autophagy 5:930–936

Previs RA, Coleman RL, Harris AL, Sood AK (2015) Molecular pathways: translational and therapeutic implications of the Notch signaling pathway in cancer. Clin Cancer Res 21:955–961

Qu X, Yu J, Bhagat G, Furuya N, Hibshoosh H, Troxel A, Rosen J, Eskelinen EL, Mizushima N, Ohsumi Y, Cattoretti G, Levine B (2003) Promotion of tumorigenesis by heterozygous disruption of the beclin 1 autophagy gene. J Clin Invest 112:1809–1820

Radogna F, Cerella C, Gaigneaux A, Christov C, Dicato M, Diederich M (2016) Cell type-dependent ROS and mitophagy response leads to apoptosis or necroptosis in neuroblastoma. Oncogene 35:3839–3853

Rangwala R, Chang YC, Hu J, Algazy KM, Evans TL, Fecher LA, Schuchter LM, Torigian DA, Panosian JT, Troxel AB, Tan KS, Heitjan DF, DeMichele AM, Vaughn DJ, Redlinger M, Alavi A, Kaiser J, Pontiggia L, Davis LE, O'Dwyer PJ, Amaravadi RK (2014) Combined MTOR and autophagy inhibition: phase I trial of hydroxychloroquine and temsirolimus in patients with advanced solid tumors and melanoma. Autophagy 10:1391–1402

Rausch V, Liu L, Apel A, Rettig T, Gladkich J, Labsch S, Kallifatidis G, Kaczorowski A, Groth A, Gross W, Gebhard MM, Schemmer P, Werner J, Salnikov AV, Zentgraf H, Buchler MW, Herr I (2012) Autophagy mediates survival of pancreatic tumour-initiating cells in a hypoxic microenvironment. J Pathol 227:325–335

Rebecca VW, Nicastri MC, Fennelly C, Chude CI, Barber-Rotenberg JS, Ronghe A, McAfee Q, McLaughlin NP, Zhang G, Goldman AR, Ojha R, Piao S, Noguera-Ortega E, Martorella A, Alicea GM, Lee JJ, Schuchter LM, Xu X, Herlyn M, Marmorstein R, Gimotty PA, Speicher DW, Winkler JD, Amaravadi RK (2019) PPT1 promotes tumor growth and is the molecular target of chloroquine derivatives in cancer. Cancer Discov 9:220–229

Ronan B, Flamand O, Vescovi L, Dureuil C, Durand L, Fassy F, Bachelot MF, Lamberton A, Mathieu M, Bertrand T, Marquette JP, El-Ahmad Y, Filoche-Romme B, Schio L, Garcia-Echeverria C, Goulaouic H, Pasquier B (2014) A highly potent and selective Vps34 inhibitor alters vesicle trafficking and autophagy. Nat Chem Biol 10:1013–1019

Rosenfeld MR, Ye X, Supko JG, Desideri S, Grossman SA, Brem S, Mikkelson T, Wang D, Chang YC, Hu J, McAfee Q, Fisher J, Troxel AB, Piao S, Heitjan DF, Tan KS, Pontiggia L, O'Dwyer PJ, Davis LE, Amaravadi RK (2014) A phase I/II trial of hydroxychloroquine in conjunction with radiation therapy and concurrent and adjuvant temozolomide in patients with newly diagnosed glioblastoma multiforme. Autophagy 10:1359–1368

Rothe K, Lin H, Lin KB, Leung A, Wang HM, Malekesmaeili M, Brinkman RR, Forrest DL, Gorski SM, Jiang X (2014) The core autophagy protein ATG4B is a potential biomarker and therapeutic target in CML stem/progenitor cells. Blood 123:3622–3634

Sanchez CG, Penfornis P, Oskowitz AZ, Boonjindasup AG, Cai DZ, Dhule SS, Rowan BG, Kelekar A, Krause DS, Pochampally RR (2011) Activation of autophagy in mesenchymal stem cells provides tumor stromal support. Carcinogenesis 32:964–972

Schneider L, Giordano S, Zelickson BR, Johnson M, Benavides G, Ouyang X, Fineberg N, Darley-Usmar VM, Zhang J (2011) Differentiation of SH-SY5Y cells to a neuronal phenotype changes cellular bioenergetics and the response to oxidative stress. Free Radic Biol Med 51:2007–2017

Sentelle RD, Senkal CE, Jiang W, Ponnusamy S, Gencer S, Selvam SP, Ramshesh VK, Peterson YK, Lemasters JJ, Szulc ZM, Bielawski J, Ogretmen B (2012) Ceramide targets autophagosomes to mitochondria and induces lethal mitophagy. Nat Chem Biol 8:831–838

Shao S, Li S, Qin Y, Wang X, Yang Y, Bai H, Zhou L, Zhao C, Wang C (2014) Spautin-1, a novel autophagy inhibitor, enhances imatinib-induced apoptosis in chronic myeloid leukemia. Int J Oncol 44:1661–1668

Sharif T, Martell E, Dai C, Kennedy BE, Murphy P, Clements DR, Kim Y, Lee PW, Gujar SA (2017) Autophagic homeostasis is required for the pluripotency of cancer stem cells. Autophagy 13:264–284

Shen Y, Li DD, Wang LL, Deng R, Zhu XF (2008) Decreased expression of autophagy-related proteins in malignant epithelial ovarian cancer. Autophagy 4:1067–1068

Shibue T, Weinberg RA (2017) EMT, CSCs, and drug resistance: the mechanistic link and clinical implications. Nat Rev Clin Oncol 14:611–629

Sosa MS, Bragado P, Aguirre-Ghiso JA (2014) Mechanisms of disseminated cancer cell dormancy: an awakening field. Nat Rev Cancer 14:611–622

Stein EM, Tallman MS (2016) Emerging therapeutic drugs for AML. Blood 127:71–78

Sui X, Kong N, Wang X, Fang Y, Hu X, Xu Y, Chen W, Wang K, Li D, Jin W, Lou F, Zheng Y, Hu H, Gong L, Zhou X, Pan H, Han W (2014) JNK confers 5-fluorouracil resistance in p53-deficient and mutant p53-expressing colon cancer cells by inducing survival autophagy. Sci Rep 4:4694

Takamura A, Komatsu M, Hara T, Sakamoto A, Kishi C, Waguri S, Eishi Y, Hino O, Tanaka K, Mizushima N (2011) Autophagy-deficient mice develop multiple liver tumors. Genes Dev 25:795–800

Takeda M, Koseki J, Takahashi H, Miyoshi N, Nishida N, Nishimura J, Hata T, Matsuda C, Mizushima T, Yamamoto H, Ishii H, Doki Y, Mori M, Haraguchi N (2019) Disruption of endolysosomal RAB5/7 efficiently eliminates colorectal cancer stem cells. Cancer Res 79:1426–1437

Trocoli A, Djavaheri-Mergny M (2011) The complex interplay between autophagy and NF-kappaB signaling pathways in cancer cells. Am J Cancer Res 1:629–649

Turcios L, Chacon E, Garcia C, Eman P, Cornea V, Jiang J, Spear B, Liu C, Watt DS, Marti F, Gedaly R (2019) Autophagic flux modulation by Wnt/beta-catenin pathway inhibition in hepatocellular carcinoma. PLoS One 14:e0212538

Viale A, Pettazzoni P, Lyssiotis CA, Ying H, Sanchez N, Marchesini M, Carugo A, Green T, Seth S, Giuliani V, Kost-Alimova M, Muller F, Colla S, Nezi L, Genovese G, Deem AK, Kapoor A, Yao W, Brunetto E, Kang Y, Yuan M, Asara JM, Wang YA, Heffernan TP,

Kimmelman AC, Wang H, Fleming JB, Cantley LC, DePinho RA, Draetta GF (2014) Oncogene ablation-resistant pancreatic cancer cells depend on mitochondrial function. Nature 514:628–632

Vitale I, Manic G, Dandrea V, De Maria R (2015) Role of autophagy in the maintenance and function of cancer stem cells. Int J Dev Biol 59:95–108

Warr MR, Binnewies M, Flach J, Reynaud D, Garg T, Malhotra R, Debnath J, Passegue E (2013) FOXO3A directs a protective autophagy program in haematopoietic stem cells. Nature 494:323–327

White E, DiPaola RS (2009) The double-edged sword of autophagy modulation in cancer. Clin Cancer Res 15:5308–5316

Wolf J, Dewi DL, Fredebohm J, Muller-Decker K, Flechtenmacher C, Hoheisel JD, Boettcher M (2013) A mammosphere formation RNAi screen reveals that ATG4A promotes a breast cancer stem-like phenotype. Breast Cancer Res 15:R109

Xiong L, Liu Z, Ouyang G, Lin L, Huang H, Kang H, Chen W, Miao X, Wen Y (2017) Autophagy inhibition enhances photocytotoxicity of Photosan-II in human colorectal cancer cells. Oncotarget 8:6419–6432

Yan C, Luo L, Guo CY, Goto S, Urata Y, Shao JH, Li TS (2017) Doxorubicin-induced mitophagy contributes to drug resistance in cancer stem cells from HCT8 human colorectal cancer cells. Cancer Lett 388:34–42

Yang Z, Zhang L, Ma A, Liu L, Li J, Gu J, Liu Y (2011) Transient mTOR inhibition facilitates continuous growth of liver tumors by modulating the maintenance of CD133+ cell populations. PLoS One 6:e28405

Yao JC, Phan AT, Jehl V, Shah G, Meric-Bernstam F (2013) Everolimus in advanced pancreatic neuroendocrine tumors: the clinical experience. Cancer Res 73:1449–1453

Yu X, Shi W, Zhang Y, Wang X, Sun S, Song Z, Liu M, Zeng Q, Cui S, Qu X (2017) CXCL12/CXCR4 axis induced miR-125b promotes invasion and confers 5-fluorouracil resistance through enhancing autophagy in colorectal cancer. Sci Rep 7:42226

Yue Z, Jin S, Yang C, Levine AJ, Heintz N (2003) Beclin 1, an autophagy gene essential for early embryonic development, is a haploinsufficient tumor suppressor. Proc Natl Acad Sci U S A 100:15077–15082

Zhang Z, Duan Q, Zhao H, Liu T, Wu H, Shen Q, Wang C, Yin T (2016) Gemcitabine treatment promotes pancreatic cancer stemness through the Nox/ROS/NF-kappaB/STAT3 signaling cascade. Cancer Lett 382:53–63

Zhou J, Li G, Zheng Y, Shen HM, Hu X, Ming QL, Huang C, Li P, Gao N (2015) A novel autophagy/mitophagy inhibitor liensinine sensitizes breast cancer cells to chemotherapy through DNM1L-mediated mitochondrial fission. Autophagy 11:1259–1279

Zhu H, Wang D, Liu Y, Su Z, Zhang L, Chen F, Zhou Y, Wu Y, Yu M, Zhang Z, Shao G (2013) Role of the hypoxia-inducible factor-1 alpha induced autophagy in the conversion of non-stem pancreatic cancer cells into CD133+ pancreatic cancer stem-like cells. Cancer Cell Int 13:119

The Autophagy Conundrum in Cancer Development, Progression and Therapeutics

11

Siddavaram Nagini, Palrasu Manikandan, and Rama Rao Malla

Abstract

Autophagy is an evolutionarily conserved process that delivers intracellular constituents to the lysosomes for degradation and recycling. Autophagy plays a central role in diverse physiological processes and has been implicated in the pathogenesis of various diseases including cancer. The role of autophagy in cancer is complex and largely context-dependent. Accumulating evidence indicates that autophagy facilitates tumorigenesis by enabling acquisition of cancer hallmarks. Autophagy manipulation has emerged as a promising strategy in cancer treatment. In this chapter, we provide an overview of the autophagic process, highlight the autophagy conundrum in cancer, examine the complex and conflicting reports on autophagy in tumour suppression and tumour promotion, as well as the role of autophagy in the acquisition of cancer hallmarks. Finally, from the clinical perspective, we summarise the evidence for autophagy-related genes and proteins as reliable markers of disease severity and prognosis and analyse the efficacy of autophagy manipulation in improving cancer treatment outcomes and circumventing chemoresistance.

Keywords

Cancer hallmarks · Autophagy · Tumorigenesis · Drug resistance · Therapeutics

---

S. Nagini (✉)
Department of Biochemistry and Biotechnology, Annamalai University, Annamalai Nagar, Chidambaram, Tamil Nadu, India

P. Manikandan
Department of Surgery, University of Miami Miller School of Medicine, Miami, FL, USA

R. R. Malla
Cancer Biology Lab, Department of Biochemistry and Bioinformatics, Institute of Science, GITAM (Deemed to be University), Visakhapatnam, Andhra Pradesh, India

© Springer Nature Singapore Pte Ltd. 2020
S. K. Bhutia (ed.), *Autophagy in Tumor and Tumor Microenvironment*,
https://doi.org/10.1007/978-981-15-6930-2_11

11.1 Introduction

Autophagy is an evolutionarily conserved process by which aberrant, unwanted proteins and damaged organelles are sequestered into double-membraned vesicles called autophagosomes and subsequently delivered to the lysosomes for degradation to maintain cellular homeostasis (Bishop and Bradshaw 2018). The term '*autophagy*', coined by Christian de Duve in 1963, is derived from the Greek words, 'auto' meaning "self" and 'phagein' meaning "to eat" (Klionsky 2008).

Autophagy is categorised into three distinct types based on the mechanism of cargo delivery to the lysosomes for degradation, microautophagy, chaperone-mediated autophagy (CMA) and macroautophagy. Microautophagy seen in yeast involves the sequestration of small cargoes by protrusion or invagination of endolysosomal membranes. CMA mediates the degradation of soluble proteins in the lysosomes with the help of molecular chaperones and lysosome-associated membrane protein 2A (LAMP2A). Macroautophagy (henceforth referred to as autophagy), the best-characterised and evolutionarily conserved type of autophagy, requires the formation of double-membrane structures termed autophagosomes for the delivery of cargoes to the lysosomes. Macroautophagy may be further classified into selective autophagy, characterised by high cargo specificity, and non-selective (bulk) autophagy which lacks cargo specificity (Allen and Baehrecke 2020; Parzych and Klionsky 2014).

Autophagy is intricately involved in health and disease. It plays a vital role in cellular turnover, development, differentiation, tissue remodelling and cell death. Autophagy is believed to function as a double-edged sword in disease processes and may have a causative or protective role. Autophagy has been implicated in ageing, infections, neurodegenerative disorders and cancer (Shintani and Klionsky 2004). Yoshinori Ohsumi was awarded the Nobel Prize for Physiology or Medicine in 2016 for his seminal work on autophagy that led to a new paradigm in understanding physiological processes such as the adaptation to starvation as well as diseases such as cancer (https://www.nobelprize.org/prizes/medicine/2016/press-release/).

11.2 Physiological Functions of Autophagy

Autophagy is essential at every stage during the development of various organisms and mediates a plethora of diverse cellular processes. Autophagy plays a critical role in the maintenance of cellular homeostasis. Under basal conditions, autophagy is involved in housekeeping functions such as removal of damaged organelles, misfolded proteins and protein aggregates. On the other hand, during starvation, autophagy promotes bioenergetic homeostasis by breaking down cellular macromolecules to generate ATP for cellular functions (Klionsky 2020; Mowers et al. 2017). Besides nutrient deprivation, autophagy is also induced to mitigate stress due to hypoxia and reactive oxygen species (ROS). During embryogenesis, autophagy catalyses the removal of paternal mitochondria. Autophagy is required for mediating immune and inflammatory response, defence against microbial infections, cell-fate determination, tissue remodelling, preservation of organelle function, recycling of intracellular proteins, prevention of toxic build-up of waste products and gene silencing. Autophagy also protects cells from undergoing programmed cell death by apoptosis (Allen and Baehrecke 2020; Singh et al. 2018).

11.3 The Autophagic Process and Components

Autophagy occurs at a basal level in all cells and can be induced by various types of stress including nutrient deprivation, hypoxia, ROS, damaged cell organelles and as a part of the DNA damage response (DDR) (Singh et al. 2018). Autophagy is divided into five stages: initiation, nucleation of the initial sequestering compartment termed the phagophore, expansion and elongation of the phagophore to form the double-membrane structure called the autophagosome, fusion of the outer membrane of the autophagosome with the lysosome to form the autolysosome and cargo degradation and recycling (Hansen et al. 2018). Each stage of autophagy has potential therapeutic targets for intervention (Fig. 11.1) (Mulcahy Levy and Thorburn 2020).

The process of autophagy is mediated by the highly conserved autophagy-related genes (ARGs), (Allen and Baehrecke 2020; Singh et al. 2018). Autophagy is initiated in response to various cellular signals by the Unc-51-like autophagy activating kinase (ULK1) complex comprising ULK1, ULK2, Atg13, Atg101 and the scaffolding protein RB1 inducible coiled-coil 1 (RBCC1) also known as FAK family kinase-interacting protein of 200 kDa (FIP200). This is followed by membrane nucleation and formation of the phagophore that requires synthesis of phosphatidylinositol-3-phosphate by activation of a class III phosphoinositide 3-kinase (PI3K) complex, composed of a PI3K, ATG14L, vacuolar protein sorting-associated proteins 15 and 34 (VPS15 and VPS34) and Beclin-1. The ATG9 trafficking system (ATG2A/ATG2B, WDR45/WIP14 and ATG9A) is responsible for elongation of the phagophore. The phagophore expands by acquisition of lipids promoted by two ubiquitin-like conjugation systems, the ATG5–ATG12–ATG16 complex and microtubule-associated protein light chain 3 (LC3) to form the autophagosome. Formation of the ATG5–ATG12–ATG16 complex is followed by conversion of the cytosolic LC3-I to the lipidated LC3-II that conjugates to phosphatidylethanolamine and incorporated into the phagophore membrane. The adaptor protein p62/sequestosome 1 (SQSTM1) binds to LC3-II during autophagosome formation and facilitates the degradation of ubiquitinated proteins (Bishop and Bradshaw 2018; Marinkovic et al. 2018). The autophagosome then fuses with a lysosome, to form an autolysosome in a process requiring small G-protein Rab7, soluble N-ethylmaleimide-sensitive factor attachment proteins (SNAREs), syntaxin17 (Stx17) and the membrane tethering complex HOPS (Dikic and Elazar 2018; Zhi et al. 2018). The autophagic process is completed within the autolysosomes by enzymatic degradation of the cargo and recycling of nutrients (Fig. 11.1).

Autophagy is regulated by the mammalian target of rapamycin (mTOR) and AMP-activated protein kinase (AMPK) signalling pathways. mTORC1 which is

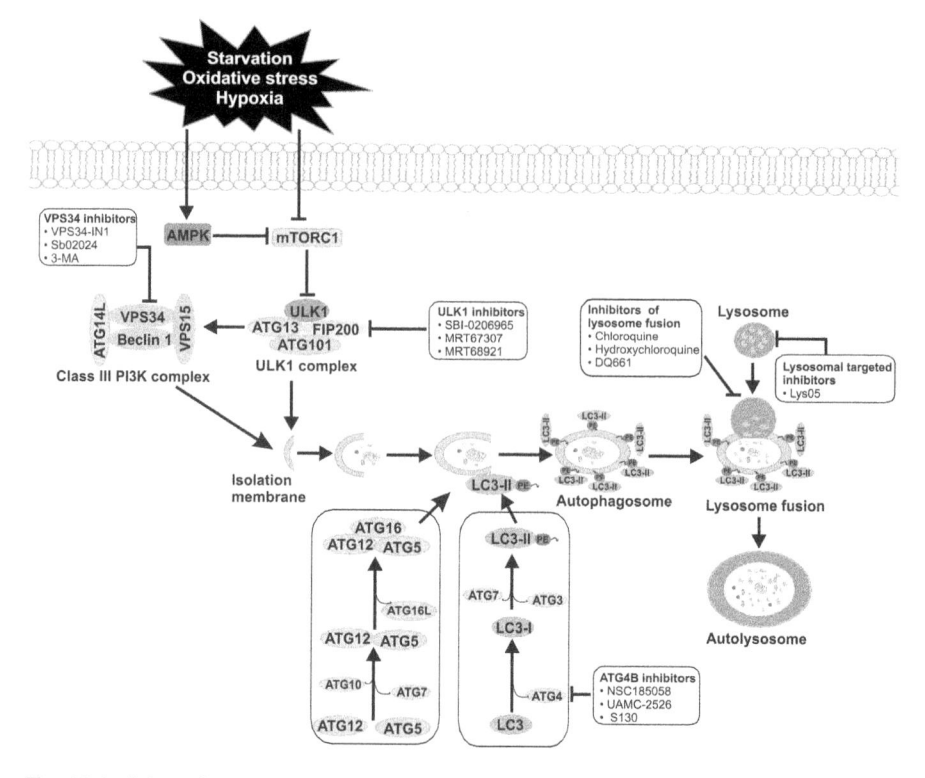

Fig. 11.1 Schematic representation of the mechanism and regulation of autophagy. The mTOR kinase is the key signalling molecule involved in the regulation of autophagy. In an unstressed state, activated mTORC1 phosphorylates and inactivates autophagy-related proteins and inhibits the ULK/FIP200/ATG13 complex with consequent inhibition of autophagy. Induction of autophagy by starvation, oxidative stress and hypoxia, inhibits mTORC1 that in turn releases and activates the ULK/FIP200/ATG13 complex. This leads to activation of the class III phosphoinositide 3-kinase (PI3K) complex comprised of Vps34, p150, Beclin-1, Atg14L and Autophagy and Beclin1 Regulator 1 (AMBRA1), which then drives the nucleation of the isolation membrane. Expansion and elongation of the isolation membrane involve conversion of cytoplasmic LC3-I to the lipidated LC3-II, followed by conjugation of phosphatidylethanolamine (PE) to LC3-II mediated by ATG4B and ATG7. Localisation of ATG5-ATG12/ATG16L complex helps in elongation by recruitment of LC3-II to the membrane. The ends of the isolation membrane fuse to form the autophagosome, which fuses with the lysosomes to form autolysosomes. The cargo is degraded in the autolysosomes by lysosomal enzymes and biomolecules recycled back to the cytoplasm

activated by nutrients and growth factors at the lysosomes induces phosphorylation of the ULK complex with repression of autophagy. On the other hand, nutrient deprivation inactivates mTORC1 leading to activation of the ULK complex and autophagy induction (Fig. 11.1). Recent evidence indicates that autophagy is also regulated by epigenetic mechanisms including histone modifications, DNA methylation and by noncoding RNAs (ncRNAs) (Baek and Kim 2017; Hu 2019). Epigenetic changes influence ARGs as well as the signalling molecules and pathways that regulate autophagy. While the co-activator-associated arginine methyltransferase

1 (CARM1) enhances transcriptional activation of ARGs, EZH2, a methyltransferase is reported to silence autophagy-activating promoters by methylation. Shin et al. (2016a) found CARM1-mediated arginine methylation (H3R17me2) as a critical epigenetic mark in autophagic induction.

11.4 Role of Autophagy in Cancer

The role of autophagy in cancer is complex and bidirectional. Autophagy has been documented to suppress or promote tumour development based on the context and the stage of tumorigenesis. Autophagy has been documented to be low in premalignant lesions and enhanced in advanced cancers (Galluzzi et al. 2015; Mulcahy Levy and Thorburn 2020).

11.4.1 Tumour Suppressive Effects of Autophagy

Autophagy prevents carcinogenesis by virtue of its ability to remove aggregated, misfolded and oncogenic proteins. Additionally, autophagy also exerts tumour-suppressive effects by stimulating the immune response. Decreased autophagy was shown to be associated with infiltration of regulatory T cells, leading to diminished immunosurveillance that facilitates tumour development (Parzych and Klionsky 2014). The tumour preventive role of autophagy has also been attributed to be mediated via scavenging endogenous sources of ROS and maintaining genomic stability (Galluzzi et al. 2015). Although genetic alterations in several ARGs have been extensively documented, a large-scale human genomic analysis of somatic mutations in ATG genes across 11 cancer types revealed that the core autophagy machinery, which plays a critical role in maintaining genomic stability does not undergo genetic alterations (Lebovitz et al. 2015).

Monoallelic deletion of Beclin-1, a haploinsufficient tumour suppressor gene, has been reported in breast, ovarian and prostate cancers (Delaney et al. 2020; Qu et al. 2003). The loss of Beclin-1 was associated with reduced autophagy and increased proliferation (Lee and Wu 2012; Zhang et al. 2018). However, biallelic Beclin-1 mutations that could cause embryonic lethality do not occur in cancer. This implies that monoallelic Beclin-1 is adequate to facilitate the requirement of functional autophagy necessary for neoplastic transformation (Yue et al. 2003). The Vps34-binding domain of Beclin-1 was shown to be essential for its tumour suppressor activity (Furuya et al. 2005). The tumour suppressor functions of Beclin-1 are also mediated through UVRAG and Bax-interacting factor-1 (Bif-1), which increase binding of Beclin-1 to Vps34 (Takahashi et al. 2007). Monoallelic deletion or mutations of UVRAG as well as downregulation of Bif-1 have been documented in diverse malignancies (Kung et al. 2011).

In addition to Beclin-1, several components of the core autophagy machinery were also found to display tumour suppressor functions. Loss-of-function mutations in ATG2B, ATG5, ATG9B and ATG12 leading to truncated ATG proteins were identified in gastric and colorectal cancers with microsatellite instability (Kang et al. 2009). Mice with deficiency of Atg5 and Atg7 showed mitochondrial damage, oxidative stress and propensity to develop liver tumours (Takamura et al. 2011). Loss of ATG4C, involved in processing LC3/ATG8 during autophagosome formation, was reported in chemically induced murine fibrosarcomas (Kimmelman 2011). Somatic mutations of ATG5 coupled with overexpression of ATG16L2 observed in various tumours prevented the interaction of ATG5 with ATG16L1, with consequent proteasomal degradation of ATG12 and ATG16L1, resulting in inhibition of autophagy (Wible et al. 2019).

p62/SQSTM1, an autophagy receptor and selective substrate for autophagy, accumulates when autophagy is inhibited with fall in levels when autophagy is induced. It thus serves as a reliable marker of autophagic flux (Mathew et al. 2009). Aberrant accumulation of p62/SQSTM1 has been reported in gastrointestinal cancer (Su et al. 2005), prostate cancer (Kitamura et al. 2006), hepatocellular carcinoma (Umemura et al. 2016), breast cancer (Li et al. 2017) and lung adenocarcinoma (Inoue et al. 2012), suggesting that autophagy inhibits tumorigenesis by decreasing p62 accumulation (Li et al. 2020).

There is growing evidence to indicate that autophagy is stimulated by well-established tumour suppressors such as TP53 and phosphatase and tensin homolog (PTEN). In HT-29 colon cancer cells, PTEN was found to promote autophagy, whereas loss-of-function mutations in PTEN suppressed autophagy (Errafiy et al. 2013). Taken together, these findings underscore the anti-tumour effects of autophagy (Fig.11.2).

11.4.2 Tumour-Promoting Effects of Autophagy

Although autophagy is reported to suppress the development and progression of tumours, substantial evidence indicates that autophagy facilitates tumorigenesis. Autophagy is a strategy that enables acquisition of cancer hallmarks and survives tumour microenvironmental stress. Several studies have demonstrated the key role of autophagy in providing essential metabolites to meet the growing demands of proliferating tumour cells (Kocaturk et al. 2019; Mulcahy Levy and Thorburn 2020; Singh et al. 2018; Yang and Klionsky 2020). Autophagy fuels enhanced metabolic and energy needs of cancer cells by mediating the degradation of macromolecules to their constituent monomer units. In addition, autophagy promotes tumour survival by enhancing tolerance to oxidative and genotoxic stress as well as stress induced by increased metabolic rate and hypoxia (Fig.11.2).

RAS are small GTPases involved in important signal pathways for proliferation, survival and metabolism. Cancers driven by the K-Ras oncogene rely heavily on autophagy even in the absence of external stressors, a phenomenon known as *'autophagy addiction'* that helps in evasion of metabolic stress and cell death (Kim et al. 2011b). Several studies have reported a correlation between RAS-mediated autophagy and the development of various human malignancies, including cancers of the lung, colon and pancreas, suggesting that autophagy plays

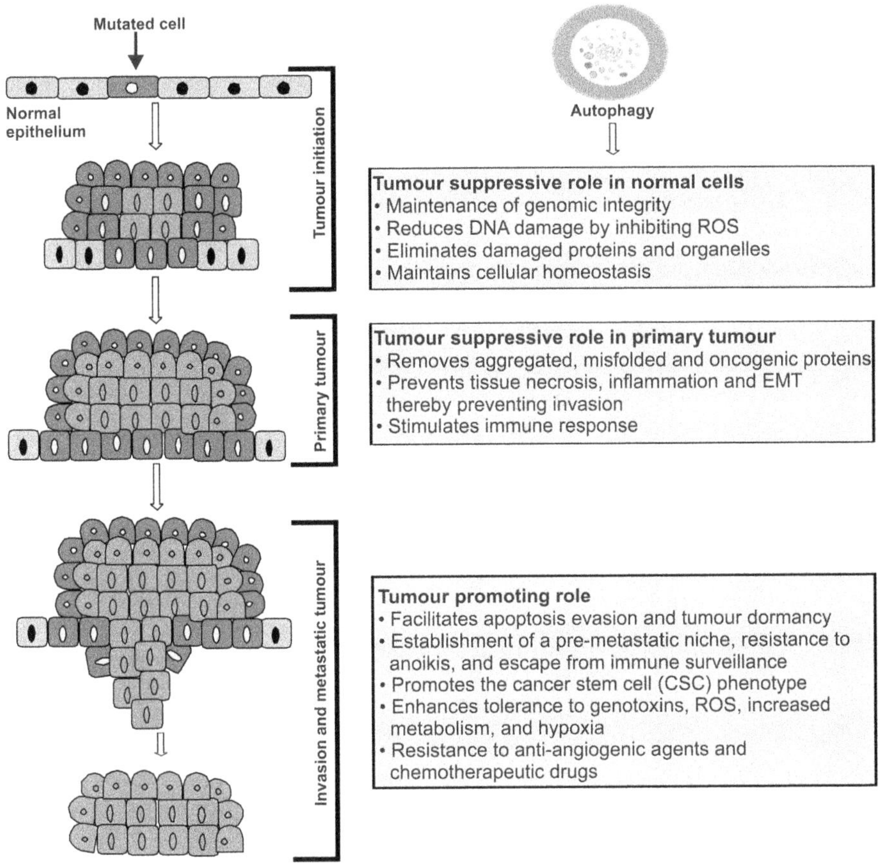

Fig. 11.2 The dual role of autophagy in tumorigenesis

an important role in survival and growth of various tumours that depend on RAS activation (Goel et al. 2015; Guo et al. 2011; Kim et al. 2011a). High rates of KRAS mutations are seen in pancreatic ductal adenocarcinomas (PDACs) that are believed to depend on autophagy to fuel tumour metabolism (Guo et al. 2011; Mulcahy Levy and Thorburn 2020; Yang et al. 2011). The tumour-promoting potential of autophagy is believed to be mediated by suppression of TP53 induction and by maintenance of mitochondrial function (Guo et al. 2013b; Mancias and Kimmelman 2011).

Cancer stem cells (CSCs) that display self-renewal and malignant transformation showed higher levels of autophagy (Nazio et al. 2019). The influence of autophagy on CSCs is rather complex and based on several factors such as origin and differentiation status. Inhibition of autophagy in CSCs induced death of CD34+ progenitor cells in chronic myeloid leukaemia, whereas in acute myeloid leukaemia, it caused expansion of progenitor cells in haematopoietic stem cells (Auberger and Puissant

2017). Conflicting findings have been reported on the effect of silencing ARGs on CSCs. While silencing of Beclin-1 or ATG genes such as ATG7, ATG12 or LC3 inhibited proliferation of CSCs, ATG7 deficiency in KRAS-driven tumours had no effect (Cufi et al. 2011; Eng et al. 2016; Gong et al. 2013).

11.5 Autophagy and Cancer Hallmarks

Tumorigenesis involves the acquisition of ten essential alterations that enable the growth and functional abilities of cancer cells to survive, proliferate, invade and disseminate, collectively denoted as *hallmarks of cancer*. These include self-sufficiency in growth signals, insensitivity to growth-inhibitory signals, evasion of programmed cell death, limitless replicative potential, sustained angiogenesis, tissue invasion and metastasis, reprogramming of energy metabolism, evading immune destruction, genome instability and inflammation (Sasahira and Kirita 2018). Several studies have unravelled the role of autophagy in the acquisition of cancer hallmarks, some of which (sustained cell proliferation, invasion, metastasis, apoptosis evasion and drug resistance) are discussed below.

11.5.1 Cell Proliferation and Autophagy

There are conflicting reports on the role of autophagy in tumour cell proliferation (Singh et al. 2018). High levels of autophagy have been documented to be essential for the growth of cancers with KRAS or BRAF mutations such as PDACs (Yang et al. 2011). In a BRAF-driven lung cancer model, Atg7 deletion resulted in tumour regression providing proof-of-concept for the involvement of autophagy in the proliferation of these tumours (Guo et al. 2013a). Other studies found a correlation between low levels of autophagy and high rate of proliferation in cancer that could be attributed to dysregulated PI3K/Akt/mTOR pathway and deletion of the tumour suppressor PTEN. Further, rapamycin, an mTOR inhibitor and autophagy inducer was shown to cause cell cycle arrest and inhibits proliferation of mantle cell lymphoma and MDA-MB-231 breast cancer cells (Chatterjee et al. 2015; Yazbeck et al. 2008). Collectively, these findings indicate that autophagy-mediated regulation of cell proliferation is context-dependent.

11.5.2 Interplay Between Autophagy and Apoptosis

Although apoptosis and autophagy are distinct forms of cell death that maintain cellular homeostasis, they are intricately interconnected by protein networks (Nikoletopoulou et al. 2013; Vijayarathna et al. 2015). Autophagy is a cytoprotective survival mechanism that tumour cells employ to evade apoptosis (Mulcahy Levy and Thorburn 2020). Understanding the mechanisms by which autophagy circumvents apoptosis in tumours will enable the development of successful therapeutic strategies. Inefficient mitochondrial outer membrane permeabilisation (MOMP) that enables tumours to recover from apoptosis and regain the ability to proliferate has been suggested as the mechanism underlying autophagy-mediated apoptosis avoidance (Ichim et al. 2015).

The BCL2 family proteins that regulate apoptosis are also involved in autophagy initiation (Fitzwalter and Thorburn 2015). The proapoptotic BH3-only proteins such as PUMA, NOXA, NIX, BID and BNIP3 disrupt the Beclin 1/BCL2 complex releasing Beclin 1 that complexes with VPS34 to stimulate autophagy. The anti-apoptotic BCL2 proteins on the other hand inhibit Beclin-1 by binding to its BH3 domain (Pattingre et al. 2005). Death-associated protein kinase (DAPK) has been shown to induce autophagy by phosphorylating Beclin 1. Upon phosphorylation, Beclin 1 dissociates from BCL-2 and binds to VPS34. In addition, DAPK also activates VPS34 via a second kinase, protein kinase D (PKD) (Eisenberg-Lerner and Kimchi 2012; Zalckvar et al. 2009).

c-jun N-terminal kinase (JNK), involved in a vast array of cellular processes, has been demonstrated to disrupt the Beclin 1-BCL-2 complex by phosphorylating BCL-2. This leads to release of Beclin 1 and formation of an active Beclin 1–VPS34 complex resulting in induction of autophagy. Wei et al. (2008) proposed a model on the dual role of JNK1-mediated BCL2 phosphorylation in regulating autophagy and apoptosis. They speculated that JNK1 initially phosphorylates BCl-2 to stimulate autophagy. However, once autophagy is unable to sustain cell survival, Bcl-2 phosphorylation inactivates its anti-apoptotic function and apoptosis is initiated.

The tumour suppressor protein TP53 also plays a dual role in autophagy based on its activation status and intracellular localisation. Cytosolic p53 inhibits autophagy by interacting with FIP200 and interfering with the ULK1 complex activity (Morselli et al. 2011; Tasdemir et al. 2008). However, under conditions of cellular stress, p53 localises to the nucleus and binds to the promoter region of multiple pro-autophagic genes, including AMPK, DRAM1, sestrin 1, sestrin 2 and PTEN, as well as pro-apoptotic genes of the BCL-2 family and p53 upregulated modulator of apoptosis (PUMA) (Budanov and Karin 2008; Gao et al. 2011; Kenzelmann Broz et al. 2013; Riley et al. 2008). Under certain conditions, p53 also induces both mitophagy and apoptosis by triggering MOMP (Youle and Narendra 2011).

The transcription factor FOXO3/FOXO3A (forkhead box O3), which confers apoptosis sensitisation by transactivating PUMA, reciprocally regulates autophagy (Warr et al. 2013). Elevated PUMA prevents the interaction between BCL2 and BAX/BAK with release of BAX/BAK MOMP and cell death by apoptosis. Fitzwalter and Thorburn (2018) postulated that FOXO3 functions as a cell surveillance mechanism to rectify perturbations in autophagy and induces apoptosis if autophagy regulation fails.

BH-3 only proteins that function at the crossroads of apoptosis and autophagy have emerged as attractive therapeutic targets in cancer. Several BH3 mimetics which are inhibitors of the anti-apoptotic BCL2 proteins have been developed. Venetoclax, a BH3-mimetic small-molecule inhibitor of BCL-2, is used in the treatment of chronic lymphocytic leukaemia (CLL) and small lymphocytic lymphoma. In acute myeloid leukaemia (AML), overexpression of vacuole membrane protein (VMP1) increased autophagic flux, protected against oxidative stress, reduced the response to venetoclax-induced MOMP and apoptotic cell death (Folkerts et al. 2019).

11.5.3 Angiogenesis and Autophagy

Angiogenesis, the formation of new blood vessels from existing vasculature, facilitates tumour invasion and metastasis. With increasing growth of a malignant tumour, the centre of the tumour is deprived of oxygen and nutrients due to decreased perfusion. Autophagy has been suggested to enable tumour cells to thrive under avascular and hypoxic conditions. In the tumour microenvironment (TME), autophagy flux induces migration of ECs and angiogenesis (Vion et al. 2017). Resistance to anti-angiogenic therapy has been attributed to high levels of autophagy in tumours. Anti-angiogenesis treatment in concert with administration of an autophagy inhibitor was found to exhibit greater efficacy besides stimulating apoptosis of tumours (Ramakrishnan et al. 2007). However, enhanced autophagy in neuroblastomas was demonstrated to block angiogenesis via degradation of pro-angiogenic gastrin-releasing peptide (GRP) (Kim et al. 2013).

Matrix glycoproteins that regulate the interplay between autophagy and angiogenesis in the tumour microenvironment are considered to be critical determinants of the fate of cancer cells. Decorin and Perlecan, matrix proteoglycans have been envisaged to influence the crosstalk between angiogenesis and autophagy signalling in endothelial cells. In a recent study, decorin, a small leucine-rich proteoglycan, was demonstrated to evoke the autophagic clearance of vascular endothelial growth factor A (VEGFA) by functioning as a partial agonist of vascular endothelial growth factor 2 (VEGFR2) in a process that requires the energy-sensing protein, AMPK and the autophagic regulator, paternally expressed gene 3 (PEG3). Further, pharmacological depletion of ATG5 led to intracellular accumulation of VEGFA, indicating that VEGFA is a substrate for autophagy. These findings underscore the therapeutic potential of decorin as a next-generation anticancer agent (Neill et al. 2020).

11.5.4 Tissue Invasion, Metastasis and Autophagy

Autophagy has a complex role in tumour invasion. In a primary tumour, autophagy prevents tissue necrosis and inflammation, thereby preventing invasion (Kenific et al. 2010). Autophagy also inhibits epithelial–mesenchymal transition (EMT) by degradation of p62/SQSTM1 as well as its cargo TWIST1 that is known to stimulate EMT (Qiang et al. 2014). However, once the tumour becomes invasive and progresses, autophagy affords protection against apoptosis and facilitates tumour dormancy. Autophagy has been implicated in various features of invasion such as cell motility, epithelial–mesenchymal transition (EMT), quiescence, stem cell phenotype and drug resistance (Mowers et al. 2017). Autophagy was found to be essential for secretion of factors critical for tumour invasion such as interleukin-6, matrix metalloproteinase-2 (MMP-2) and WNT-5A (Lock et al. 2014). Interestingly, hypoxia and transforming growth factor beta (TGFβ) are known to induce EMT also induce autophagy (Kiyono et al. 2009; Li et al. 2013). MicroRNA-mediated suppression of Smad2 was found to interrupt autophagy, resulting in inhibition of cell survival and invasive potential (Zhai et al. 2015). Conversely, ULK2, which promotes autophagy enhanced EMT and invasiveness (Kim et al. 2016). Autophagy was reported to induce EMT via SPHK1-TRAF2-Beclin-1-CDH1 signal cascades in hepatocellular carcinoma cells (Liu et al. 2017a).

Emerging evidence indicates the involvement of autophagy in metastasis (Dower et al. 2018). Several steps in the metastatic cascade are believed to be autophagy-dependent, including establishment of a pre-metastatic niche, tumour cell dormancy, resistance to anoikis and escape from immune surveillance (Kenific et al. 2010; Mowers et al. 2017). Autophagy also plays an important role in preventing tumour cells that detach from the ECM from dying by the process of anoikis, thereby promoting metastasis (Lock and Debnath 2008). Autophagy is induced by the same factors that promote metastasis such as hypoxia. Interestingly, several features of autophagy, such as mesenchymal characteristics, escape from immune surveillance and stem cell-like phenotype, are shared by metastasis. Increased staining for the autophagy marker, microtubule-associated light chain B (LC3B), is a common feature in solid tumours that is associated with metastasis (Lazova et al. 2012). Increased autophagy and EMT promote the cancer stem cell (CSC) phenotype that drives metastasis (May et al. 2011). In breast ductal carcinoma in situ (DCIS), high levels of autophagy were observed in subpopulations of cells that displayed tumour-invasive potential and stem cell phenotype (Espina et al. 2010).

The tumour microenvironment (TME), which interacts with the malignant tumour, profoundly influences tumour progression as well as therapeutic response. Autophagy is documented to promote migration and invasion of tumour cells, maintain tumour cell stemness and drug-resistance phenotypes and influence the crosstalk between the tumour and the TME (Mowers et al. 2018). In the TME, autophagy facilitates polarisation of macrophages into tumour-associated macrophages (TAMs) (Chen et al. 2014; Wen et al. 2018), and differentiation of fibroblasts into cancer-associated fibroblasts (CAFs) (Ngabire and Kim 2017; Peiris-Pages et al. 2015; Wang et al. 2017) and myeloid-derived suppressor cells (MDSCs) (Dong et al. 2017; Ostrand-Rosenberg et al. 2020).

The interplay between autophagy and exosomes is increasingly recognised to influence the TME. Exosomes, cargo-laden vesicles secreted by various cell types, establish intercellular communication to transfer their contents such as RNA and proteins to other cells, which may impact autophagy. Both exosomes and autophagy exert influence on the TME and metastasis and reciprocally regulate each other (Lin et al. 2019; Ruivo et al. 2017). The interaction of autophagy and exosomes is also mediated by autophagy-related proteins. ATG5 silencing significantly attenuated the release of exosomes as well as exosome-mediated lipidation of LC3B, a central protein of the autophagy pathway (Xu et al. 2018a).

11.5.5 Drug Resistance and Autophagy

There is growing evidence to indicate the involvement of autophagy in resistance to chemotherapeutic agents. The anticancer drug 5-fluorouracil (5-FU) used in the treatment of solid tumours such as breast, pancreatic and colorectal cancers, inhibits thymidylate synthetase, an enzyme essential for DNA synthesis. Induction of cytoprotective autophagy that results in chemoresistance is a major limitation of this drug. Autophagy induction by 5-FU has been attributed to overexpression of beclin-1, followed by conversion of LC3I to LC3II, JNK-mediated protective autophagy and BCL2-mediated autophagic flux. The DNA-damaging chemotherapeutic drug, cisplatin, is also documented to induce autophagy and chemoresistance. Several mechanisms have been suggested for enhanced autophagy induced by cisplatin. These include modulation of ERK pathway and upregulation of beclin 1 with consequent conversion of LC3 proteins, increase of ATG7 expression and downregulation of miR-199a-5p (Xu et al. 2012). However, combined administration of cisplatin and an autophagy inhibitor induced tumour cell death. In mitoxantone-resistant breast cancer cells, miR-181a targets *Atg5* and impedes autophagy by targeting breast cancer–resistance protein (Jiao et al. 2013). Likewise, miR-874 inhibits autophagy and sensitises gastric cancer cells to chemotherapy via the target gene ATG16L1 (Huang et al. 2018). Shuhua et al. (2015) observed a positive correlation between the expressions of the ARGs Raptor, Rictor and Beclin1 and the multidrug resistance (MDR) gene in colorectal cancer (CRC) patients. Targeting autophagy by modulating *Atgs* such as *Beclin1* (Eum and Lee 2011), *Atg5* (Ge et al. 2014), *Atg7* (Singh et al. 2012) and *Atg12* (An et al. 2015) sensitised MDR cells to therapeutic agents. Taken together, these findings imply that chemoresistance can be circumvented by targeting autophagy.

11.6 ARGs as Prognostic Markers

There is substantial evidence to indicate that ARGs are reliable markers of disease severity and prognosis (Bortnik and Gorski 2017; Yang and Klionsky 2020). The expression levels of ATG genes vary based on the site of the tumour and stage of the disease. In colon cancer, ATG16L2, CAPN2 and TP63 were upregulated, whereas SIRT1, RPS6KB1, PEX3, UVRAG and NAF1 were downregulated and associated with disease recurrence (Mo et al. 2019). On the other hand, in gastric cancer, ULK1, Beclin-1, ATG3 and ATG10 were identified as favourable prognostic markers (Cao et al. 2016). An eight-gene autophagy-related signature (*BLOC1S1, IL24, NRG4, PDK4, PEX3, PRKG1, SIRT2* and *WDR45L*) was identified as an independent and accurate predictor for the prognosis of serous ovarian cancer (An et al. 2018). Recently, Mao et al. (2020) showed that ATGs are crucial factors in the progression of HCC and could serve as potential prognostic markers for diagnosis and treatment. An autophagy score signature was validated to classify CRC patients into low and high risk of early relapse to predict post-operative survival (Zhou et al. 2019). Gene expression microarray data obtained from TCGA was used to develop ARG expression signature as a predictive tool for overall survival (OS) and disease-free survival (DFS) in prostate cancer patients. Five OS-related and 22 DFS-related ARG signatures were identified that could function as promising prognostic biomarkers of prostate cancer (Hu et al. 2020). Despite these studies, correlation between the ARG signature and the cancer type still remains obscure. The moonlighting functions of ATG proteins are believed to be responsible for the lack of correlation. Many ATG proteins are multifunctional and exert their influence beyond autophagy on diverse signalling pathways and cellular processes.

11.7 Autophagy Manipulation in Cancer Therapeutics

Autophagy manipulation has emerged as a promising strategy in cancer treatment. However, the paradoxical role of autophagy in cancer merits attention while designing therapeutic strategies. While enhancing autophagy is an option in premalignant lesions, and in some malignant tumours, inhibiting autophagy appears to be effective in many tumours, especially in advanced cancers. Several clinical trials are underway to target autophagy in cancer with more emphasis on the discovery and development of drugs that inhibit autophagy (Towers and Thorburn 2016).

11.7.1 Autophagy Induction

Several chemotherapeutic drugs are known to induce autophagy. The mToR inhibitor rapamycin has been successfully used to inhibit angiogenesis by preventing the synthesis of VEGF and downstream signalling events. Temsirolimus and everolimus, water-soluble analogues of rapamycin administered alone or in combination with chemotherapeutic drugs inhibited proliferation and induced autophagic cell death in mantle cell lymphoma and acute lymphoblastic leukaemia (Crazzolara et al. 2009; Yazbeck et al. 2008). Significant improvement in progression-free survival (PFS) was evident with everolimus treatment in patients with advanced neuroendocrine tumours in the Phase III RAD001 in Advanced Neuroendocrine Tumours (RADIANT)-3 and RADIANT-4 studies, respectively (Gajate et al. 2017). Everolimus in combination with exemestane, an aromatase inhibitor was found to be an important treatment option for patients with hormone receptor-positive (HR+) and human epidermal growth factor receptor 2- (HER2-) metastatic breast cancer (Riccardi et al. 2018). Combination chemotherapy with the autophagy inducers temozolomide and dasatinib was effective in killing glioblastoma cells resistant to apoptosis (Milano et al. 2009).

11.7.2 Autophagy Inhibition

There is substantial evidence to indicate that autophagy enhances tumour development and progression as well as chemoresistance in a wide variety of neoplasms.

Ample evidence from cell-based in vitro studies, genetically engineered mouse models (GEMMs) and patient-derived xenograft (PDX) mouse models demonstrate that autophagy inhibition by anti-cancer drugs enhances tumour cell death (Levy and Thorburn 2011; Mulcahy Levy and Thorburn 2020). Autophagy inhibition has been demonstrated to sensitise tumour cells to chemotherapeutic agents and potentiate apoptosis (Amaravadi et al. 2016).

Autophagy inhibition as a treatment modality for cancer may be tumour-specific or systemic. Tumour-specific autophagy inhibition causes perturbations in tumour cell metabolism, impairment in redox and energy homeostasis, mitochondrial dysfunction and reduced nucleotide pools eventually leading to tumour cell death. Systemic inhibition of autophagy, on the other hand, causes changes in the tumour microenvironment (Kimmelman and White 2017).

Several strategies have been used to inhibit autophagy in malignant tumours including small molecule inhibitors, genetic ablation of ARGs such as beclin-1, ATG5 or ATG7, and repurposed drugs such as chloroquine (Mulcahy Levy and Thorburn 2020). Current clinical efforts have explored the different stages of autophagy as potential therapeutic targets to maximise benefit in cancer treatment (Mulcahy Levy and Thorburn 2020). The serine/threonine kinases ULK1 and ULK2 are prime targets to block autophagy in the early stages. The selective ATP competitive inhibitor of ULK1 kinase, SBI-0206965 (SBI) was found to induce apoptosis in lung cancer during nutrient deprivation (Egan et al. 2015). Preclinical results using inhibitors of VPS34 (VPS34-IN1 and SB02024), ATG4B (NSC185058, UAMC-2526 and S130) are encouraging (Dyczynski et al. 2018; Fu et al. 2019).

Autophagy inhibition both alone and in combination with anticancer drugs is emerging as a promising option in cancer therapy. The autophagy inhibitor 3-methyladenine (3-MA) when used in concert with tratsuzumab increased chemotherapeutic efficacy in HER2-positive breast cancer cells (Jain et al. 2013). Treatment with 3-MA or deletion of beclin-1 induced chemosensitisation of hepatocellular carcinoma cells (Song et al. 2009). Knockdown of ARGs was found to overcome resistance to tamoxifen in ER-positive breast cancer cells (Cook et al. 2011). In cisplatin-resistant ovarian cancer cells, Atg5 deletion induced apoptosis (Wang and Wu 2014).

11.7.2.1 Chloroquine and Hydroxycloroquine

The antimalarial drug chloroquine (7-chloro-4-(4-diethylamino-1-methylbutylamino)-quinoline, CQ) has attracted significant attention as a promising anticancer agent, a classic example of drug repurposing. Both CQ and hydroxychloroquine (HCQ) have been approved by the Food and Drug Administration (FDA) for clinical trials in cancer. CQ is a small molecule that is unprotonated at physiological pH. Being lipophilic, it traverses the cell membrane and accumulates in acidic compartments such as the lysosomes (Weyerhauser et al. 2018). CQ inhibits autophagy by preventing the fusion of autophagosome with the lysososome (Yang et al. 2013). CQ treatment reverted resistance to chemotherapeutic and anti-angiogenesis drugs (Selvakumaran et al. 2013).

Addition of a hydroxyl group to CQ lowered toxicity of CQ while retaining the efficacy. A large number of clinical trials have revealed the adjuvant effects of CQ/HCQ for diverse neoplasms. Following the identification of CQ as an autophagy inhibitor by Murakami et al. (1998), CQ was demonstrated to significantly improve clinical outcomes in patients with glioblastoma (Briceno et al. 2003). Subsequently, CQ/HCQ was reported to exhibit anti-neoplastic properties on a wide range of tumours (Xu et al. 2018b). CQ treatment is recognised to sensitise colorectal cancer cells to anti-angiogenesis treatment, DNA damaging chemotherapeutic drugs and photosan-II-mediated photodynamic therapy (PS-PDT) (Xiong et al. 2017). The ability of CQ to sensitise malignant tumours to radiation and chemotherapy was impaired by pharmacological inhibition or siRNA ablation of Beclin-1. CQ acts on a wide spectrum of molecular targets such as p53, NF–κB and ATM kinase, reflecting its functional pleiotropy. CQ has been hypothesised to play a dual role by activating DNA damage response (DDR) and suppressing DNA repair, thereby shifting the balance towards cell death (Weyerhauser et al. 2018). Recent research has provided evidence that CQ exerts anticancer effects independent of its ability to inhibit autophagy (Eng et al. 2016).

11.7.2.2 Lysosome-Targeted Inhibitors

Although CQ/HCQ showed positive results in GBM and pancreatic tumours, clinical efficacy was not encouraging in other tumours. Several lysosomal targeted inhibitors that are potent and selective are being developed as potential alternatives to CQ/HCQ (Mulcahy Levy and Thorburn 2020). Lys05, a bisaminoquinoline and DQ661, a dimeric quinacrine that concurrently inhibits lysosomes by deacidification and impairs lysosomal recruitment of mTOR were successful as single agents in mouse models of melanoma and CRC. DQ661 displayed greater efficacy relative to HCQ and Lys05 especially in acidic tumours, because it is able to maintain its activity in acidic media. Additionally, DQ661 was also found to be promising in combination with gemcitabine in PDAC (McAfee et al. 2012; Pellegrini et al. 2014; Rebecca et al. 2017).

11.7.2.3 Epigenetic Modulation of Autophagy

Given the importance of epigenetic players in regulating autophagy, epigenetic modifiers that influence autophagy through histone acetylation, methylation of CpG islands and by ncRNAs have been used to manipulate autophagy. Several natural products have been documented to target autophagy via epigenetic modification (Vidoni et al. 2019). Curcumin was demonstrated to inhibit autophagy by restoring the expression of miR-143 and induce apoptosis of prostate cancer cells exposed to radiation (Liu et al. 2017b). Ellagic acid, a naturally occurring polyphenol abundantly found in fruits and vegetables that exerts antiproliferative effects has been reported to inhibit CARM1-mediated H3R17 methylation, thereby suppressing autophagy (Shin et al. 2016b). Studies from this laboratory demonstrated that gedunin and nimbolide, limonoids from the neem tree (*Azadirachta indica*) exert their antiproliferative effects by inhibiting cytoprotective autophagy and inducing apoptosis in oral cancer cell lines and in the hamster buccal pouch model of oral oncogenesis (Sophia et al. 2018; Tanagala et al. 2018). While gedunin mediated its effects via downregulation of the oncomiR, miR-21, nimbolide augmented apoptosis by overcoming the shielding effects of autophagy through modulation of the PI3K/Akt/GSK-3β signalling axis as well as the ncRNAs miR-126 and HOTAIR. Autophagy modulators are thus a valuable addition to the armamentarium of compounds that offer promise in cancer therapeutics.

11.7.2.4 Pitfalls of Autophagy Inhibition

There are several concerns in using autophagy manipulation as a therapeutic strategy in cancer. Many of the autophagy inhibitors including CQ/HCQ are not autophagy specific and affect other essential signalling pathways. For instance, in dormant murine breast cancer stem cells autophagy inhibition induced aberrant expression of 6-phosphofructo2-kinase/fructose-2,6-biphosphatase 3 (PFKFB3) leading to proliferation and recurrent metastatic disease (Yang et al. 2013). The cytotoxicity due to global autophagy inhibition induced by some compounds is another concern although it may be circumvented by therapy breaks or by using agents that cause incomplete autophagy inhibition such as CQ. The uptake of HCQ is pH-dependent which limits its effectiveness in solid tumours that show differences in pH between central and peripheral regions (Pellegrini et al. 2014).

Autophagy inhibition has been reported to cause side effects such as inflammation and tissue damage. This can be overcome by intermittent dosing of autophagy inhibitors. An inducible dominant-negative ATG4BC74A mutant mouse model that mimics a pharmacological inhibitor by reversibly manipulating autophagy without a complete blockade been developed (Yang et al. 2018). The interplay between autophagy and apoptosis lends credence to the development of intermittent autophagy inhibitors. However, the appropriate dose of autophagy inhibitors remains to be standardised.

Treatment outcomes may also depend on the concept of autophagy addiction. RAS-driven tumours such as PDACs may respond better to autophagy inhibition compared to autophagy-independent tumours providing a rationale for initiating clinical trials targeting autophagy addiction. Autophagy inhibition decreased tumour growth in xenograft models of PDAC and improved surgical outcomes in PDAC patients who were pre-operatively treated with gemcitabine, nab-paclitaxel and HCQ (Boone et al. 2015; La Belle Flynn et al. 2019). Autophagy inhibition in combination with direct targeting of MEK or ERK was found to be beneficial and clinical trials have been developed for NRAS melanoma and PDAC respectively (Kinsey et al. 2019). In addition to RAS, mutations in other genes have also been used to identify autophagy-dependence as well as to predict response to autophagy inhibition such as the epidermal growth factor receptor (EGFR) that regulates pathways influencing autophagy. GBM tumours expressing EGFR variant III (EGFRvIII), as well as head and neck squamous cell carcinoma (HNSCC) are autophagy-dependent and respond to autophagy inhibition (Jutten et al. 2018). Clinical trials have been carried out on autophagy inhibition in NSCLC and GBM patients with overexpressed or mutant EGFR (Massachusetts General Hospital 2019, https://ClinicalTrials.gov/show/

NCT00977470; Maastricht Radiation Oncology 2020, https://ClinicalTrials.gov/show/NCT02378532).

11.8 Conclusion

The role of autophagy in cancer is highly complex and paradoxical. While autophagy has suppressive effects on some tumours, in most cases, autophagy is a survival pathway that enables tumour proliferation and progression. In particular, the interplay between autophagy and apoptosis is intriguing and has implications for cancer therapy. Autophagy is a therapeutically targetable process, although there are many factors that need to be considered to maximise benefit. It is increasingly important to weigh options such as targeting the early or late stages of the pathway, stage of the disease that will respond best to intervention, whether to use an autophagy inducer or inhibitor and whether to administer the autophagy modulator as a single agent or in combination. Patient selection is critical in delineating positive findings as well as to identify non-responders. Rationally based interventions are therefore essential to effectively maximise therapeutic benefit and minimise adverse outcomes.

Acknowledgements Financial support from the University Grants Commission, Basic Science Research (No.F.18-1/2011(BSR)) and the Science and Engineering Research Board (#EMR/2016/001984) of Department of Science and Technology New Delhi, India to Siddavaram Nagini is gratefully acknowledged.

References

Allen EA, Baehrecke EH (2020) Autophagy in animal development. Cell Death Differ 27:903–918

Amaravadi R, Kimmelman AC, White E (2016) Recent insights into the function of autophagy in cancer. Genes Dev 30:1913–1930

An Y, Zhang Z, Shang Y, Jiang X, Dong J, Yu P, Nie Y, Zhao Q (2015) miR-23b-3p regulates the chemoresistance of gastric cancer cells by targeting ATG12 and HMGB2. Cell Death Dis 6: e1766

An Y, Bi F, You Y, Liu X, Yang Q (2018) Development of a novel autophagy-related prognostic signature for serous ovarian cancer. J Cancer 9:4058–4071

Auberger P, Puissant A (2017) Autophagy, a key mechanism of oncogenesis and resistance in leukemia. Blood 129:547–552

Baek SH, Kim KI (2017) Epigenetic control of autophagy: nuclear events gain more attention. Mol Cell 65:781–785

Bishop E, Bradshaw TD (2018) Autophagy modulation: a prudent approach in cancer treatment? Cancer Chemother Pharmacol 82:913–922

Boone BA, Bahary N, Zureikat AH, Moser AJ, Normolle DP, Wu WC, Singhi AD, Bao P, Bartlett DL, Liotta LA, Espina V, Loughran P, Lotze MT, Zeh HJ 3rd (2015) Safety and biologic response of pre-operative autophagy inhibition in combination with gemcitabine in patients with pancreatic adenocarcinoma. Ann Surg Oncol 22:4402–4410

Bortnik S, Gorski SM (2017) Clinical applications of autophagy proteins in cancer: from potential targets to biomarkers. Int J Mol Sci 18:1496

Briceno E, Reyes S, Sotelo J (2003) Therapy of glioblastoma multiforme improved by the antimutagenic chloroquine. Neurosurg Focus 14:e3

Budanov AV, Karin M (2008) p53 target genes sestrin1 and sestrin2 connect genotoxic stress and mTOR signaling. Cell 134:451–460

Cao QH, Liu F, Yang ZL, Fu XH, Yang ZH, Liu Q, Wang L, Wan XB, Fan XJ (2016) Prognostic value of autophagy related proteins ULK1, Beclin 1, ATG3, ATG5, ATG7, ATG9, ATG10, ATG12, LC3B and p62/SQSTM1 in gastric cancer. Am J Transl Res 8:3831–3847

Chatterjee A, Mukhopadhyay S, Tung K, Patel D, Foster DA (2015) Rapamycin-induced G1 cell cycle arrest employs both TGF-beta and Rb pathways. Cancer Lett 360:134–140

Chen P, Cescon M, Bonaldo P (2014) Autophagy-mediated regulation of macrophages and its applications for cancer. Autophagy 10:192–200

Cook KL, Shajahan AN, Clarke R (2011) Autophagy and endocrine resistance in breast cancer. Expert Rev Anticancer Ther 11:1283–1294

Crazzolara R, Cisterne A, Thien M, Hewson J, Baraz R, Bradstock KF, Bendall LJ (2009) Potentiating effects of RAD001 (Everolimus) on vincristine therapy in childhood acute lymphoblastic leukemia. Blood 113:3297–3306

Cufi S, Vazquez-Martin A, Oliveras-Ferraros C, Martin-Castillo B, Vellon L, Menendez JA (2011) Autophagy positively regulates the CD44(+) CD24(−/low) breast cancer stem-like phenotype. Cell Cycle 10:3871–3885

Delaney JR, Patel CB, Bapat J, Jones CM, Ramos-Zapatero M, Ortell KK, Tanios R, Haghighiabyaneh M, Axelrod J, DeStefano JW, Tancioni I, Schlaepfer DD, Harismendy O, La Spada AR, Stupack DG (2020) Autophagy gene haploinsufficiency drives chromosome instability, increases migration, and promotes early ovarian tumors. PLoS Genet 16:e1008558

Dikic I, Elazar Z (2018) Mechanism and medical implications of mammalian autophagy. Nat Rev Mol Cell Biol 19:349–364

Dong G, Si C, Zhang Q, Yan F, Li C, Zhang H, Ma Q, Dai J, Li Z, Shi H, Wang B, Zhang J, Ming J, Hu Y, Geng S, Zhang Y, Li L, Xiong H (2017) Autophagy regulates accumulation and functional activity of granulocytic myeloid-derived suppressor cells via STAT3 signaling in endotoxin shock. Biochim Biophys Acta Mol Basis Dis 1863:2796–2807

Dower CM, Wills CA, Frisch SM, Wang HG (2018) Mechanisms and context underlying the role of autophagy in cancer metastasis. Autophagy 14:1110–1128

Dyczynski M, Yu Y, Otrocka M, Parpal S, Braga T, Henley AB, Zazzi H, Lerner M, Wennerberg K, Viklund J, Martinsson J, Grander D, De Milito A, Pokrovskaja Tamm K (2018) Targeting autophagy by small molecule inhibitors of vacuolar protein sorting 34 (Vps34) improves the sensitivity of breast cancer cells to Sunitinib. Cancer Lett 435:32–43

Egan DF, Chun MG, Vamos M, Zou H, Rong J, Miller CJ, Lou HJ, Raveendra-Panickar D, Yang CC, Sheffler DJ, Teriete P, Asara JM, Turk BE, Cosford ND, Shaw RJ (2015) Small molecule inhibition of the autophagy kinase ULK1 and identification of ULK1 substrates. Mol Cell 59:285–297

Eisenberg-Lerner A, Kimchi A (2012) PKD is a kinase of Vps34 that mediates ROS-induced autophagy downstream of DAPk. Cell Death Differ 19:788–797

Eng CH, Wang Z, Tkach D, Toral-Barza L, Ugwonali S, Liu S, Fitzgerald SL, George E, Frias E, Cochran N, De Jesus R, McAllister G, Hoffman GR, Bray K, Lemon L, Lucas J, Fantin VR, Abraham RT, Murphy LO, Nyfeler B (2016) Macroautophagy is dispensable for growth of KRAS mutant tumors and chloroquine efficacy. Proc Natl Acad Sci U S A 113:182–187

Errafiy R, Aguado C, Ghislat G, Esteve JM, Gil A, Loutfi M, Knecht E (2013) PTEN increases autophagy and inhibits the ubiquitin-proteasome pathway in glioma cells independently of its lipid phosphatase activity. PLoS One 8:e83318

Espina V, Mariani BD, Gallagher RI, Tran K, Banks S, Wiedemann J, Huryk H, Mueller C, Adamo L, Deng J, Petricoin EF, Pastore L, Zaman S, Menezes G, Mize J, Johal J, Edmiston K, Liotta LA (2010) Malignant precursor cells pre-exist in human breast DCIS and require autophagy for survival. PLoS One 5:e10240

Eum KH, Lee M (2011) Targeting the autophagy pathway using ectopic expression of Beclin 1 in combination with rapamycin in drug-resistant v-Ha-ras-transformed NIH 3T3 cells. Mol Cells 31:231–238

Fitzwalter BE, Thorburn A (2015) Recent insights into cell death and autophagy. FEBS J 282:4279–4288

Fitzwalter BE, Thorburn A (2018) FOXO3 links autophagy to apoptosis. Autophagy 14:1467–1468

Folkerts H, Wierenga AT, van den Heuvel FA, Woldhuis RR, Kluit DS, Jaques J, Schuringa JJ, Vellenga E (2019) Elevated VMP1 expression in acute myeloid leukemia amplifies autophagy and is protective against venetoclax-induced apoptosis. Cell Death Dis 10:421

Fu Y, Hong L, Xu J, Zhong G, Gu Q, Gu Q, Guan Y, Zheng X, Dai Q, Luo X, Liu C, Huang Z, Yin XM, Liu P, Li M (2019) Discovery of a small molecule targeting autophagy via ATG4B inhibition and cell death of colorectal cancer cells in vitro and in vivo. Autophagy 15:295–311

Furuya N, Yu J, Byfield M, Pattingre S, Levine B (2005) The evolutionarily conserved domain of Beclin 1 is required for Vps34 binding, autophagy and tumor suppressor function. Autophagy 1:46–52

Gajate P, Martinez-Saez O, Alonso-Gordoa T, Grande E (2017) Emerging use of everolimus in the treatment of neuroendocrine tumors. Cancer Manag Res 9:215–224

Galluzzi L, Pietrocola F, Bravo-San Pedro JM, Amaravadi RK, Baehrecke EH, Cecconi F, Codogno P, Debnath J, Gewirtz DA, Karantza V, Kimmelman A, Kumar S, Levine B, Maiuri MC, Martin SJ, Penninger J, Piacentini M, Rubinsztein DC, Simon HU, Simonsen A, Thorburn AM, Velasco G, Ryan KM, Kroemer G (2015) Autophagy in malignant transformation and cancer progression. EMBO J 34:856–880

Gao W, Shen Z, Shang L, Wang X (2011) Upregulation of human autophagy-initiation kinase ULK1 by tumor suppressor p53 contributes to DNA-damage-induced cell death. Cell Death Differ 18:1598–1607

Ge J, Chen Z, Huang J, Chen J, Yuan W, Deng Z, Chen Z (2014) Upregulation of autophagy-related gene-5 (ATG-5) is associated with chemoresistance in human gastric cancer. PLoS One 9: e110293

Goel S, Huang J, Klampfer L (2015) K-Ras, intestinal homeostasis and colon cancer. Curr Clin Pharmacol 10:73–81

Gong C, Bauvy C, Tonelli G, Yue W, Delomenie C, Nicolas V, Zhu Y, Domergue V, Marin-Esteban V, Tharinger H, Delbos L, Gary-Gouy H, Morel AP, Ghavami S, Song E, Codogno P, Mehrpour M (2013) Beclin 1 and autophagy are required for the tumorigenicity of breast cancer stem-like/progenitor cells. Oncogene 32:2261–2272, 2272e.1–11

Guo JY, Chen HY, Mathew R, Fan J, Strohecker AM, Karsli-Uzunbas G, Kamphorst JJ, Chen G, Lemons JM, Karantza V, Coller HA, Dipaola RS, Gelinas C, Rabinowitz JD, White E (2011) Activated Ras requires autophagy to maintain oxidative metabolism and tumorigenesis. Genes Dev 25:460–470

Guo JY, Karsli-Uzunbas G, Mathew R, Aisner SC, Kamphorst JJ, Strohecker AM, Chen G, Price S, Lu W, Teng X, Snyder E, Santanam U, Dipaola RS, Jacks T, Rabinowitz JD, White E (2013a) Autophagy suppresses progression of K-ras-induced lung tumors to oncocytomas and maintains lipid homeostasis. Genes Dev 27:1447–1461

Guo JY, Xia B, White E (2013b) Autophagy-mediated tumor promotion. Cell 155:1216–1219

Hansen M, Rubinsztein DC, Walker DW (2018) Autophagy as a promoter of longevity: insights from model organisms. Nat Rev Mol Cell Biol 19:579–593

Hu LF (2019) Epigenetic regulation of autophagy. Adv Exp Med Biol 1206:221–236

Hu D, Jiang L, Luo S, Zhao X, Hu H, Zhao G, Tang W (2020) Development of an autophagy-related gene expression signature for prognosis prediction in prostate cancer patients. J Transl Med 18:160

Huang H, Tang J, Zhang L, Bu Y, Zhang X (2018) miR-874 regulates multiple-drug resistance in gastric cancer by targeting ATG16L1. Int J Oncol 53:2769–2779

Ichim G, Lopez J, Ahmed SU, Muthalagu N, Giampazolias E, Delgado ME, Haller M, Riley JS, Mason SM, Athineos D, Parsons MJ, van de Kooij B, Bouchier-Hayes L, Chalmers AJ,

Rooswinkel RW, Oberst A, Blyth K, Rehm M, Murphy DJ, Tait SWG (2015) Limited mitochondrial permeabilization causes DNA damage and genomic instability in the absence of cell death. Mol Cell 57:860–872

Inoue D, Suzuki T, Mitsuishi Y, Miki Y, Suzuki S, Sugawara S, Watanabe M, Sakurada A, Endo C, Uruno A, Sasano H, Nakagawa T, Satoh K, Tanaka N, Kubo H, Motohashi H, Yamamoto M (2012) Accumulation of p62/SQSTM1 is associated with poor prognosis in patients with lung adenocarcinoma. Cancer Sci 103:760–766

Jain K, Paranandi KS, Sridharan S, Basu A (2013) Autophagy in breast cancer and its implications for therapy. Am J Cancer Res 3:251–265

Jiao X, Zhao L, Ma M, Bai X, He M, Yan Y, Wang Y, Chen Q, Zhao X, Zhou M, Cui Z, Zheng Z, Wang E, Wei M (2013) MiR-181a enhances drug sensitivity in mitoxantone-resistant breast cancer cells by targeting breast cancer resistance protein (BCRP/ABCG2). Breast Cancer Res Treat 139:717–730

Jutten B, Keulers TG, Peeters HJM, Schaaf MBE, Savelkouls KGM, Compter I, Clarijs R, Schijns O, Ackermans L, Teernstra OPM, Zonneveld MI, Colaris RME, Dubois L, Vooijs MA, Bussink J, Sotelo J, Theys J, Lammering G, Rouschop KMA (2018) EGFRvIII expression triggers a metabolic dependency and therapeutic vulnerability sensitive to autophagy inhibition. Autophagy 14:283–295

Kang MR, Kim MS, Oh JE, Kim YR, Song SY, Kim SS, Ahn CH, Yoo NJ, Lee SH (2009) Frameshift mutations of autophagy-related genes ATG2B, ATG5, ATG9B and ATG12 in gastric and colorectal cancers with microsatellite instability. J Pathol 217:702–706

Kenific CM, Thorburn A, Debnath J (2010) Autophagy and metastasis: another double-edged sword. Curr Opin Cell Biol 22:241–245

Kenzelmann Broz D, Spano Mello S, Bieging KT, Jiang D, Dusek RL, Brady CA, Sidow A, Attardi LD (2013) Global genomic profiling reveals an extensive p53-regulated autophagy program contributing to key p53 responses. Genes Dev 27:1016–1031

Kim JH, Kim HY, Lee YK, Yoon YS, Xu WG, Yoon JK, Choi SE, Ko YG, Kim MJ, Lee SJ, Wang HJ, Yoon G (2011a) Involvement of mitophagy in oncogenic K-Ras-induced transformation: overcoming a cellular energy deficit from glucose deficiency. Autophagy 7:1187–1198

Kim MJ, Woo SJ, Yoon CH, Lee JS, An S, Choi YH, Hwang SG, Yoon G, Lee SJ (2011b) Involvement of autophagy in oncogenic K-Ras-induced malignant cell transformation. J Biol Chem 286:12924–12932

Kim KW, Paul P, Qiao J, Lee S, Chung DH (2013) Enhanced autophagy blocks angiogenesis via degradation of gastrin-releasing peptide in neuroblastoma cells. Autophagy 9:1579–1590

Kim YH, Baek SH, Kim EK, Ha JM, Jin SY, Lee HS, Ha HK, Song SH, Kim SJ, Shin HK, Yong J, Kim DH, Kim CD, Bae SS (2016) Uncoordinated 51-like kinase 2 signaling pathway regulates epithelial-mesenchymal transition in A549 lung cancer cells. FEBS Lett 590:1365–1374

Kimmelman AC (2011) The dynamic nature of autophagy in cancer. Genes Dev 25:1999–2010

Kimmelman AC, White E (2017) Autophagy and tumor metabolism. Cell Metab 25:1037–1043

Kinsey CG, Camolotto SA, Boespflug AM, Guillen KP, Foth M, Truong A, Schuman SS, Shea JE, Seipp MT, Yap JT, Burrell LD, Lum DH, Whisenant JR, Gilcrease GW 3rd, Cavalieri CC, Rehbein KM, Cutler SL, Affolter KE, Welm AL, Welm BE, Scaife CL, Snyder EL, McMahon M (2019) Protective autophagy elicited by RAF→MEK→ERK inhibition suggests a treatment strategy for RAS-driven cancers. Nat Med 25:620–627

Kitamura H, Torigoe T, Asanuma H, Hisasue SI, Suzuki K, Tsukamoto T, Satoh M, Sato N (2006) Cytosolic overexpression of p62 sequestosome 1 in neoplastic prostate tissue. Histopathology 48:157–161

Kiyono K, Suzuki HI, Matsuyama H, Morishita Y, Komuro A, Kano MR, Sugimoto K, Miyazono K (2009) Autophagy is activated by TGF-beta and potentiates TGF-beta-mediated growth inhibition in human hepatocellular carcinoma cells. Cancer Res 69:8844–8852

Klionsky DJ (2008) Autophagy revisited: a conversation with Christian de Duve. Autophagy 4:740–743

Klionsky DJ (2020) Autophagy participates in, well, just about everything. Cell Death Differ 27:831–832

Kocaturk NM, Akkoc Y, Kig C, Bayraktar O, Gozuacik D, Kutlu O (2019) Autophagy as a molecular target for cancer treatment. Eur J Pharm Sci 134:116–137

Kung CP, Budina A, Balaburski G, Bergenstock MK, Murphy M (2011) Autophagy in tumor suppression and cancer therapy. Crit Rev Eukaryot Gene Expr 21:71–100

La Belle Flynn A, Calhoun BC, Sharma A, Chang JC, Almasan A, Schiemann WP (2019) Autophagy inhibition elicits emergence from metastatic dormancy by inducing and stabilizing Pfkfb3 expression. Nat Commun 10:3668

Lazova R, Camp RL, Klump V, Siddiqui SF, Amaravadi RK, Pawelek JM (2012) Punctate LC3B expression is a common feature of solid tumors and associated with proliferation, metastasis, and poor outcome. Clin Cancer Res 18:370–379

Lebovitz CB, Robertson AG, Goya R, Jones SJ, Morin RD, Marra MA, Gorski SM (2015) Cross-cancer profiling of molecular alterations within the human autophagy interaction network. Autophagy 11:1668–1687

Lee JG, Wu R (2012) Combination erlotinib-cisplatin and Atg3-mediated autophagy in erlotinib resistant lung cancer. PLoS One 7:e48532

Levy JM, Thorburn A (2011) Targeting autophagy during cancer therapy to improve clinical outcomes. Pharmacol Ther 131:130–141

Li J, Yang B, Zhou Q, Wu Y, Shang D, Guo Y, Song Z, Zheng Q, Xiong J (2013) Autophagy promotes hepatocellular carcinoma cell invasion through activation of epithelial-mesenchymal transition. Carcinogenesis 34:1343–1351

Li SS, Xu LZ, Zhou W, Yao S, Wang CL, Xia JL, Wang HF, Kamran M, Xue XY, Dong L, Wang J, Ding XD, Bella L, Bugeon L, Xu J, Zheng FM, Dallman MJ, Lam EWF, Liu Q (2017) p62/SQSTM1 interacts with vimentin to enhance breast cancer metastasis. Carcinogenesis 38:1092–1103

Li X, He S, Ma B (2020) Autophagy and autophagy-related proteins in cancer. Mol Cancer 19:12

Lin J, Lu X, Liao S, Chen X, Wang S, Zhao C, Li X, Xu YZ, Liu HF, Pan Q (2019) Cross-regulation between exosomal and autophagic pathways: promising therapy targets in disease. Discov Med 27:201–210

Liu H, Ma Y, He HW, Zhao WL, Shao RG (2017a) SPHK1 (sphingosine kinase 1) induces epithelial-mesenchymal transition by promoting the autophagy-linked lysosomal degradation of CDH1/E-cadherin in hepatoma cells. Autophagy 13:900–913

Liu J, Li M, Wang Y, Luo J (2017b) Curcumin sensitizes prostate cancer cells to radiation partly via epigenetic activation of miR-143 and miR-143 mediated autophagy inhibition. J Drug Target 25:645–652

Lock R, Debnath J (2008) Extracellular matrix regulation of autophagy. Curr Opin Cell Biol 20:583–588

Lock R, Kenific CM, Leidal AM, Salas E, Debnath J (2014) Autophagy-dependent production of secreted factors facilitates oncogenic RAS-driven invasion. Cancer Discov 4:466–479

Maastricht Radiation Oncology (2020) The addition of chloroquine to chemoradiation for glioblastoma. https://ClinicalTrials.gov/show/NCT02378532

Mancias JD, Kimmelman AC (2011) Targeting autophagy addiction in cancer. Oncotarget 2:1302–1306

Mao D, Zhang Z, Zhao X, Dong X (2020) Autophagy-related genes prognosis signature as potential predictive markers for immunotherapy in hepatocellular carcinoma. Peer J 8:e8383

Marinkovic M, Sprung M, Buljubasic M, Novak I (2018) Autophagy modulation in cancer: current knowledge on action and therapy. Oxidative Med Cell Longev 2018:8023821

Massachusetts General Hospital (2019) Erlotinib with or without hydroxychloroquine in chemo-naive advanced NSCLC and (EGFR) mutations. https://ClinicalTrials.gov/show/NCT00977470

Mathew R, Karp CM, Beaudoin B, Vuong N, Chen G, Chen HY, Bray K, Reddy A, Bhanot G, Gelinas C, Dipaola RS, Karantza-Wadsworth V, White E (2009) Autophagy suppresses tumorigenesis through elimination of p62. Cell 137:1062–1075

May CD, Sphyris N, Evans KW, Werden SJ, Guo W, Mani SA (2011) Epithelial-mesenchymal transition and cancer stem cells: a dangerously dynamic duo in breast cancer progression. Breast Cancer Res 13:202

McAfee Q, Zhang Z, Samanta A, Levi SM, Ma XH, Piao S, Lynch JP, Uehara T, Sepulveda AR, Davis LE, Winkler JD, Amaravadi RK (2012) Autophagy inhibitor Lys05 has single-agent antitumor activity and reproduces the phenotype of a genetic autophagy deficiency. Proc Natl Acad Sci U S A 109:8253–8258

Milano V, Piao Y, LaFortune T, de Groot J (2009) Dasatinib-induced autophagy is enhanced in combination with temozolomide in glioma. Mol Cancer Ther 8:394–406

Mo S, Dai W, Xiang W, Li Y, Feng Y, Zhang L, Li Q, Cai G (2019) Prognostic and predictive value of an autophagy-related signature for early relapse in stages I-III colon cancer. Carcinogenesis 40:861–870

Morselli E, Shen S, Ruckenstuhl C, Bauer MA, Marino G, Galluzzi L, Criollo A, Michaud M, Maiuri MC, Chano T, Madeo F, Kroemer G (2011) p53 inhibits autophagy by interacting with the human ortholog of yeast Atg17, RB1CC1/FIP200. Cell Cycle 10:2763–2769

Mowers EE, Sharifi MN, Macleod KF (2017) Autophagy in cancer metastasis. Oncogene 36:1619–1630

Mowers EE, Sharifi MN, Macleod KF (2018) Functions of autophagy in the tumor microenvironment and cancer metastasis. FEBS J 285:1751–1766

Mulcahy Levy JM, Thorburn A (2020) Autophagy in cancer: moving from understanding mechanism to improving therapy responses in patients. Cell Death Differ 27:843–857

Murakami N, Oyama F, Gu Y, McLennan IS, Nonaka I, Ihara Y (1998) Accumulation of tau in autophagic vacuoles in chloroquine myopathy. J Neuropathol Exp Neurol 57:664–673

Nazio F, Bordi M, Cianfanelli V, Locatelli F, Cecconi F (2019) Autophagy and cancer stem cells: molecular mechanisms and therapeutic applications. Cell Death Differ 26:690–702

Neill T, Chen CG, Buraschi S, Iozzo RV (2020) Catabolic degradation of endothelial VEGFA via autophagy. J Biol Chem 295:6064–6079

Ngabire D, Kim GD (2017) Autophagy and inflammatory response in the tumor microenvironment. Int J Mol Sci 18:2016

Nikoletopoulou V, Markaki M, Palikaras K, Tavernarakis N (2013) Crosstalk between apoptosis, necrosis and autophagy. Biochim Biophys Acta 1833:3448–3459

Ohsumi Y (2016) MLA style: press release. NobelPrize.org. Nobel Media AB 2020. 11 May 2020. https://www.nobelprize.org/prizes/medicine/2016/press-release/

Ostrand-Rosenberg S, Beury DW, Parker KH, Horn LA (2020) Survival of the fittest: how myeloid-derived suppressor cells survive in the inhospitable tumor microenvironment. Cancer Immunol Immunother 69:215–221

Parzych KR, Klionsky DJ (2014) An overview of autophagy: morphology, mechanism, and regulation. Antioxid Redox Signal 20:460–473

Pattingre S, Tassa A, Qu X, Garuti R, Liang XH, Mizushima N, Packer M, Schneider MD, Levine B (2005) Bcl-2 antiapoptotic proteins inhibit Beclin 1-dependent autophagy. Cell 122:927–939

Peiris-Pages M, Smith DL, Gyorffy B, Sotgia F, Lisanti MP (2015) Proteomic identification of prognostic tumour biomarkers, using chemotherapy-induced cancer-associated fibroblasts. Aging (Albany NY) 7:816–838

Pellegrini P, Strambi A, Zipoli C, Hagg-Olofsson M, Buoncervello M, Linder S, De Milito A (2014) Acidic extracellular pH neutralizes the autophagy-inhibiting activity of chloroquine: implications for cancer therapies. Autophagy 10:562–571

Qiang L, Zhao B, Ming M, Wang N, He TC, Hwang S, Thorburn A, He YY (2014) Regulation of cell proliferation and migration by p62 through stabilization of Twist1. Proc Natl Acad Sci U S A 111:9241–9246

Qu X, Yu J, Bhagat G, Furuya N, Hibshoosh H, Troxel A, Rosen J, Eskelinen EL, Mizushima N, Ohsumi Y, Cattoretti G, Levine B (2003) Promotion of tumorigenesis by heterozygous disruption of the beclin 1 autophagy gene. J Clin Invest 112:1809–1820

Ramakrishnan S, Nguyen TM, Subramanian IV, Kelekar A (2007) Autophagy and angiogenesis inhibition. Autophagy 3:512–515

Rebecca VW, Nicastri MC, McLaughlin N, Fennelly C, McAfee Q, Ronghe A, Nofal M, Lim CY, Witze E, Chude CI, Zhang G, Alicea GM, Piao S, Murugan S, Ojha R, Levi SM, Wei Z, Barber-Rotenberg JS, Murphy ME, Mills GB, Lu Y, Rabinowitz J, Marmorstein R, Liu Q, Liu S, Xu X, Herlyn M, Zoncu R, Brady DC, Speicher DW, Winkler JD, Amaravadi RK (2017) A unified approach to targeting the lysosome's degradative and growth signaling roles. Cancer Discov 7:1266–1283

Riccardi F, Colantuoni G, Diana A, Mocerino C, Carteni G, Lauria R, Febbraro A, Nuzzo F, Addeo R, Marano O, Incoronato P, De Placido S, Ciardiello F, Orditura M (2018) Exemestane and Everolimus combination treatment of hormone receptor positive, HER2 negative metastatic breast cancer: a retrospective study of 9 cancer centers in the Campania Region (Southern Italy) focused on activity, efficacy and safety. Mol Clin Oncol 9:255–263

Riley T, Sontag E, Chen P, Levine A (2008) Transcriptional control of human p53-regulated genes. Nat Rev Mol Cell Biol 9:402–412

Ruivo CF, Adem B, Silva M, Melo SA (2017) The biology of cancer exosomes: insights and new perspectives. Cancer Res 77:6480–6488

Sasahira T, Kirita T (2018) Hallmarks of cancer-related newly prognostic factors of oral squamous cell carcinoma. Int J Mol Sci 19:2413

Selvakumaran M, Amaravadi RK, Vasilevskaya IA, O'Dwyer PJ (2013) Autophagy inhibition sensitizes colon cancer cells to antiangiogenic and cytotoxic therapy. Clin Cancer Res 19:2995–3007

Shin HJ, Kim H, Oh S, Lee JG, Kee M, Ko HJ, Kweon MN, Won KJ, Baek SH (2016a) AMPK-SKP2-CARM1 signalling cascade in transcriptional regulation of autophagy. Nature 534:553–557

Shin HR, Kim H, Kim KI, Baek SH (2016b) Epigenetic and transcriptional regulation of autophagy. Autophagy 12:2248–2249

Shintani T, Klionsky DJ (2004) Autophagy in health and disease: a double-edged sword. Science 306:990–995

Shuhua W, Chenbo S, Yangyang L, Xiangqian G, Shuang H, Tangyue L, Dong T (2015) Autophagy-related genes Raptor, Rictor, and Beclin1 expression and relationship with multi-drug resistance in colorectal carcinoma. Hum Pathol 46:1752–1759

Singh BN, Kumar D, Shankar S, Srivastava RK (2012) Rottlerin induces autophagy which leads to apoptotic cell death through inhibition of PI3K/Akt/mTOR pathway in human pancreatic cancer stem cells. Biochem Pharmacol 84:1154–1163

Singh SS, Vats S, Chia AY, Tan TZ, Deng S, Ong MS, Arfuso F, Yap CT, Goh BC, Sethi G, Huang RY, Shen HM, Manjithaya R, Kumar AP (2018) Dual role of autophagy in hallmarks of cancer. Oncogene 37:1142–1158

Song J, Qu Z, Guo X, Zhao Q, Zhao X, Gao L, Sun K, Shen F, Wu M, Wei L (2009) Hypoxia-induced autophagy contributes to the chemoresistance of hepatocellular carcinoma cells. Autophagy 5:1131–1144

Sophia J, Kowshik J, Dwivedi A, Bhutia SK, Manavathi B, Mishra R, Nagini S (2018) Nimbolide, a neem limonoid inhibits cytoprotective autophagy to activate apoptosis via modulation of the PI3K/Akt/GSK-3beta signalling pathway in oral cancer. Cell Death Dis 9:1087

Su Y, Qian H, Zhang J, Wang S, Shi P, Peng X (2005) The diversity expression of p62 in digestive system cancers. Clin Immunol 116:118–123

Takahashi Y, Coppola D, Matsushita N, Cualing HD, Sun M, Sato Y, Liang C, Jung JU, Cheng JQ, Mule JJ, Pledger WJ, Wang HG (2007) Bif-1 interacts with Beclin 1 through UVRAG and regulates autophagy and tumorigenesis. Nat Cell Biol 9:1142–1151

Takamura A, Komatsu M, Hara T, Sakamoto A, Kishi C, Waguri S, Eishi Y, Hino O, Tanaka K, Mizushima N (2011) Autophagy-deficient mice develop multiple liver tumors. Genes Dev 25:795–800

Tanagala KKK, Baba AB, Kowshik J, Reddy GB, Nagini S (2018) Gedunin, a neem limonoid in combination with epalrestat inhibits cancer hallmarks by attenuating aldose reductase-driven oncogenic signaling in SCC131 oral cancer cells. Anti Cancer Agents Med Chem 18:2042–2052

Tasdemir E, Maiuri MC, Galluzzi L, Vitale I, Djavaheri-Mergny M, D'Amelio M, Criollo A, Morselli E, Zhu C, Harper F, Nannmark U, Samara C, Pinton P, Vicencio JM, Carnuccio R, Moll UM, Madeo F, Paterlini-Brechot P, Rizzuto R, Szabadkai G, Pierron G, Blomgren K, Tavernarakis N, Codogno P, Cecconi F, Kroemer G (2008) Regulation of autophagy by cytoplasmic p53. Nat Cell Biol 10:676–687

Towers CG, Thorburn A (2016) Therapeutic targeting of autophagy. EBioMedicine 14:15–23

Umemura A, He F, Taniguchi K, Nakagawa H, Yamachika S, Font-Burgada J, Zhong Z, Subramaniam S, Raghunandan S, Duran A, Linares JF, Reina-Campos M, Umemura S, Valasek MA, Seki E, Yamaguchi K, Koike K, Itoh Y, Diaz-Meco MT, Moscat J, Karin M (2016) p62, upregulated during preneoplasia, induces hepatocellular carcinogenesis by maintaining survival of stressed HCC-initiating cells. Cancer Cell 29:935–948

Vidoni C, Ferraresi A, Secomandi E, Vallino L, Dhanasekaran DN, Isidoro C (2019) Epigenetic targeting of autophagy for cancer prevention and treatment by natural compounds. Semin Cancer Biol. https://doi.org/10.1016/j.semcancer.2019.04.006

Vijayarathna S, Gothai S, Jothy SL, Chen Y, Kanwar JR, Sasidharan S (2015) Can cancer therapy be achieved by bridging apoptosis and autophagy: a method based on microRNA-dependent gene therapy and phytochemical targets. Asian Pac J Cancer Prev 16:7435–7439

Vion AC, Kheloufi M, Hammoutene A, Poisson J, Lasselin J, Devue C, Pic I, Dupont N, Busse J, Stark K, Lafaurie-Janvore J, Barakat AI, Loyer X, Souyri M, Viollet B, Julia P, Tedgui A, Codogno P, Boulanger CM, Rautou PE (2017) Autophagy is required for endothelial cell alignment and atheroprotection under physiological blood flow. Proc Natl Acad Sci U S A 114:E8675–E8684

Wang J, Wu GS (2014) Role of autophagy in cisplatin resistance in ovarian cancer cells. J Biol Chem 289:17163–17173

Wang Y, Gan G, Wang B, Wu J, Cao Y, Zhu D, Xu Y, Wang X, Han H, Li X, Ye M, Zhao J, Mi J (2017) Cancer-associated fibroblasts promote irradiated cancer cell recovery through autophagy. EBioMedicine 17:45–56

Warr MR, Binnewies M, Flach J, Reynaud D, Garg T, Malhotra R, Debnath J, Passegue E (2013) FOXO3A directs a protective autophagy program in haematopoietic stem cells. Nature 494:323–327

Wei Y, Sinha S, Levine B (2008) Dual role of JNK1-mediated phosphorylation of Bcl-2 in autophagy and apoptosis regulation. Autophagy 7:949–951

Wen ZF, Liu H, Gao R, Zhou M, Ma J, Zhang Y, Zhao J, Chen Y, Zhang T, Huang F, Pan N, Zhang J, Fox BA, Hu HM, Wang LX (2018) Tumor cell-released autophagosomes (TRAPs) promote immunosuppression through induction of M2-like macrophages with increased expression of PD-L1. J Immunother Cancer 6:151

Weyerhauser P, Kantelhardt SR, Kim EL (2018) Re-purposing chloroquine for glioblastoma: potential merits and confounding variables. Front Oncol 8:335

Wible DJ, Chao HP, Tang DG, Bratton SB (2019) ATG5 cancer mutations and alternative mRNA splicing reveal a conjugation switch that regulates ATG12-ATG5-ATG16L1 complex assembly and autophagy. Cell Discov 5:42

Xiong L, Liu Z, Ouyang G, Lin L, Huang H, Kang H, Chen W, Miao X, Wen Y (2017) Autophagy inhibition enhances photocytotoxicity of Photosan-II in human colorectal cancer cells. Oncotarget 8:6419–6432

Xu N, Zhang J, Shen C, Luo Y, Xia L, Xue F, Xia Q (2012) Cisplatin-induced downregulation of miR-199a-5p increases drug resistance by activating autophagy in HCC cell. Biochem Biophys Res Commun 423:826–831

Xu J, Camfield R, Gorski SM (2018a) The interplay between exosomes and autophagy—partners in crime. J Cell Sci 131:jcs215210

Xu R, Ji Z, Xu C, Zhu J (2018b) The clinical value of using chloroquine or hydroxychloroquine as autophagy inhibitors in the treatment of cancers: a systematic review and meta-analysis. Medicine (Baltimore) 97:e12912

Yang Y, Klionsky DJ (2020) Autophagy and disease: unanswered questions. Cell Death Differ 27:858–871

Yang S, Wang X, Contino G, Liesa M, Sahin E, Ying H, Bause A, Li Y, Stommel JM, Dell'antonio G, Mautner J, Tonon G, Haigis M, Shirihai OS, Doglioni C, Bardeesy N, Kimmelman AC (2011) Pancreatic cancers require autophagy for tumor growth. Genes Dev 25:717–729

Yang YP, Hu LF, Zheng HF, Mao CJ, Hu WD, Xiong KP, Wang F, Liu CF (2013) Application and interpretation of current autophagy inhibitors and activators. Acta Pharmacol Sin 34:625–635

Yang A, Herter-Sprie G, Zhang H, Lin EY, Biancur D, Wang X, Deng J, Hai J, Yang S, Wong KK, Kimmelman AC (2018) Autophagy sustains pancreatic cancer growth through both cell-autonomous and nonautonomous mechanisms. Cancer Discov 8:276–287

Yazbeck VY, Buglio D, Georgakis GV, Li Y, Iwado E, Romaguera JE, Kondo S, Younes A (2008) Temsirolimus downregulates p21 without altering cyclin D1 expression and induces autophagy and synergizes with vorinostat in mantle cell lymphoma. Exp Hematol 36:443–450

Youle RJ, Narendra DP (2011) Mechanisms of mitophagy. Nat Rev Mol Cell Biol 12:9–14

Yue Z, Jin S, Yang C, Levine AJ, Heintz N (2003) Beclin 1, an autophagy gene essential for early embryonic development, is a haploinsufficient tumor suppressor. Proc Natl Acad Sci U S A 100:15077–15082

Zalckvar E, Berissi H, Eisenstein M, Kimchi A (2009) Phosphorylation of Beclin 1 by DAP-kinase promotes autophagy by weakening its interactions with Bcl-2 and Bcl-XL. Autophagy 5:720–722

Zhai H, Fesler A, Ba Y, Wu S, Ju J (2015) Inhibition of colorectal cancer stem cell survival and invasive potential by hsa-miR-140-5p mediated suppression of Smad2 and autophagy. Oncotarget 6:19735–19746

Zhang Y, Li F, Liu L, Jiang H, Jiang X, Ge X, Cao J, Wang Z, Zhang L, Wang Y (2018) Salinomycin-induced autophagy blocks apoptosis via the ATG3/AKT/mTOR signaling axis in PC-3 cells. Life Sci 207:451–460

Zhi X, Feng W, Rong Y, Liu R (2018) Anatomy of autophagy: from the beginning to the end. Cell Mol Life Sci 75:815–831

Zhou Z, Mo S, Dai W, Ying Z, Zhang L, Xiang W, Han L, Wang Z, Li Q, Wang R, Cai G (2019) Development and validation of an autophagy score signature for the prediction of post-operative survival in colorectal cancer. Front Oncol 9:878

Targeting Autophagy in Cancer: Therapeutic Implications

12

Dipanwita Das Mukherjee, Satabdi Datta Choudhury, and Gopal Chakrabarti

Abstract

Autophagy is an evolutionary conserved self-degradation process that occurs ubiquitously in eukaryotes. It plays an important role in maintenance of cellular homeostasis by balancing the energy resources or through removal of misfolded proteins and damaged organelles. During autophagy, the recycling of the long-lived proteins or organelles is executed through their engulfment into double-membrane autophagosome followed by their lysosomal degradation via formation of autophagolysosome. Interestingly, autophagy is under tight regulation by a group of genes called autophagy-related genes (ATG) in association with various signalling pathways. Literature review suggests that autophagy is implicated in numerous developmental and other physiological processes such as cell differentiation, cell survival, cell death, nutrient starvation response and its dysregulation, often, leads to many pathological conditions including cancer. Generally, under normal physiological conditions, basal autophagy occurs in all cells but it is induced only in response to specific intra- or extra-cellular stimuli. In cancer, depending on the context, autophagy can be paradoxical in nature

---

D. Das Mukherjee
Department of Biotechnology and Dr. B. C. Guha Centre for Genetic Engineering and Biotechnology, University of Calcutta, Kolkata, West Bengal, India

Division of Molecular Medicine, Bose Institute, Kolkata, West Bengal, India

S. Datta Choudhury
Department of Biotechnology and Dr. B. C. Guha Centre for Genetic Engineering and Biotechnology, University of Calcutta, Kolkata, West Bengal, India

Department of Zoology, Sri Guru Granth Sahib World University, Fatehgarh Sahib, Punjab, India

G. Chakrabarti (✉)
Department of Biotechnology and Dr. B. C. Guha Centre for Genetic Engineering and Biotechnology, University of Calcutta, Kolkata, West Bengal, India
e-mail: gcbcg@caluniv.ac.in

© Springer Nature Singapore Pte Ltd. 2020
S. K. Bhutia (ed.), *Autophagy in Tumor and Tumor Microenvironment*,
https://doi.org/10.1007/978-981-15-6930-2_12

(i.e. tumour-suppressive or tumour-promoting) and has also been documented to have the remarkable role in development of chemoresistance, thus, justifying the effectiveness of cancer therapeutic intervention through stimulation or inhibition of autophagy. Henceforth, in this chapter, we have summarised the autophagy in a nutshell, with focus on its mechanism, monitoring methods, regulation and context-dependent role in cancer and explored how the manipulation of autophagy could be beneficial towards improved cancer cure, as evident from the numerous in vitro and in vivo studies as well as clinical trials.

Keywords

Autophagy · Cancer therapy · Chemoresistance · Autophagy regulated genes · Clinical trials

12.1 Introduction

Autophagy is a dynamic process responsible for degradation and turnover of cellular proteins and organelle. It is accomplished through sequestration of target cellular constituents or organelle into double-membrane vesicles called autophagosome, which in turn fuses with lysosome to form autophagolysosome, wherein they are degraded by lysosomal proteases for recycling. It is an evolutionary conserved ubiquitous process occurring in eukaryotes (Klionsky and Emr 2000; Levine and Klionsky 2004). In recent times, autophagy has gained immense attention in clinical research owing to their versatile role in diverse physiological and pathophysiological conditions, amongst which cancer is of particular importance. Autophagy occurs at basal level in most of the cells. However, it can also be induced in response to specific stimuli, wherein autophagy is context-dependent (Mizushima 2007; White 2012; Amaravadi et al. 2016). However, the differences between basal autophagy and stimuli-induced autophagy and their relevance are not yet well-understood.

The Greek term 'autophagy' meaning 'self-eating' was coined by Christian de Duve in 1963 based on the electron microscopic studies displaying single or double-membrane vesicles containing parts of sequestered cytoplasm with variable degree of disintegrated organelles, especially mitochondria and other intracellular structures. Remarkable progress in understanding autophagy has been reported in the last few decades by decoding its molecular mechanism and significance in various physiological processes (Klionsky 2007; Levy et al. 2017). The breakthrough discovery of the detailed mechanism of regulation and execution of autophagy at molecular level in yeast *Saccharomyces cerevisiae* by Yoshinori Ohsumi has been awarded the 2016 Nobel prize in Physiology and Medicine.

12.2 Types of Autophagy

Autophagy can be categorised into three different types: macro-autophagy, micro-autophagy and chaperon-mediated autophagy. Macroautophagy involves engulfment of cytoplasmic proteins and organelle into double-membrane bound vesicles called autophagosome, which is trafficked to lysosome to form autolysosome for degradation by lysosomal proteases. In contrast, microautophagy is characterised by the internalization of the substrate through invagination of lysosomal or endosomal membrane followed by their lysosomal degradation (Li et al. 2012). However, in chaperon-mediated autophagy (CMA), the cargo protein contains KFERQ-like pentapeptide motif, which is recognised by cytosolic chaperone protein called heat shock cognate 70 (HSP-70) for their translocation to the lysosomal lumen through interaction with lysosomal-associated membrane protein 2A (LAMP 2A) receptor (Kaushik et al. 2011). Although both micro and macro-autophagy are capable of targeting large structures through selective and non-selective mechanism, CMA is constitutively selective in nature and thereby, restricted to turnover of specific protein with well-defined KFERQ motif. It is important to note that non-selective autophagy involves the direct engulfment of the cytoplasm and its components into the autophagosome (in macroautophagy) or through invagination of the lysosomal membrane (in microautophagy). While, in contrast, selective autophagy is mediated by specific targeting of the cargo, either cellular proteins or organelles, hallmarked with degradation signal (most commonly, ubiquitin in mammals) through interaction with autophagy cargo receptor, which serves as molecular bridge, for their degradation by autophagy (Kaur and Debnath 2015; Levy et al. 2017).

12.3 Mechanism of Autophagy

Autophagy is a complex, multi-step process under the intricate control of a set of 30 evolutionary conserved, autophagy-regulated genes (ATG), which were identified in yeast and mostly, have well-recognised mammalian orthologue. It divided into three distinct stages: autophagosome biogenesis, fusion with lysosome and lysosomal degradation of intravesicular constituents (Fig. 12.1). The autophagosome formation is initiated at the phagophore assembly point through the activation of ULK (UNC 51-like kinase) complex comprising of ULK1, ULK2 and ATG13, FIP200 (FAK family kinase interacting protein of 200 kDa) and ATG 101. This is followed by the nucleation stage when the ULK complex targets class III PI3 kinase complex—consisting of Beclin 1 (Atg6 in yeast), VPS34 (vacuolar protein sorting 34; also known as PIK3C3), ATG14, UVRAG (UV radiation resistance-associated gene protein; also known as p63) and AMBRA1 (activating molecule in BECN1-regulated autophagy protein 1)—promotes production of autophagosome-specific phosphatidyl-inositol-3-phosphate. Finally, the ATG5–ATG12–ATG16 complex along with ATG4B–ATG7 complex facilitates the expansion of the autophagosome membrane through lipidation of the microtubule-associated protein light chain 1 (LC3I), which is the mammalian homologue of yeast Atg8, and GABARAP (γ-aminobutyric type A

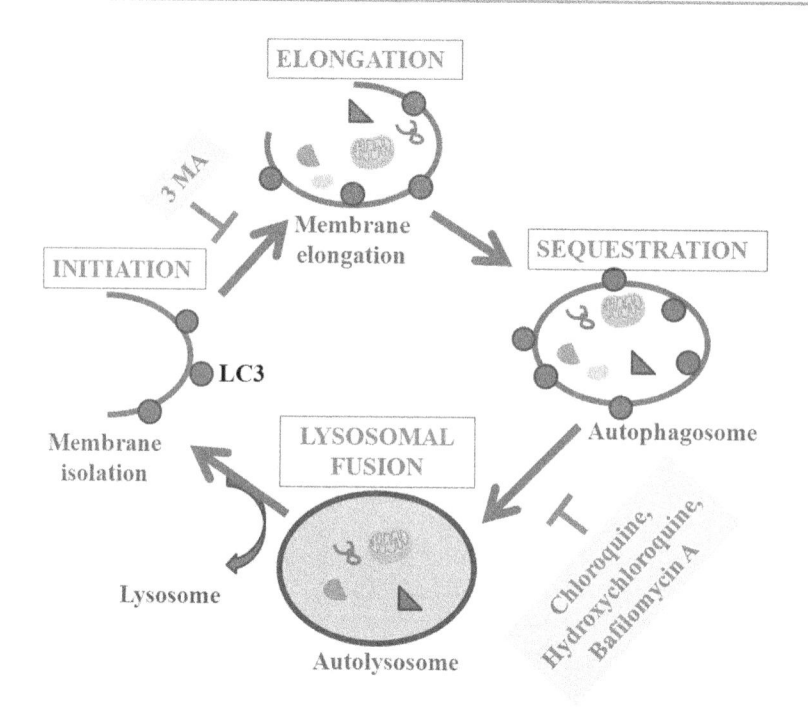

Fig. 12.1 The mechanism of autophagy. Autophagy is a multistep cellular process comprising of autophagosome initiation, elongation of the autophagosomal membrane, sequestration of the cargo and fusion of the autophagosome with the lysosome for degradation of the constituents. It can be inhibited at particular steps by specific inhibitors (such as 3MA—early phase autophagy inhibitor that inhibits autophagosome formation and chloroquine, hydroxychloroquine, bafilomycin A—late phase autophagy inhibitors that prevents the fusion of autophagosome with lysosome)

(GABAA)-receptor associated protein) with phosphatidylethanolamine (PE) to form LC3II and GABARAP-II, which in turn co-localise followed by their recruitment to the membrane. Interestingly, LC3BII, the well-known autophagosome marker found on the autophagosomal membrane, has been reported to facilitate the hemifusion of membranes and cargo selection for degradation, possibly through regulation of variable protein–protein interaction (Mizushima 2007; Levy et al. 2017; Meijer and Codogno 2004; Mizushima et al. 2011; Onorati et al. 2018). However, the significance of LCB-related molecules in autophagy needs further investigation (Fig. 12.1).

12.4 Methods of Monitoring Autophagy

In present-day autophagy research, the detection and quantification of autophagosome along with biochemical validation of the autophagic markers comprises the principal methods of monitoring autophagy. The electron microscopy is the most conventional and oldest method that enables the visualisation of the autophagosome at the ultrastructural level. It is of immense interest to note that in

1950s, the autophagy was first discovered by electron microscopic study of the lysosome (Klionsky 2007). However, advancement of autophagy research called for the formulation of easier and more accessible assays of autophagy detection. The advent of LC3B as the signature of autophagosome has simplified the revelation of autophagy by the light microscopic detection of LC3B or GFP–LC3B puncta. Finally, the conversion of LC3I to LC3 II by immunoblotting with LC3 antibody is the widely employed biochemical assay to confirm autophagy. Furthermore, immunoblot depicting the turnover of p62 is also used to expose autophagy (Mizushima 2004; Mizushima et al. 2010; Yoshii and Mizushima 2017).

Nonetheless, Levine et al. have highlighted the misconception of the direct correlation of the number of autophagosome with the autophagic activity (Mizushima et al. 2010). Owing to the dynamic nature, at any point of time, the number of autophagosome is the function of the balance between their formation rate and fusion rate with the lysosome. Henceforth, the autophagosome accumulation represents either induction of autophagy or suppression of the downstream pathway necessitating the measurement of the autophagic flux, in absence and presence of pharmacological inhibitors and activators, as an essential parameter for uncovering the status of autophagy. The commonly used pharmacological inhibitors include PI3-kinase inhibitors (such as wortmannin, 3-MA and LY294002), microtubule-disrupting agents (e.g. nocodazole), etc. while rapamycin and its analogue, CCI-779, BH3 mimetics (ABT737) and many others are used as autophagy activators. Further, manipulation of the autophagy by knockdown or knockout and over-expression of the ATG genes are also adopted to analyse autophagic flux. The methods used to measure autophagy comprises of LC3 turnover assay, degradation of LC3 and other selective targets, specifically p62 as well as radiolabelled long-lived protein and mRFP-GFP-LC3 assay. The mRFP–GFP–LC3 assay is an interesting test, which exploits the principle of lysosomal stability of RFP versus the quenching of GFP in acidic lysosomal compartment and thus, ascertains the localisation of LC3 depending on their fluorescence properties (Mizushima et al. 2010; Mizushima 2004; Yoshii and Mizushima 2017). Owing to the limitation of each of these assays, combination of the independent experimental methods is usually recommended as the most appropriate technique to estimate autophagy.

12.5 Regulation of Autophagy

Numerous signalling pathways have been involved in up and down-regulation of autophagy. However, the lack of information to understand the detailed molecular mechanism of the autophagy regulation in both cancer and normal cells calls for further investigation (Fig. 12.2).

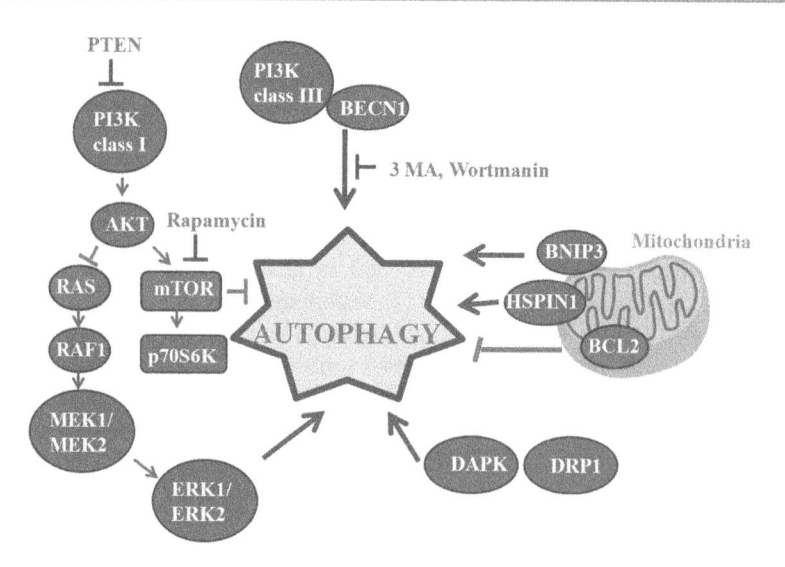

Fig. 12.2 Regulation of autophagy. The different molecular signalling cascade involved in modulation of autophagy. The green arrows indicate activation and the red indicates inhibition of autophagy

12.5.1 The PI3K–AKT–mTOR Signalling Pathway

According to recent reports, phosphatidylinositol 3-kinase (PI3K)–AKT–mTOR pathway is important for negative regulation of autophagy. AKT, a serine-threonine kinase, activates mTOR, a TOR kinase, which leads to suppression of autophagy. Studies in yeast have demonstrated that TOR kinase, which lies upstream of the autophagy-related genes, serves as the guard in autophagy initiation (Schmelzle and Hall 2000). Moreover, in mammalian cells, mTOR integrates with growth factor signalling cascade, thereby, regulating autophagy. It is interesting to note that class I and III PI3K have opposing role in regulation of autophagy: Class I PI3K, which is activated through growth factor receptor, inhibits autophagy while activation of class III PI3K facilitates autophagy by promoting sequestration of cytoplasmic cargo (Petiot et al. 2000). The tumour suppressor genes, like oncogenic RAS and phosphatase and tensin homologue (*PTEN*), also regulate autophagy through PI3K–AKT–mTOR pathway. Oncogenic RAS activates class I PI3K while PTEN deactivates class I PI3K, thereby, suppressing and initiating autophagy, respectively, through modulation of AKT (Arico et al. 2001). In addition, mutation of PTEN, located on chromosome 10q23, in various cancers activates AKT and thus, inhibits autophagy. PI3K–AKT–mTOR signalling pathway is dependent on nutrient availability like nitrogen or amino acids, which leads to transcriptional and translational regulation by p70s6 kinase and 4E binding protein 1 (Wang and Klionsky 2003).

12.5.2 Beclin 1 (BECN1) and Other Pathways

Beclin 1 is a coiled coil protein which is a BCL-2 interacting gene product. It is the first reported molecule to directly link tumourigenesis with autophagy. Previous reports have indicated significant role of class III PI3K in regulation of autophagosome formation and also, promotion of transport of the lysosomal enzymes from trans-golgi network (TGN) to the lysosome. Hence, BECN1 binds to class III PI3K to form BECN1–PI3K complex, which localises in TGN and presumably, facilitates sorting of putative autophagosomal components followed by autophagy induction (Liang et al. 1999).

Other molecules implicated in regulation of autophagy in cancer cells include BCL2 and its family members (BNIP3 and HSPIN1), death-associated protein kinase (DAPK), death-associated related protein kinase 1 (DRP1) and mitogen-activated kinases. BNIP3 (BCL2–adenovirus E1B 19-kDa-interacting protein 3) and HSPIN1 (a human homologue of the *Drosophila melanogaster* spin gene product) are the member of the BCL2 homology 3 (BH3)-only subfamily of the BCL2 family proteins (Vande Velde et al. 2000). They have been reported to induce caspase-independent autophagic cell death in various cancer cell lines. In addition, literature survey has documented that bone marrow-derived cells from BAX and BAK-deficient mice or murine embryonic fibroblast (MEF) are apoptosis resistant but susceptible to autophagy induction upon withdrawal of growth factor or exposure to the chemotherapeutic agent, etoposide (Lum et al. 2005). These, collectively, strengthens the relevance of BNIP3 and HSPIN1 in regulation of autophagy. The DAPK, DRP1 and mitogen-activated protein kinases belong to the family of serine–threonine kinases that regulate a plethora of cellular responses including autophagy. For example, DAPK and DRP1, which are regulated by Ca^{2+}–calmodulin, induce autophagy in MCF7 and HeLa cell (Inbal et al. 2002). While the stimulation of extracellular signal-regulated kinases ERK1 and ERK2, by the RAS–RAF1–mitogen-activated protein kinase (MEK) signalling cascade, induces autophagy in HT-29 colon cancer cell (Ogier-Denis et al. 2000) and buffers the metabolic stress (Degenhardt et al. 2006). For instance, during nutrient starvation, autophagy serves as the alternative energy reservoir whereas it also expedites the adaptation of cancer cells to cellular damage by removing the damaged proteins and organelles (Mizushima 2007).

12.6 Autophagy: The Double-Edged Sword

Autophagy has versatile role in diverse cellular processes and diseases. Basal autophagy occurs constitutively and performs its homeostatic function in conjugation with proteasome degradation pathway to facilitate protein and organelle quality control (Mathew et al. 2007; Mizushima 2007; Ravikumar et al. 2002). It has also been reported to help in elimination of pathogens and apoptotic bodies (Colombo 2007; Qu et al. 2007).

Mounting evidences suggest that autophagy has a pivotal role in cancer, although, its role in sustaining cell survival or inducing cell death is paradoxical (Baehrecke 2005). Autophagy is a well-conserved survival mechanism in several tumour types, which is rendered by protecting the cancer cells from undergoing programmed cell death. It is the most widely used mechanism of the cancer cells to survive. Therefore, inhibition of the autophagy is often exploited as the most feasible approach to sensitise the tumour cells to apoptosis and forms the basis of numerous cancer clinical trials. Nonetheless, in some situations, autophagy can also induce cell death, which is called programmed cell death type II (PCD II) or lethal autophagy. However, the autophagic cell death and apoptosis can be distinguished based on morphological and biochemical features. For instance, in contrast to apoptosis, autophagic cell death is caspase-independent and characterised by degradation of Golgi apparatus, polyribosome and endoplasmic reticulum prior to nuclear destruction (Bursch et al. 2000). Interestingly, during anti-cancer treatment, protective autophagy is initially triggered at the early stage by sequestering the damaged organelle and protein. But once the cellular damage crosses a certain threshold, lethal autophagy or death-inducing is activated to remove the damaged cells from the tissue (Kondo et al. 2005). Although apoptosis and autophagy are interconnected but little is known about the crosstalk between them. Recently, prothymosin-α, inhibitor of apoptosome formation in neuron, has been identified as plausible candidate for modulating the switch between apoptosis and autophagy (Kondo et al. 2005). Intriguingly, autophagy is dependent on multiple factors such as the nature and duration of stimulus, cell type, etc. For example, arsenic oxide (As_2O_3)-induced autophagy in glioma cells while in leukaemia cells, it triggered apoptotic cell death. Similarly, in contrast to the DNA alkylating agent, cisplatins, temozolomide (TMZ) induced autophagy, instead of apoptosis, in several cancer cell lines (Pelicano et al. 2003; Kanzawa et al. 2004). Moreover, it is interesting to document that while tamoxifen induced apoptosis in some cells, it also induced autophagy in other and both apoptosis and autophagy in the rest of breast cancer cells (Bursch et al. 1996). Henceforth, the modern cancer researchers have focused on investigation of the intricate regulation of autophagy and deciphering the interlink between the apoptosis and autophagy.

12.7 Role of Autophagy in Chemoresistance

A large number of recent studies suggests autophagy plays pivotal role in development of chemoresistance (Datta et al. 2017; Hu et al. 2012); in addition, various articles provide increasing evidences that inhibition of autophagy, in combination with various anticancer drugs can augment cytotoxicity on cancer cells leading to attenuation of chemoresistance development and metastasis process (Datta et al. 2019; Follo et al. 2018; Levy et al. 2014) (Fig. 12.3).

Epirubicin, one of the leading drugs used for breast cancer treatment, has shown evidences of autophagy induction in MCF7 breast cancer cell lines, which leads to cytoprotection of the cells from the chemotherapeutic stress induced by this drug. Similarly, autophagy inhibition has also shown elevated cytotoxic effect of various

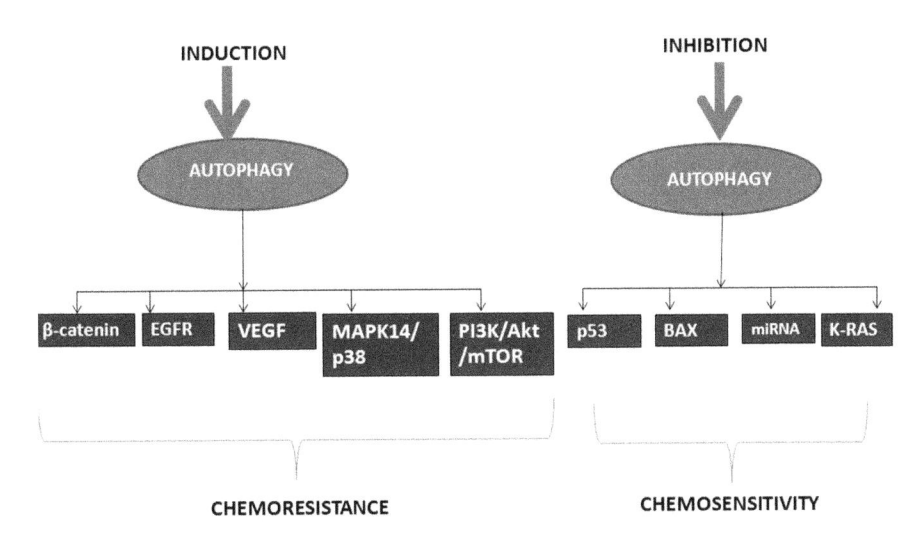

Fig. 12.3 The modulation of chemoresistance and chemosensitivity by autophagy. Schematic diagram depicting the activation of different cellular signalling pathways in cancer cells through autophagy induction or inhibition, leading to chemoresistance or chemosensitivity, respectively

chemotherapeutic drugs like 5-fluorouracil, irinotecan in colorectal cancer, oesophageal cancer, etc. (Chen et al. 2011; Sasaki et al. 2010). Likewise, in human hepatocarcinoma cell lines, autophagy level gets elevated with oxaliplatin treatment, and suppression of autophagy enhances oxaliplatin-induced cell death (Guo et al. 2013). Some of the leading drugs for treatment of lung cancers, like topetocan and paclitaxel, also shows elevated autophagy levels in lung cancer cells, which may ultimately aid in development of chemoresistance against these drugs and inhibition of autophagy has shown promising role in prevention of chemoresistance development against these drugs in lung cancer cells (Datta et al. 2019; Goldberg et al. 2012).

There are varieties of molecular mechanisms via which autophagy induction may lead to chemoresistance development in various cancers; epidermal growth factor is a key regulatory factor for cell survival. Through its binding to cell surface receptors, EGF can induce the activation of three signalling pathways that aids in cancer development and progression, Ras/MAPK, PI3K/Akt and JAK/STATs (Henson and Gibson 2006). In malignant peripheral nerve sheath tumour (MPNST) PD168393, an EGFR-TKI, may induce autophagy as a cytostatic but not a cytotoxic response in malignant peripheral nerve sheath tumour (MPNST) cells that was accompanied by suppression of Akt and mTOR activation. The aberrant expression of PI3K/AKt pathway may also aid in chemoresistance development and PI3K/Akt inhibitors may also lead to increased cytotoxicity of chemotherapeutic drugs against cancer cells by autophagy blockage. In many pre-clinical models autophagy inhibition has shown increased cytotoxic effect, by elevated p53 activity. Vascular endothelial growth factor-C (VEGF-C) is a secreted growth factor involved in many oncogenic processes, which shows autophagy promoting activities in many cancer cells and VEGF-C inhibitors have been reported to increase cytotoxic effect of anticancer drugs by downregulation of cellular autophagy. Activation of MAPK14/p38 also triggers survival-promoting autophagy to protect tumour cells against the cytotoxic effects of chemotherapeutic drugs. In addition, various micro-RNAs may also play key role in chemoresistance development by either inhibition or up-regulation of cellular autophagy, for example inhibition of miR30a (a potent autophagic inhibitor) may lead to chemoresistance development and elevated expression of miR30a may aggravate cytotoxicity of cancer cells by inhibition of autophagy; similarly, miR-199a-5p (an autophagic inducer) may lead to chemoresistance development to cisplatin and vice versa. Moreover, recent reports suggest that some paclitaxel resistant cell lines also show reduced expression of miR16 and 17, which usually exhibits inhibitory effects on beclin-1 expression and elevated expression of these miRNAs may increase sensitivity of these resistant cell lines towards paclitaxel by down-regulation of autophagy (Chatterjee et al. 2015).

However, in spite of its clear prosurvival role, autophagy has also shown to have a prodeath role under certain circumstances, following treatment with a specific set of chemotherapeutic agents, either by enhancing the induction of apoptosis or mediating 'autophagic cell death' by K-RAS, ERK pathways.

12.8 Autophagy Inhibitors

The autophagy inhibitors, whose effectiveness in in vivo and safety in clinical trials have been approved by the FDA, are the antimalarial drugs chloroquine (CQ) and its derivative hydroxychloroquine (HCQ); these are lysomotrophic drugs which raise the lysosomal pH, thereby preventing fusion of lysosomes with autophagosomes and thus, preventing autophagosomal degradation (Fox 1993; Mauthe et al. 2018). Both CQ and HCQ have been investigated in preclinical studies or clinical trials. In addition to antimalarial drugs, inhibition of autophagy by either pharmacological approaches or via genetic silencing of autophagy regulatory genes such as Beclin 1, ATG6, ATG5, ATG7 or ATG12 (Table 12.1) also results in sensitisation of cancer cells to a variety of chemotherapeutic drugs. Different autophagy inhibitors block autophagy at different well-defined stages. For example, another antimalarial drug bafilomycin A1 can inhibit autophagosome fusion with lysosomes and autophagosome degradation in the final stage of autophagy. Class III PI3K inhibitors (3-methyladenine (3-MA), LY294002 and Wortmannin) or knockdown of autophagy regulatory genes are involved in the initiation/expansion stage of autophagy (Liu et al. 2013; Zhao et al. 2012) (Table 12.1).

Although some previous articles have linked autophagy with cell death (Acharya et al. 2011; Lin and Baehrecke 2015; Paul et al. 2020), increasing number of recent research articles have also displayed the promising role of autophagy in cancer cell survival, wherein autophagy inhibition enhanced the chemo-sensitivity of cancer cells towards a wide range of chemotherapeutic drugs (Bhattacharya et al. 2016; Cournoyer et al. 2019; Dyczynski et al. 2018; Ganguli et al. 2014; Pagotto et al.

Table 12.1 Strategies used for inhibition of autophagy

Drug	Target	Effect
I. Pharmacological agents		
Chloroquine	Lysosomal pH	Inhibit autophagosome fusion with lysosomes and autophagosome degradation
Hydroxychloroquine	Lysosomal pH	Inhibit autophagosome fusion with lysosomes and autophagosome degradation
Monensin	Change endocytic and lysosomal pH	Inhibit the initiation/expansion stage of autophagy
Bafilomycin A1	Class III PI3K inhibitor	Inhibit the initiation/expansion stage of autophagy
3-Methyladenine	Class III PI3K inhibitor	Inhibit the initiation/expansion stage of autophagy
Wortmannin	Class III PI3K inhibitor	Inhibit the initiation/expansion stage of autophagy
LY294002	Class III PI3K inhibitor	Inhibit the initiation/expansion stage of autophagy
Pyrvinium	Class III PI3K inhibitor	Inhibit the initiation/expansion stage of autophagy
II. Genetic silencing of autophagy regulatory genes		
A. miRNA		
miR-140	ATG12 inhibition	Autophagy inhibition
miR-502	RAB1B	Autophagy inhibition
miR106a/b	ATG16L and ATG12	Autophagy inhibition
miR-183	UVRAG	Autophagy inhibition
miR-22	BTG1	Autophagy inhibition
miR-4093p	Beclin-1	Autophagy inhibition
B. si-RNA		
ATG12-siRNA	ATG12	Autophagy inhibition
ATG5-siRNA	ATG5	Autophagy inhibition
Beclin1-siRNA	Beclin1	Autophagy inhibition
ATG7-SiRNA	ATG7	Autophagy inhibition
ATG6-siRNA	ATG6	Autophagy inhibition

2017). In addition, many reports also suggest that autophagy inhibition may prevent chemo-resistance development in many cancer cell lines (Belounis et al. 2016; Datta et al. 2019). Hence, literature review has established the differential role of autophagy under different conditions. Therefore, finding the exact role of autophagy in a given cancer type, under a given condition is the key factor in determining the clinical approach for apt cancer chemotherapy.

12.9 Clinical Trials

Owing to the opposing, context-dependent role of autophagy in cancer, several studies have proposed that manipulation of autophagy, by stimulation or inhibition, could enhance the efficacy of multiple cancer therapies. However, till date,

Table 12.2 Autophagy inhibitors and their clinical application in different types of cancers

Autophagy inhibitor	Tumour	Additional treatment	Clinical trial phase
Hydroxychloroquine (HCQ)	1. Solid tumours and melanoma	Temsirolimus	I
	2. Malignant solid tumours and colorectal cancer	Vorinostat	I
	3. Non-Hodgkin's lymphoma	Doxorubicin	I
	4. Glioblastoma	Temozolomide and radiation	I/II
	5. Refractory myeloma	Bortezomib	I
	6. Pancreatic adenocarcinoma	Gemcitabine or capecitabine	I/II
	7. Non-small cell lung cancer	Erlotinib	I
	8. Adult solid neoplasm	Sunitinib malate	I
	9. Advanced cancers	MK-2206 (protein kinase B (Akt inhibitor))	I
	10. Small cell lung cancer	Gemcitabine/ carboplatin	I/II
	11. Renal cell carcinoma	IL-2	I/II
	12. Estrogen receptor-positive breast cancer and prostate cancer	None	I and II
Chloroquine	1. Glioblastoma	Temozolomide and radiation	
	2. Brain metastases: Non-small cell lung cancer, small cell lung cancer and ovarian cancer	Radiation	II and Pilot

chloroquine (CQ) and hydroxychloroquine (HCQ) are the only clinically approved and available drugs to inhibit autophagy in clinical models. As tabulated in Table 12.2, mounting preclinical evidences have documented that the inhibition of autophagy with CQ or HCQ alone or in combination with other drugs or radiation caused significant improvement in clinical outcome in cancer patients (Barnard et al. 2014; Briceno et al. 2003; Chude and Amaravadi 2017; Eldredge et al. 2013; Levy et al. 2017; Mahalingam et al. 2014; Rangwala et al. 2014; Rojas-Puentes et al. 2013; Vogl et al. 2014).

12.10 Conclusions

The significance of autophagy in tumourigenesis and cancer treatment makes it an important target for therapeutic intervention. However, till date, autophagy and its role in cancer are poorly understood. Therefore, the attempt to manipulate autophagy should be designed depending on its specific role in that particular scenario of malignancy. The two different and competing approaches of autophagy modulation are generally adopted towards improvement in cancer therapy. Firstly, cancer cells undergoing lethal autophagy could be exposed to mTOR inhibitors such as rapamycin and its derivatives: CCI-779, RAD001 and AP23573 in order to aggravate autophagic cell death culminating in suppression of a broad range of tumours (Chan 2004). Secondly, in contrast to the above strategy, inhibition of protective autophagy, with autophagy inhibitors such as chloroquine, hydroxychloroquine, bafilomycin A, etc. enhances the therapeutic potential of cancer therapeutics through sensitisation of the cancer cells to apoptotic cell death, as supported by mounting number of clinical trials (Barnard et al. 2014; Chude and Amaravadi 2017; Kanzawa et al. 2003; Levy et al. 2017; Mahalingam et al. 2014; Rangwala et al. 2014). However, both of these attempts in modulation of autophagy yield best outcome when combined with conventional cancer therapies.

Presently, numerous research groups throughout the globe have focused on delineating the detailed mechanism and signalling network of autophagy and understanding its intricate role in various types and stages of cancer. Henceforth, these extensive studies could enlighten new strategies of enhancing the efficacy of the currently available therapeutic options towards successful cancer cure.

Acknowledgement D.D.M. is thankful to "DBT-Research Associateship program in Biotechnology and Life Sciences" under Department of Biotechnology (Govt. of India) for the financial support

References

Acharya BR, Bhattacharyya S, Choudhury D, Chakrabarti G (2011) The microtubule depolymerizing agent naphthazarin induces both apoptosis and autophagy in A549 lung cancer cells. Apoptosis 16:924–939

Amaravadi R, Kimmelman AC, White E (2016) Recent insights into the function of autophagy in cancer. Genes Dev 30:1913–1930

Arico S, Petiot A, Bauvy C, Dubbelhuis PF, Meijer AJ, Codogno P, Ogier-Denis E (2001) The tumor suppressor PTEN positively regulates macroautophagy by inhibiting the phosphatidylinositol 3-kinase/protein kinase B pathway. J Biol Chem 276:35243–35246

Baehrecke EH (2005) Autophagy: dual roles in life and death? Nat Rev Mol Cell Biol 6:505–510

Barnard RA, Wittenburg LA, Amaravadi RK, Gustafson DL, Thorburn A, Thamm DH (2014) Phase I clinical trial and pharmacodynamic evaluation of combination hydroxychloroquine and doxorubicin treatment in pet dogs treated for spontaneously occurring lymphoma. Autophagy 10:1415–1425

Belounis A, Nyalendo C, Le Gall R, Imbriglio TV, Mahma M, Teira P, Beaunoyer M, Cournoyer S, Haddad E, Vassal G, Sartelet H (2016) Autophagy is associated with chemoresistance in neuroblastoma. BMC Cancer 16:891

Bhattacharya S, Das A, Datta S, Ganguli A, Chakrabarti G (2016) Colchicine induces autophagy and senescence in lung cancer cells at clinically admissible concentration: potential use of colchicine in combination with autophagy inhibitor in cancer therapy. Tumour Biol 37:10653–10664

Briceno E, Reyes S, Sotelo J (2003) Therapy of glioblastoma multiforme improved by the antimutagenic chloroquine. Neurosurg Focus 14:e3

Bursch W, Ellinger A, Kienzl H, Torok L, Pandey S, Sikorska M, Walker R, Hermann RS (1996) Active cell death induced by the anti-estrogens tamoxifen and ICI 164 384 in human mammary carcinoma cells (MCF-7) in culture: the role of autophagy. Carcinogenesis 17:1595–1607

Bursch W, Ellinger A, Gerner C, Frohwein U, Schulte-Hermann R (2000) Programmed cell death (PCD). Apoptosis, autophagic PCD, or others? Ann N Y Acad Sci 926:1–12

Chan S (2004) Targeting the mammalian target of rapamycin (mTOR): a new approach to treating cancer. Br J Cancer 91:1420–1424

Chatterjee A, Chattopadhyay D, Chakrabarti G (2015) MiR-16 targets Bcl-2 in paclitaxel-resistant lung cancer cells and overexpression of miR-16 along with miR-17 causes unprecedented sensitivity by simultaneously modulating autophagy and apoptosis. Cell Signal 27:189–203

Chen YS, Song HX, Lu Y, Li X, Chen T, Zhang Y, Xue JX, Liu H, Kan B, Yang G, Fu T (2011) Autophagy inhibition contributes to radiation sensitization of esophageal squamous carcinoma cells. Dis Esophagus 24:437–443

Chude CI, Amaravadi RK (2017) Targeting autophagy in cancer: update on clinical trials and novel inhibitors. Int J Mol Sci 18:1279

Colombo MI (2007) Autophagy: a pathogen driven process. IUBMB Life 59:238–242

Cournoyer S, Addioui A, Belounis A, Beaunoyer M, Nyalendo C, Le Gall R, Teira P, Haddad E, Vassal G, Sartelet H (2019) GX15-070 (Obatoclax), a Bcl-2 family proteins inhibitor engenders apoptosis and pro-survival autophagy and increases chemosensitivity in neuroblastoma. BMC Cancer 19:1018

Datta S, Choudhury D, Das A, Das Mukherjee D, Das N, Roy SS, Chakrabarti G (2017) Paclitaxel resistance development is associated with biphasic changes in reactive oxygen species, mitochondrial membrane potential and autophagy with elevated energy production capacity in lung cancer cells: a chronological study. Tumour Biol 39. https://doi.org/10.1177/1010428317694314

Datta S, Choudhury D, Das A, Mukherjee DD, Dasgupta M, Bandopadhyay S, Chakrabarti G (2019) Autophagy inhibition with chloroquine reverts paclitaxel resistance and attenuates metastatic potential in human nonsmall lung adenocarcinoma A549 cells via ROS mediated modulation of beta-catenin pathway. Apoptosis 24:414–433

Degenhardt K, Mathew R, Beaudoin B, Bray K, Anderson D, Chen G, Mukherjee C, Shi Y, Gelinas C, Fan Y, Nelson DA, Jin S, White E (2006) Autophagy promotes tumor cell survival and restricts necrosis, inflammation, and tumorigenesis. Cancer Cell 10:51–64

Dyczynski M, Yu Y, Otrocka M, Parpal S, Braga T, Henley AB, Zazzi H, Lerner M, Wennerberg K, Viklund J, Martinsson J, Grander D, De Milito A, Pokrovskaja Tamm K (2018) Targeting autophagy by small molecule inhibitors of vacuolar protein sorting 34 (Vps34) improves the sensitivity of breast cancer cells to Sunitinib. Cancer Lett 435:32–43

Eldredge HB, Denittis A, Duhadaway JB, Chernick M, Metz R, Prendergast GC (2013) Concurrent whole brain radiotherapy and short-course chloroquine in patients with brain metastases: a pilot trial. J Radiat Oncol 2:315

Follo C, Cheng Y, Richards WG, Bueno R, Broaddus VC (2018) Inhibition of autophagy initiation potentiates chemosensitivity in mesothelioma. Mol Carcinog 57:319–332

Fox RI (1993) Mechanism of action of hydroxychloroquine as an antirheumatic drug. Semin Arthritis Rheum 23:82–91

Ganguli A, Choudhury D, Datta S, Bhattacharya S, Chakrabarti G (2014) Inhibition of autophagy by chloroquine potentiates synergistically anti-cancer property of artemisinin by promoting ROS dependent apoptosis. Biochimie 107(Pt B):338–349

Goldberg SB, Supko JG, Neal JW, Muzikansky A, Digumarthy S, Fidias P, Temel JS, Heist RS, Shaw AT, McCarthy PO, Lynch TJ, Sharma S, Settleman JE, Sequist LV (2012) A phase I study of erlotinib and hydroxychloroquine in advanced non-small-cell lung cancer. J Thorac Oncol 7:1602–1608

Guo XL, Li D, Sun K, Wang J, Liu Y, Song JR, Zhao QD, Zhang SS, Deng WJ, Zhao X, Wu MC, Wei LX (2013) Inhibition of autophagy enhances anticancer effects of bevacizumab in hepatocarcinoma. J Mol Med (Berl) 91:473–483

Henson ES, Gibson SB (2006) Surviving cell death through epidermal growth factor (EGF) signal transduction pathways: implications for cancer therapy. Cell Signal 18:2089–2097

Hu YL, Jahangiri A, Delay M, Aghi MK (2012) Tumor cell autophagy as an adaptive response mediating resistance to treatments such as antiangiogenic therapy. Cancer Res 72:4294–4299

Inbal B, Bialik S, Sabanay I, Shani G, Kimchi A (2002) DAP kinase and DRP-1 mediate membrane blebbing and the formation of autophagic vesicles during programmed cell death. J Cell Biol 157:455–468

Kanzawa T, Kondo Y, Ito H, Kondo S, Germano I (2003) Induction of autophagic cell death in malignant glioma cells by arsenic trioxide. Cancer Res 63:2103–2108

Kanzawa T, Germano IM, Komata T, Ito H, Kondo Y, Kondo S (2004) Role of autophagy in temozolomide-induced cytotoxicity for malignant glioma cells. Cell Death Differ 11:448–457

Kaur J, Debnath J (2015) Autophagy at the crossroads of catabolism and anabolism. Nat Rev Mol Cell Biol 16:461–472

Kaushik S, Bandyopadhyay U, Sridhar S, Kiffin R, Martinez-Vicente M, Kon M, Orenstein SJ, Wong E, Cuervo AM (2011) Chaperone-mediated autophagy at a glance. J Cell Sci 124:495–499

Klionsky DJ (2007) Autophagy: from phenomenology to molecular understanding in less than a decade. Nat Rev Mol Cell Biol 8:931–937

Klionsky DJ, Emr SD (2000) Autophagy as a regulated pathway of cellular degradation. Science 290:1717–1721

Kondo Y, Kanzawa T, Sawaya R, Kondo S (2005) The role of autophagy in cancer development and response to therapy. Nat Rev Cancer 5:726–734

Levine B, Klionsky DJ (2004) Development by self-digestion: molecular mechanisms and biological functions of autophagy. Dev Cell 6:463–477

Levy JM, Thompson JC, Griesinger AM, Amani V, Donson AM, Birks DK, Morgan MJ, Mirsky DM, Handler MH, Foreman NK, Thorburn A (2014) Autophagy inhibition improves chemosensitivity in BRAF(V600E) brain tumors. Cancer Discov 4:773–780

Levy JMM, Towers CG, Thorburn A (2017) Targeting autophagy in cancer. Nat Rev Cancer 17:528–542

Li WW, Li J, Bao JK (2012) Microautophagy: lesser-known self-eating. Cell Mol Life Sci 69:1125–1136

Liang XH, Jackson S, Seaman M, Brown K, Kempkes B, Hibshoosh H, Levine B (1999) Induction of autophagy and inhibition of tumorigenesis by beclin 1. Nature 402:672–676

Lin L, Baehrecke EH (2015) Autophagy, cell death, and cancer. Mol Cell Oncol 2:e985913

Liu F, Liu D, Yang Y, Zhao S (2013) Effect of autophagy inhibition on chemotherapy-induced apoptosis in A549 lung cancer cells. Oncol Lett 5:1261–1265

Lum JJ, Bauer DE, Kong M, Harris MH, Li C, Lindsten T, Thompson CB (2005) Growth factor regulation of autophagy and cell survival in the absence of apoptosis. Cell 120:237–248

Mahalingam D, Mita M, Sarantopoulos J, Wood L, Amaravadi RK, Davis LE, Mita AC, Curiel TJ, Espitia CM, Nawrocki ST, Giles FJ, Carew JS (2014) Combined autophagy and HDAC inhibition: a phase I safety, tolerability, pharmacokinetic, and pharmacodynamic analysis of hydroxychloroquine in combination with the HDAC inhibitor vorinostat in patients with advanced solid tumors. Autophagy 10:1403–1414

Mathew R, Karantza-Wadsworth V, White E (2007) Role of autophagy in cancer. Nat Rev Cancer 7:961–967

Mauthe M, Orhon I, Rocchi C, Zhou X, Luhr M, Hijlkema KJ, Coppes RP, Engedal N, Mari M, Reggiori F (2018) Chloroquine inhibits autophagic flux by decreasing autophagosome-lysosome fusion. Autophagy 14:1435–1455

Meijer AJ, Codogno P (2004) Regulation and role of autophagy in mammalian cells. Int J Biochem Cell Biol 36:2445–2462

Mizushima N (2004) Methods for monitoring autophagy. Int J Biochem Cell Biol 36:2491–2502

Mizushima N (2007) Autophagy: process and function. Genes Dev 21:2861–2873

Mizushima N, Yoshimori T, Levine B (2010) Methods in mammalian autophagy research. Cell 140:313–326

Mizushima N, Yoshimori T, Ohsumi Y (2011) The role of Atg proteins in autophagosome formation. Annu Rev Cell Dev Biol 27:107–132

Ogier-Denis E, Pattingre S, El Benna J, Codogno P (2000) Erk1/2-dependent phosphorylation of Galpha-interacting protein stimulates its GTPase accelerating activity and autophagy in human colon cancer cells. J Biol Chem 275:39090–39095

Onorati AV, Dyczynski M, Ojha R, Amaravadi RK (2018) Targeting autophagy in cancer. Cancer 124:3307–3318

Pagotto A, Pilotto G, Mazzoldi EL, Nicoletto MO, Frezzini S, Pasto A, Amadori A (2017) Autophagy inhibition reduces chemoresistance and tumorigenic potential of human ovarian cancer stem cells. Cell Death Dis 8:e2943

Paul S, Chakrabarty S, Ghosh S, Nag D, Das A, Dastidar DG, Dasgupta M, Dutta N, Kumari M, Pal M, Chakrabarti G (2020) Targeting cellular microtubule by phytochemical apocynin exhibits autophagy-mediated apoptosis to inhibit lung carcinoma progression and tumorigenesis. Phytomedicine 67:153152

Pelicano H, Feng L, Zhou Y, Carew JS, Hileman EO, Plunkett W, Keating MJ, Huang P (2003) Inhibition of mitochondrial respiration: a novel strategy to enhance drug-induced apoptosis in human leukemia cells by a reactive oxygen species-mediated mechanism. J Biol Chem 278:37832–37839

Petiot A, Ogier-Denis E, Blommaart EF, Meijer AJ, Codogno P (2000) Distinct classes of phosphatidylinositol 3′-kinases are involved in signaling pathways that control macroautophagy in HT-29 cells. J Biol Chem 275:992–998

Qu X, Zou Z, Sun Q, Luby-Phelps K, Cheng P, Hogan RN, Gilpin C, Levine B (2007) Autophagy gene-dependent clearance of apoptotic cells during embryonic development. Cell 128:931–946

Rangwala R, Leone R, Chang YC, Fecher LA, Schuchter LM, Kramer A, Tan KS, Heitjan DF, Rodgers G, Gallagher M, Piao S, Troxel AB, Evans TL, DeMichele AM, Nathanson KL, O'Dwyer PJ, Kaiser J, Pontiggia L, Davis LE, Amaravadi RK (2014) Phase I trial of hydroxychloroquine with dose-intense temozolomide in patients with advanced solid tumors and melanoma. Autophagy 10:1369–1379

Ravikumar B, Duden R, Rubinsztein DC (2002) Aggregate-prone proteins with polyglutamine and polyalanine expansions are degraded by autophagy. Hum Mol Genet 11:1107–1117

Rojas-Puentes LL, Gonzalez-Pinedo M, Crismatt A, Ortega-Gomez A, Gamboa-Vignolle C, Nunez-Gomez R, Dorantes-Gallareta Y, Arce-Salinas C, Arrieta O (2013) Phase II randomized, double-blind, placebo-controlled study of whole-brain irradiation with concomitant chloroquine for brain metastases. Radiat Oncol 8:209

Sasaki K, Tsuno NH, Sunami E, Tsurita G, Kawai K, Okaji Y, Nishikawa T, Shuno Y, Hongo K, Hiyoshi M, Kaneko M, Kitayama J, Takahashi K, Nagawa H (2010) Chloroquine potentiates the anti-cancer effect of 5-fluorouracil on colon cancer cells. BMC Cancer 10:370

Schmelzle T, Hall MN (2000) TOR, a central controller of cell growth. Cell 103:253–262

Vande Velde C, Cizeau J, Dubik D, Alimonti J, Brown T, Israels S, Hakem R, Greenberg AH (2000) BNIP3 and genetic control of necrosis-like cell death through the mitochondrial permeability transition pore. Mol Cell Biol 20:5454–5468

Vogl DT, Stadtmauer EA, Tan KS, Heitjan DF, Davis LE, Pontiggia L, Rangwala R, Piao S, Chang YC, Scott EC, Paul TM, Nichols CW, Porter DL, Kaplan J, Mallon G, Bradner JE, Amaravadi RK (2014) Combined autophagy and proteasome inhibition: a phase 1 trial of hydroxychloroquine and bortezomib in patients with relapsed/refractory myeloma. Autophagy 10:1380–1390

Wang CW, Klionsky DJ (2003) The molecular mechanism of autophagy. Mol Med 9:65–76

White E (2012) Deconvoluting the context-dependent role for autophagy in cancer. Nat Rev Cancer 12:401–410

Yoshii SR, Mizushima N (2017) Monitoring and measuring autophagy. Int J Mol Sci 18:1865

Zhao S, Ma CM, Liu CX, Wei W, Sun Y, Yan H, Wu YL (2012) Autophagy inhibition enhances isobavachalcone-induced cell death in multiple myeloma cells. Int J Mol Med 30:939–944

Mechanistic Insights into Autophagosome-Lysosome Fusion in Cancer Therapeutics

Kewal Kumar Mahapatra, Soumya Ranjan Mishra, Bishnu Prasad Behera, Prakash Priyadarshi Praharaj, Debasna Pritimanjari Panigrahi, Chandra Sekhar Bhol, Srimanta Patra, and Sujit Kumar Bhutia

Abstract

Autophagy is a highly conserved metabolic process to degrade cytoplasmic materials, damaged organelles and aggregated milieus through formation of autolysosome with the help of lysosomal acid hydrolases. This multistep event has a dual role in the regulation of cancer. The termination step of autophagy involving fusion of autophagosome and lysosome degrades autopahgic cargo to establish the cellular homeostasis. This is the point which defines the cytoprotective nature of autophagy in some cases and on the contrary, it also acts as tumor suppressor step due to accumulation of autophagic vacuoles or excessive autopahgic flux in most of the cancers. In this setting, many different drugs have been developed and discovered to combat the various type of cancers through the modulation of autophagy in the cancer cells either by its activation or suppression mechanism. Interestingly, targeting the genes of stage-specific regulation of autophagy could be a novel approach for the directive suppression and prevention of cancer. Moreover, the endocytosis–autophagy crosstalk gives an edge to develop drugs which specifically acts upon the cancer cells and modulate the autophagy process. In addition, the drugs modulating the fusion step of autophagy may be the novel molecule for the cancer therapeutics which needs to be investigated further for future cancer treatment.

Keywords

Autophagy · Autophagosome · Autolysosome · Cancer

K. K. Mahapatra · S. R. Mishra · B. P. Behera · P. P. Praharaj · D. P. Panigrahi · C. S. Bhol · S. Patra · S. K. Bhutia (✉)
Department of Life Science, National Institute of Technology Rourkela, Rourkela, Odisha, India
e-mail: sujitb@nitrkl.ac.in

13.1 Introduction

Autophagy is a conserved self-digestion process to maintain the cellular integrity and homeostasis in the cells through vesicular trafficking pathways. The genes regulating the process of autophagy are called ATGs and until now 41 ATGs have been discovered in the case of yeasts (Mahapatra et al. 2019). This process helps in nutrient recycling and removes damaged organelles, proteins to prevent many disorders. It plays a pivotal role in the regulation of cancer by playing a dual role in cancer regulation, i.e. by acting as a cytoprotective and cell death phenomenon. Initially, it supports tumor progression by protecting the cancer cells but the excessive events of autophagy lead to cell death in the case of cancer cells (Eskelinen 2011; Singh et al. 2018).

To date, three major types of autophagy have been reported, i.e. microautophagy, macroautophagy, and chaperone-mediated autophagy but among them, macroautophagy has been extensively studied and is referred to as autophagy from here on in this chapter (Parzych and Klionsky 2014). It may be selective (e.g., mitophagy, pexophagy) or bulk degradation process and subdivided into four distinct stages, i.e., (1) nucleation and initiation of the phagophore, (2) formation of the matured autophagosomes, (3) fusion events with endosomes and/or lysosomes, and (4) degradation of entrapped contents in autolysosomes or autophagic flux (Deretic 2006). This stage-specific regulation of autophagy has a vital role in the regulation of various metabolic pathways and although, all the steps have their own significance, the fusion events play a crucial role in case of the regulation of various diseases. It helps in the regulation of various physiological homeostatic responses like nutrient recycling which helps in the attainment of the nutrient need of the cells during stressed conditions. It decides the fate of the cargos entrapped inside and hence, it has the major role to play in the clearance of the milieus inside the cell to keep it clean and healthy (Levine 2007). Moreover, the recovery of the depleted lysosomal pool by autophagic lysosome reformation also depends upon the fusion events and hence, it regulates the promotion or healing of many physiological disorders which are solely based on vesicular trafficking therapeutics (Chen and Yu 2018).

In this connection, we have described the role of stage-specific regulation of autophagy in response to various conditions which specifically emphasized on the fusion events and cancer progression in case of mammals. Moreover, we have tried to establish a link between autophagic fusion with that of disease progression in response to several genes. Then the role of different natural and synthetic compounds has been discussed in the modulation of the stage-specific regulation of autophagy and lastly, we have discussed the roles and challenges of natural compounds in treating cancer.

13.2 Mechanism of Autophagy

Autophagy is initiated by the formation of pre-autophagosomal structures called omegasomes in the case of mammals which is the same as the phagophore assembly site or PAS in case of yeasts. PI(3)P-enriched membranes modulate the formation of omegasomes which are tethers to the endoplasmic reticulum (ER) by its FYVE domain (Axe et al. 2008; Hurley and Young 2017). Furthermore, activation of unc-51 like autophagy activating kinase-1 (ULK1) is essential for autophagy induction which is further regulated by the level of the signaling complexes like mTOR and AMPK. In case of mammals, the initiation complex comprises of ULK1, ATG13, and FIP200. Moreover, ULK1 activation leads to phosphorylation of remaining complex members, i.e., ATG13 and FIP200. Then the phosphatidylinositol is phosphorylated by phosphatidylinositol 3 kinase (PI3KC3) and hence the phagophore nucleation took place followed by the activation of Beclin-1 and Vps15/p150 in case of mammals. It helps in mobilization of effector molecules onto the PAS and results in membrane elongation (Mahapatra et al. 2019; Santana-Codina et al. 2017).

After the successful nucleation of the phagophore membrane, it is elongated to form the complete and closed double-membrane vesicular structures called autophagosomes. This undergoes a two-step process for completion which mimics like ubiquitin-like reactions. At first the E1 ubiquitin-activating enzyme, ATG7 activates the ATG12 which is linked to ATG5 by E2 ubiquitin-conjugating enzyme, i.e., ATG10. Hence formed complex ATG12–ATG5 is then formed the ternary complex by interacting with ATG16L1 which again as a whole act as E3 like ubiquitin ligase. Post formation of this ternary complex or the E3 ubiquitin ligase the second reaction is triggered by the addition of phosphatidylethanolamine to the ATG8 family proteins like LC3, GATE-16, and GABARAP. Hence, the ATG8 family members remain as a marker of the autophagy which is recovered from the cytosolic side after the successful completion of the autophagy but the proteins facing inwards or which are present inside are gets degraded by the action of the acid hydrolases of the lysosomes (Kabeya et al. 2004; Panda et al. 2015; Qiu et al. 2013; Weidberg et al. 2011). These conjugation systems result in the formation of closed vesicle autophagosomes that are available with entrapped cargos for degradation by lysosomal hydrolases (Fig. 13.1).

The matured autophagosomes are readily available for the fusion with lysosomes to form a closed structure known as autolysosomes. Autolysosomes are the organelles in which cargo degradation took place and the newly formed nutrients are recycled back for the maintenance of the homeostasis (Klionsky et al. 2014). Several genes have been reported to control these fusion events. It is mainly of two types where the mature autophagosome can directly fuse with the lysosome to form an autolysosome or first with late endosome to form an amphisome and then with the lysosome to form the autolysosomes (Sanchez-Wandelmer and Reggiori 2013). It has been reported that the insertion of SNARE followed by the recruitment of HOPS complex onto the membrane is the major regulator of the fusion process. Furthermore, STX-17 interacts with SNAP-29 and VAMP-8 to facilitate these fusion events

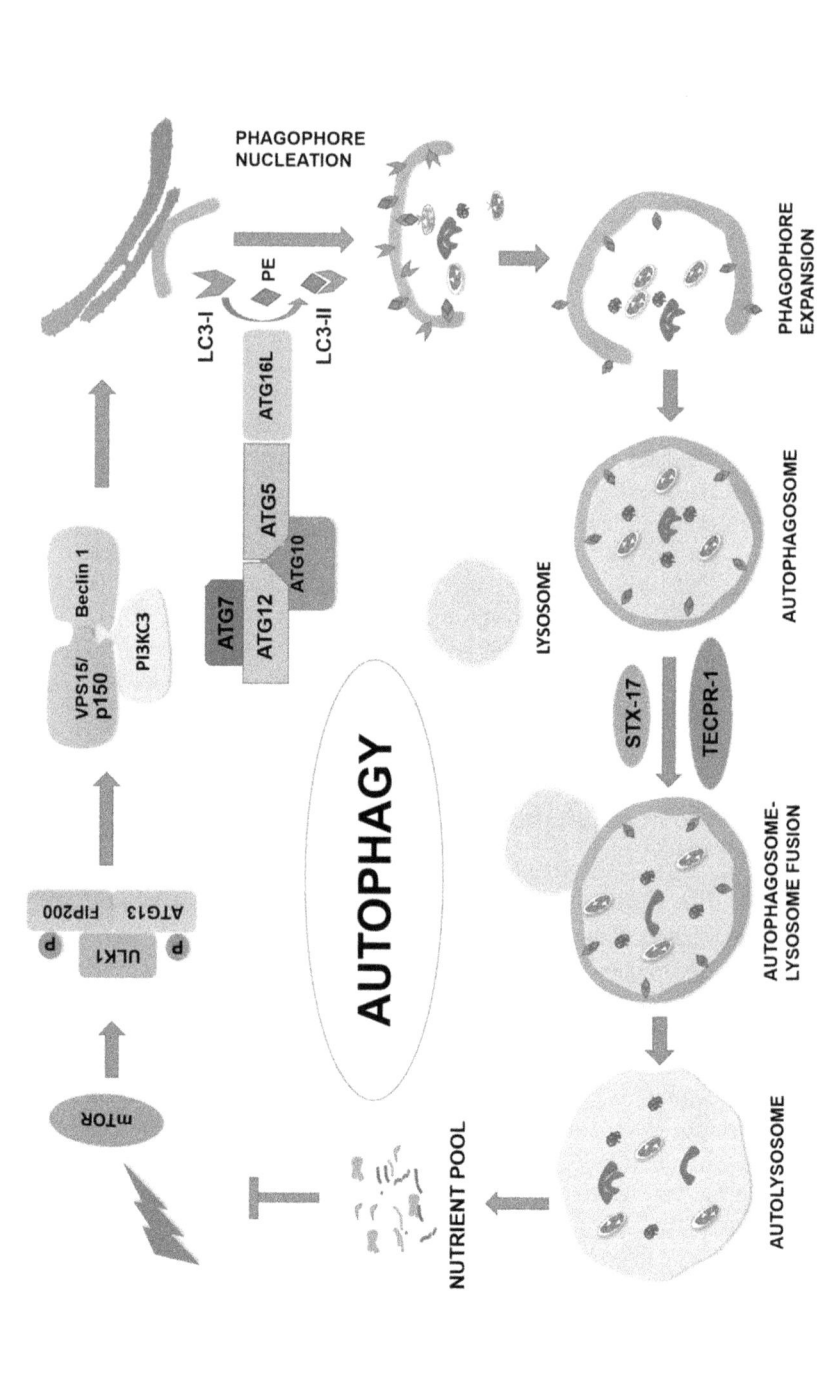

Fig. 13.1 Overall mechanism of autophagy. Autophagy is induced by depletion of mTOR followed by ULK1 activation, which leads to nucleation of the phagophore after phosphorylation of ATG13 and FIP200. Then, autophagosome formation is triggered by Beclin-1 which remains as a complex with VPS15/p150 and PI3KC3. The membrane gets elongated by the action of the ternary complex ATG12–ATG5–ATG16L1, which acts as E3 ligase and facilitates the addition of phosphatidylethanolamine to LC3-I to form LC3-II. Followed by the formation of a matured autophagosome the STX-17 and TECPR-1 helps in fusion of autophagosome and lysosome to form the autolysosome. Then the cargo entrapped within the autolysosome get degraded and nutrient is recycled inside the cells to maintain the homeostasis

(Itakura et al. 2012). Likewise, TECPR1 is able to interact with PI3P to recruit autophagosomes onto the lysosomes to modulate the fusion process (Chen et al. 2012). On successful completion of the fusion process, the cargo gets degraded and the cells are released from the stressed condition to facilitate the normal functioning.

Earlier it is believed that fusion event is the termination step of autophagy for the attainment of homeostasis in the cells. But due to autophagy, lysosomes are depleted in the cells and has to be counterbalanced to continue the autophagy and other lysosome assocaited processes. This stage is overcome by the process of formation of newly formed lysosomes from the existing autolysosomes by the phenomenon called autophagic lysosome reformation (ALR) (Chen and Yu 2013). The newly formed lysosomes may again be available for the continuation of the autophagy process in the cells. It is started by the recruitment of the KIF5B, Clathrin, and phosphatidylinositol-4,5-bisphosphate $(PI(4,5)P_2)$ onto the autolysosomes when the mTOR gets reactivated and it results in the formation of a tubule (Du et al. 2016; Liu and Klionsky 2018; Rong et al. 2012; Zhang et al. 2016). This tubule is then elongated and gets budded off to form a new vesicle called as protolysosome. It happens only after the dissociation of the Rab7 from the autolysosomes and when the lipid status is normal (Chen and Yu 2017; Guerra and Bucci 2016). These newly formed protolysosomes get matured to form the active functional lysosomes. These lysosomes are again available for the other processes in the cell to maintain the lysosomal homeostasis (Fig. 13.2).

13.3 Autophagic Fusion in the Regulation of Cancer

Autophagy plays an essential role in the regulation of cancer. It shows a dual regulation mechanism, i.e., it may act as a tumor promoter or tumor suppressor agent depending upon the circumstances. This conditional mechanism imparts a major contribution to carcinogenesis and its prevention (Yun and Lee 2018). Drugs and therapy used for the control and prevention of different cancers are mostly targeted the autophagic pathways and hence can able to regulate the cancer progression. Moreover, autophagic genes have been reported to regulate cancer progression, e.g., Beclin-1 acts as a haploinsufficient tumor suppressor gene (Edinger and Thompson 2003). Targeting various steps of autophagy could have potent anticancer effect to regulate cancer growth and progression. Out of them the last step of autophagy has the utmost significance, i.e., the fusion between autophagosomes and the lysosomes. This step decides the fate of the inner cargos and maintenance of the homeostasis inside the cells. It also plays a vital role in the regulation of lysosomal homeostasis in the cells through ALR as discussed before.

This fusion event mainly depends upon two major components, i.e., autophagosomes and lysosomes. If any of the components is incompetent then it halts the fusion between them. So, for the completion of the successful fusion; competent and active functional autophagosomes and lysosomes are the utmost need for the process (Lőrincz and Juhász 2020). Several genes have been reported to regulate these processes. Silent mating type information regulation 1 (SIRT1) has

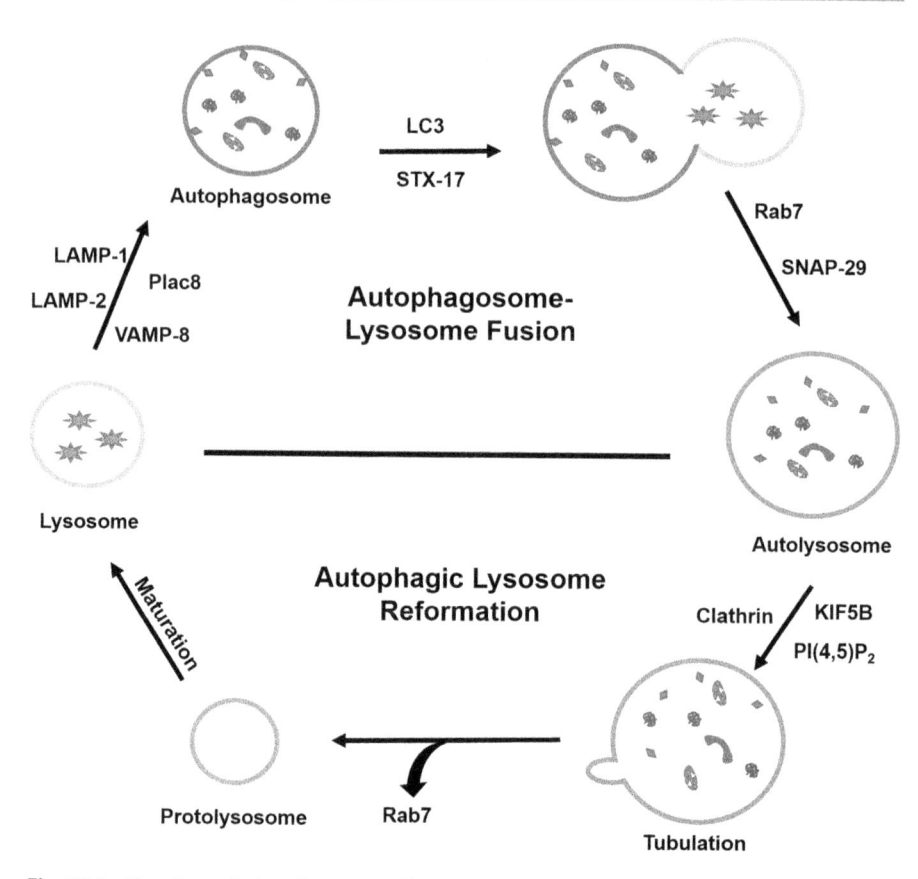

Fig. 13.2 Genetic regulation of stage-specific autophagy and autolysosome reformation. Autophagic vesicle autophagosome and the endocytic vesicle lysosomes have distinct set of genes which gets interacted to facilitate the fusion process with the help of Rab7 and SNAP-29 adaptor proteins. Then the successive autophagic fusion cycles lead to depletion in the lysosomal pool inside the cells. Hence, the protein like KIF5B, Clathrin, and PI(4,5)P$_2$ are recruited onto autolysosome to form a tubule like structure which will buds off to form a protolysosome after the dissociation of Rab7. Then it is matured to form a active functional lysosome which helps in reestablishment of the lysosomal pool inside the cells by the process of autophagic lysosome reformation

a dual role in the case of cancer progression. It deacetylates the proteins to regulate their function. It is reported to have increased expression in several cancers like gastro-esophageal junction (GEJ) cancer, breast cancer, colorectal adenocarcinoma, hepatocellular carcinoma, gastric cancer, non-small cell lung cancer, soft tissue sarcomas, ovarian cancer, prostate cancer, cervical cancer, medulloblastoma, and lymphoma. It helps in the deacetylation of ATG5, ATG7, ATG8 and promotes nuclear translocation of FoxO1, which subsequently elevates the level of Rab7, which is essential for fusion of autophagosome and lysosomes. Hence, indirectly SIRT1 promotes fusion events and modulates autophagy to promote cancer progression in gastric cancer (Qiu et al. 2015). Furthermore, Plac8 has a critical role in the progression of pancreatic cancers. It shows lysosomal localization in the case of vesicular trafficking and it has been reported that depletion of Plac8 does not lead to alteration in the lysosomal pH but results in a significant decrease in the LAMP-2 and LC3 co-localization signifying its role in the regulation of fusion events in case of cancer. It specifically targets the late stages of autophagy whereas the initial stages remain unaltered. Hence, the accumulation of autophagosomes took place inside the cells leading to cell death in the case of pancreatic cancer cells (Kinsey et al. 2014). Moreover, the endocytic protein Rab7 has several roles in the maturation of vesicles like late endosomes and fusion of autophagosomes with lysosomes. It is a major adaptor protein that establishes the crosstalk between autophagy and endocytosis. It has been observed that the dysregulation and aberration of Rab7 inside the cells may lead to various diseases including cancer. Hence, it is an effective agent regulating the fusion events in the case of autophagy and hence regulating the cancer progression (Liang and Jung 2010; Zhang et al. 2009). In addition, LAMP-3 another lysosomal marker is reported to control the lysosomal fusion process. The addition of drug tamoxifen elevated the expression of LAMP-3 and autophagy in breast cancer. When the LAMP-3 is depleted inside, the cell shows the absence of fusion events as suggested by LC3-LAMP-1 co-localization assay in the case of MCF7 breast cancer cells. Hence, LAMP-3 involves autophagy induction and shows tamoxifen resistance in the case of breast cancer cells (Nagelkerke et al. 2014). Another gene MALAT1 is found to activate autophagy and promote tumor progression in case of pancreatic cancer. It elevates the autophagy levels by regulating HuR and TIA-1. Depletion of MALAT1 results in the depletion of well-known fusion marker LAMP-2 and an increase in the expression of p62. Hence, MALAT1 might regulate the fusion events for its oncogenic response in pancreatic cancer cells. Although, it seems that depletion of MALAT1 involves a decrease in the degradation of cargos, the actual mechanism still needs to be investigated which will give a novel target for pancreatic cancer cells (Li et al. 2016). Furthermore, BAP31 has shown an antitumor mechanism and resides on the ER sites. It interacts with STX-17 in normal conditions and hence prevents the autophagic response in normal conditions. When BAP31 is depleted inside the cells, STX-17 is free to interact with ATG14L and regulate the autophagy response in cancer cells. BAP31 acts as a tumor suppressor gene and inhibits the autophagy to regulate cell death in cancer cells and also acts as a sequester molecule for STX-17. But STX-17 also has a pivotal role in fusion events. So, the role of BAP31 in regulating autophagy and STX-17 should be investigated in detail to confirm the stage-specific regulation of autophagy (Machihara and Namba 2019). Moreover, p62 an adaptor protein for autophagy shows a dual role in the regulation of cancer. Elevation in expression of p62 is a marker for the accumulation of cargo inside the vesicles and hence, it signifies the inhibition of late stages in autophagy. But a high level of p62 is also observed in tumor promotion and it is implicated to resistance towards chemotherapy in cancer cells. So, for the prognosis of cancer, it might act as a valuable marker for the scientific community (Islam et al. 2018). These genes have been involved in the regulation of cancer by modulating the autophagic response inside the cells. They might affect different stages of autophagy for the modulation of cancer pathways but the late-stage autophagic regulation has an utmost upper hand in this regard. As the late-stage autophagy inhibition is the point to decide the dual role of autophagy in cancer progression and metastasis, it is the need of the hour to target these genes to make them a potential tracker for the prognosis of various cancers and their therapeutics (Fig. 13.2).

13.4 Molecules in Promotion of Autophagic Fusion

Although the autophagy process is regulated by certain genes and the basic pathway is already well established some molecules modulate the stage-specific regulation of autophagy. It might enhance the frequency of the autophagy for the autophagic clearance inside the cells or inhibit a specific step in autophagy process. Curcumin an anticancer drug against malignant glioma, chronic myeloid leukemia, and esophageal cancer cells have the ability to induce the autophagic response inside the cells to provoke cell death and control cancer cell proliferation. It helps in increased expression of autophagic protein Beclin-1 and LC3 turnover. Hence, it plays an active role in autophagosome biogenesis. Furthermore, Reserveratol a phytoalexin shows anticancer activity against various type of cancers. It helps in the induction of autophagy by Beclin-1 independent pathway. It regulates the autophagy response by modulation of mTOR and Akt level inside the cells (Zhang et al. 2012). Drug resistance is the most common problem faced now a days for treating several cancers (Vasan et al. 2019). Due to the development of resistance against the specific drugs those are very much ineffective against the same due to the insensitivity of the cells (Munck et al. 2014). This may be due to the deregulation of apoptotic pathways in those resistant cells. But drugs like Saikosaponin-d which is a SERCA inhibitor, is found to modulate the cell death phenomenon in case of the cervical and breast cancer cells which are incompetent for undergoing apoptosis. Saikosaponin-d modulates Beclin-1 independent autophagic activity by increasing the frequency of autophagic flux. As the flux turnover increases the autophagy turnover also increases simultaneously (Wong et al. 2013). Silibinin which is a natural flavonoid has anticancer property against breast cancer cells. It has been reported that silibinin induces cell death in breast cancer cells by autophagy-dependent pathway by triggering ROS production and depletion in the ATP level. It can activate the ATG12–ATG5 axis formation for the modulation of autophagy but it was observed that in the presence of autophagic inhibitor like 3-Methyladenine (3-MA) and Bafilomycin A1 (Baf-A1) the cell death induction potential of silibinin reduces significantly. Hence, it is clear that it can activate autophagy to speed up the autophagic flux turnover and regulate cell death in breast cancer cells (Jiang et al. 2015). Pterostilbene is an effective anticancer agent against bladder cancer cells. It helps in the induction of autophagy through the inhibition of Akt/mTOR/p70S6K pathway and causes activation of the MEK/ERK1/2 pathway. It has been observed that Pterostilbene along with autophagy inhibitors like 3-MA and Baf-A1 is a much more potent anticancer drug against bladder cancer cells (Chen et al. 2010). Furthermore, Rottlerin is a potential anticancer agent against bladder cancer cells. It checks the cell proliferation by caspase-independent pathway. As per reports, the level of caspases and PARP which are the potent marker of the apoptosis remain unaltered. But it can induce autophagy to some great extent which might play an active role in regulation of cell death in bladder cancer cells or it may be said that it is an active molecule having a potent role in autophagic cell death mechanism (Qi et al. 2016). Furthermore, the isoflavonoid Genistein isolated from *Genista tinctoria* shows its cytotoxic effects in ovarian cancer. It is also a potent autophagy inducer and triggers the autophagosome biogenesis in A2780 cells by regulating the Akt expression (Gossner et al. 2007; Spagnuolo et al. 2015). Another small molecule Berberine has an active role in checking the growth of liver cancer cells. It can able to induce apoptosis by mitochondrial-dependent pathway. In addition to this, it can also activate autophagy cascade by modulating the Akt/mTOR pathway and trigger Beclin-1 dependent autophagy to form the mature autophagosomes which help in autophagic clearance of the cells (Wang et al. 2010). Furthermore, a quinine methide triterpenoid Celastrol shows anticancer activity against gastric cancer cells both in vitro and in vivo. It helps in inhibiting the cell proliferation in gastric cancer cells and simultaneously elevated the expression of autophagic marker proteins ATG5, ATG7, Beclin-1, and also the LC3 turnover. Hence, it might cause higher turnover of autophagic flux in case of gastric cancer cells, which leads to cell death (Lee et al. 2014). In addition to that, a bufadienolide from toad venom Arenobufagin shows its anticancer potential against hepatocellular carcinoma (HCC). It can able to inhibit cell proliferation by triggering apoptosis and helps in cell survival by inducing autophagic response inside the cells. It triggers the autophagic flux turnover for autophagic clearance in the cells which will restore the nutrients that supports tumor progression through PI3K/Akt/mTOR-dependent pathway. But the simultaneous addition of stage-specific autophagy inhibitors like 3-MA, Baf-A1, and chloroquine (CQ) could lead to enhanced cell death in hepatocellular carcinoma (HCC). Hence, the compound arenobufagin along with autophagy inhibitor can be a novel combined therapy compound against HCC in the case of humans (Zhang et al. 2013). Likewise, several compounds have been reported to induce autophagy in case of various cancers. But in most of the cases, autophagy plays a protective role for the progression of tumor growth and metastasis by maintaining the homeostasis within the cells. So, even if the compound activates autophagy they need another inhibitor for inducing cell death phenomenon in case of cancer to control their proliferative rate. Hence, scientists must focus to search for compounds which can inhibit the process of autophagy in stage-specific manner to control the disease progression in case of various cancer conditions (Table 13.1).

13.5 Molecules Inhibiting Autophagic Fusion in Regulation of Cancer Progression

The process of autophagy is well known for its dual role in cancer. On one hand, it inhibits cancer initiation but on the other, it helps in cancer progression in later stages. Hence therapeutic regulation of autophagy has a critical role in the war against cancer. Moreover, inhibition of protective autophagy and the promotion of

Table 13.1 Drugs targeting autophagy pathway to promote autophagosome and lysosome fusion

Drug name	Source	Autophagy target	References
Curcumin	*Curcumin longa*	Beclin-1, LC3	Zhang et al. (2012)
Reserveratol	*Polygonum cuspidatum*	Akt, mTOR	Zhang et al. (2012)
Saikosaponin-d	*Bupleurum falcatum* L.	LC3	Wong et al. (2013)
Silibinin	*Silybum marianum*	ATG12, ATG5	Jiang et al. (2015)
Pterostilbene	*Pterocarpus*, *Vitis vinifera* leaves, some berries (e.g., blueberries and cranberries), and in grapes	Akt, mTOR, p70S6K	Chen et al. (2010)
Rottlerin	Synthetic molecule also known as mallotoxin	LC3	Qi et al. (2016)
Genistein	*Genista tinctoria*	Akt	Gossner et al. (2007), Spagnuolo et al. (2015)
Berberine	*Coptidis rhizoma*	Akt, mTOR	Wang et al. (2010)
Celastrol	*Trypterigium wilfordii*	Beclin-1, ATG5, ATG7, LC3	Lee et al. (2014)
Arenobufagin	Bufadienolide from toad venom	PI3K, Akt, mTOR	Zhang et al. (2013)

autophagic cell death in cancer cells are our target to be achieved. Various compounds including a large number of natural products isolated and purified from plants are now gaining importance for targeting various key pathways in cancer therapeutics including the tumor-promoting autophagy response.

Inhibition of tumor-promoting autophagy is now considered as a novel way of targeting cancer and in this regard, several drugs are tested and found effective in inhibiting autophagy by targeting the fusion of the autophagosome with a lysosome in vitro. Among them, CQ and its derivative hydroxychloroquine (HCQ) are the two most effective drugs which are under clinical trial. The effect of CQ on autophagy inhibits the autophagic degradation process by limiting the delivery of sequestered cargo into the lysosome by blocking the fusion of the autophagosome with the lysosome and not through inhibiting the lysosomal action (Mauthe et al. 2018). Andrographolide (Andro), a diterpenoid lactone isolated from *Andrographis paniculata* is well known for its anticancer and anti-inflammatory effect but it has also been reported to be a potent inhibitor of protective autophagy by disrupting autophagosome–lysosome fusion without affecting the lysosomal function. Thus Andro treatment along with cisplatin, a DNA-damaging agent, leads to the sensitization of cancer cells towards apoptotic cell death (Zhou et al. 2012). Ginsenoside Ro (Ro) a natural product isolated from the plant *Panax ginseng* leads to autophagy inhibition by targeting autophagosome–lysosome fusion. Ro influences the lysosomal acidification and inhibits lysosomal Cathepsin activity, thus leads to an accumulation of LC3 and an increase in p62 in esophageal cancer cell lines. Ro mediated autophagic hindrance occurs through increasing ROS production by ESR2 (estrogen receptor 2)–NCF1 (neutrophil cytosolic factor 1) axis. Ro modulates the 5-FU induced protective autophagy and sensitizes chemoresistant esophageal cancer cells to cell death. Hence Ro could be used as a novel anticancer weapon in combination with others to counteract chemoresistance (Zheng et al. 2016). Berbamine (BBM), another natural compound, is reported to inhibit autophagy through inhibition of SNARE-mediated autophagosome–lysosome fusion in breast cancer cells through inhibition of the interaction between SNAP-29 and VAMP-8. Furthermore, BBM upregulates the expression of BNIP3 and induces the interaction between BNIP3 and SNAP-29. Hence such autophagy inhibitory effect of BBM could be exploited for their potential anticancer strategies through inhibition of autophagosome–lysosome fusion in different cancers (Fu et al. 2018). Oblongifolin c (Oc), extracted from *Garcinia yunnanensis hu*, is also reported to be a potent autophagy inhibitor as it inhibits autophagic flux through targeting autophagosome–lysosome fusion. Furthermore, Oc obstructs the lysosomal proteolytic activity through alteration of lysosomal acidification and reducing lysosomal cathepsin expression. Such anti-autophagic activity of Oc has been well explored as a potential strategy against cancer as it can sensitize the cancer cells to Caspase-3-dependent apoptosis in vitro. Furthermore, combined treatment with Oc and caloric restriction exaggerates anticancer efficacy of Oc in vivo (Lao et al. 2014). Withaferin A (WFA) and Withanolides isolated from the plant *Withania somnifera* were reported to block the autophagic flux and decrease lysosomal proteolytic activity, thus inhibiting the autophagosome–lysosome fusion in breast cancer cell lines. However, this action of WFA is not through alteration of the characteristic of autophagolysosome but through inhibiting the lysosomal protease activity. Furthermore, WFA is a potent AMPK inhibitor and hence in combination with 2-deoxyglucose it helps in inducing apoptosis and inhibits breast cancer growth (Muniraj et al. 2019). ECDD-S27, a compound derived from plant's arylnaphthalene lignan glycoside, i.e., cleistanthin A, was reported to have potent anticancer activity through targeting the autophagy process. ECDD-S27 treatment leads to impaired autophagic flux and the accumulation of autophagic vacuoles inside the cells. The mechanism of its action is through targeting the V-ATPase and thus hindering the lysosomal acidification (Paha et al. 2019). In addition to that, Vacuolin1 another compound which inhibits the fusion between autophagosome and lysosome and thereby hinders the autophagic process is reported to be more efficient as well as less toxic than chloroquine. Vacuoline1 targets the lysosomal acidification and calcium influx to the lysosome and thereby inhibits the lysosomal activity. Also, vacuoline1 mediates autophagosome–lysosome fusion inhibition by activation of the Rab5 GTPase activity, because the knockdown of Rab5 significantly hinders the fusion inhibition by vacuoline1 whereas the overexpression of the same enhances the fusion inhibition. However, though Rab5 activation is required for vacluolin1-mediated fusion inhibition, it is not the only target. Along with Rab5 activation, vacuoline1 alkalinize

Table 13.2 Drugs inhibiting autophagosome and lysosome fusion targeting lysosomal pathway

Drug name	Source	Mode of action	References
Chloroquine and hydroxychloroquine	Synthetic molecule	Lysosomal activity remains intact	Mauthe et al. (2018)
Andrographolide	*Andrographis paniculata*	Lysosomal activity remains intact	Zhou et al. (2012)
Ginsenoside Ro	*Panax ginseng*	Inhibits lysosomal cathepsin activity	Zheng et al. (2016)
Berbamine	*Berberis amurensis Rupr*	Affects the interaction between SNAP29 and VAMP8	Fu et al. (2018)
Oblongifolin-c	*Garcinia yunnanensis hu*	Alter lysosomal pH and decreases cathepsin activity	Lao et al. (2014)
Withaferin A	*Withania somnifera*	Decreases lysosomal protease activity	Muniraj et al. (2019)
ECDD-S27	Synthetic molecule	Affects lysosomal pH by targeting V-ATPase	Paha et al. (2019)
Vacuoline1	Small molecule	Targets lysosomal acidification and Calcium influx	Lu et al. (2014)
Pulsatilla saponin D	*Pulsatilla chinensis*	Alter lysosomal pH and decreases cathepsin activity	Wang et al. (2019)

lysosomal system, reduces calcium concentration of lysosome and altogether these affect the autophagosome–lysosome fusion process (Lu et al. 2014). Furthermore, Pulsatilla saponin D (PSD) another potent autophagy inhibitor leads to inhibition of autophagosome–lysosome fusion and thereby autophagic cargo degradation. PSD is also reported for significant inhibition of lysosomal acidification and activation of lysosomal cathepsin in breast cancer cell lines. Such anti-autophagic activity of PSD through blockage of autophagosome–lysosome fusion and lysosomal acidification can be a good anticancer strategy and the use of PSD along with camptothecin will increase its anticancer potential in breast cancer (Wang et al. 2019). So the natural products paved a nice way towards the cancer prevention and therapeutics not only by inhibiting tumor-promoting autophagy but also by sensitizing the chemoresistant tumors towards cell death (Table 13.2).

13.6 Potential Challenges and Future Perspectives

Many documents have been reported on natural and synthetic molecules to modulate autophagy and cancer in human systems for possible cancer therapeutics. Although the in vitro assay of those compounds already established in relation to autophagy and cancer, the preclinical and clinical trials are yet to be confirmed. As per now, autophagy is a multistep process having a dual role in case of the cancer progression and metastasis. The drugs discovered to date mainly targeting autophagy to various stages in it. Mostly, they can induce autophagy for the survival of the cells by promoting the autophagic flux turnover. This can generate a nutrient pool that can be reused by the cells to maintain the homeostasis (He et al. 2018). Also, excessive autophagy leads to a loss in the lysosomal pool inside the cells and can be recovered by the process of ALR which is essential to maintain the lysosomal homeostasis in the cells. Hence, the genes which are localized to lysosomes mostly drive the fusion between autophagosome and lysosomes to facilitate the autophagic response. This fusion point is also essential because of its relevance in autophagy and endocytosis crosstalk (Mahapatra et al. 2019). So, it is essential to target the fusion step of autophagy which is known to be the termination step to date. As this step decides the fate of the cells, i.e., in case of excessive autophagic flux turn over or in the absence of autophagic flux results in cell death in most of the cancer cells due to the deprivation of nutrients or the accumulation of autophagic vacuoles, respectively (Bialik et al. 2018; González-Polo et al. 2005). Whereas, the undisturbed fusion leads to cytoprotective autophagy in cancer cells. Hence, the fusion step aids the dual role to play during cancer for autophagy response inside the cells. So, it is an open challenge for the scientists to rediscover and rethink about the directive inhibition or modulation of autophagy in cancer patients to achieve the actual effective and cancer-specific drugs targeting those specific cancer cells. As the reports suggested a critical role of the endocytosis in regulation of autophagy, it is the most crucial step to redesign the drugs which can specifically act upon the cancer cells and modulate the autophagic response in those cells to get a better and specific therapeutic approach in cancer patients. Hence, it is the need of the hour to check the efficacy of drugs targeting this fusion process to find a novel therapeutic approach for cancer progression and metastasis.

13.7 Conclusion

Although science has been progressed much for a decade on the advancement of cancer therapeutics by natural components which targets stage-specific regulation of autophagy; it is the need of the time to establish preclinical and clinical approaches toward these targets. These compounds target various genes and pathways in the cells, so doing in vitro validation is not enough for the therapeutic strategies against cancer. The disease models for different cancers need to be established for the enforcement of these drugs and their validation. Moreover, natural therapeutics have an edge over classical methods like surgery, irradiation, and chemotherapy due to its less toxicity and commercial availability. It can be potentially used by many of the victims irrespective of the socio-economic status of the patients. Hence, it is the time to focus on these novel targets for the service of mankind in cancer prevention and its therapeutics.

Acknowledgments The authors declare that they have no conflict of interest in this chapter. K.K.M. is highly indebted to the Government of India, Ministry of Science and Technology, Department of Science and Technology (No. DST/INSPIRE Fellowship/2017/IF170344). Research support was partly provided by the Science and Engineering Research Board (SERB) [EMR/2016/001246], Department of Science and Technology.

References

Axe EL, Walker SA, Manifava M, Chandra P, Roderick HL, Habermann A, Griffiths G, Ktistakis NT (2008) Autophagosome formation from membrane compartments enriched in phosphatidylinositol 3-phosphate and dynamically connected to the endoplasmic reticulum. J Cell Biol 182:685–701

Bialik S, Dasari SK, Kimchi A (2018) Autophagy-dependent cell death—where, how and why a cell eats itself to death. J Cell Sci 131:jcs215152

Chen Y, Yu L (2013) Autophagic lysosome reformation. Exp Cell Res 319:142–146

Chen Y, Yu L (2017) Recent progress in autophagic lysosome reformation. Traffic (Copenhagen, Denmark) 18:358–361

Chen Y, Yu L (2018) Development of research into autophagic lysosome reformation. Mol Cells 41:45–49

Chen R-J, Ho C-T, Wang Y-J (2010) Pterostilbene induces autophagy and apoptosis in sensitive and chemoresistant human bladder cancer cells. Mol Nutr Food Res 54:1819–1832

Chen D, Fan W, Lu Y, Ding X, Chen S, Zhong Q (2012) A mammalian autophagosome maturation mechanism mediated by TECPR1 and the Atg12-Atg5 conjugate. Mol Cell 45:629–641

Deretic V (2006) Autophagy as an immune defense mechanism. Curr Opin Immunol 18:375–382

Du W, Su QP, Chen Y, Zhu Y, Jiang D, Rong Y, Zhang S, Zhang Y, Ren H, Zhang C, Wang X, Gao N, Wang Y, Sun L, Sun Y, Yu L (2016) Kinesin 1 drives autolysosome tubulation. Dev Cell 37:326–336

Edinger AL, Thompson CB (2003) Defective autophagy leads to cancer. Cancer Cell 4:422–424

Eskelinen EL (2011) The dual role of autophagy in cancer. Curr Opin Pharmacol 11:294–300

Fu R, Deng Q, Zhang H, Hu X, Li Y, Liu Y, Hu J, Luo Q, Zhang Y, Jiang X, Li L, Yang C, Gao N (2018) A novel autophagy inhibitor berbamine blocks SNARE-mediated autophagosome-lysosome fusion through upregulation of BNIP3. Cell Death Dis 9:243

González-Polo R-A, Boya P, Pauleau A-L, Jalil A, Larochette N, Souquère S, Eskelinen E-L, Pierron G, Saftig P, Kroemer G (2005) The apoptosis/autophagy paradox: autophagic vacuolization before apoptotic death. J Cell Sci 118:3091–3102

Gossner G, Choi M, Tan L, Fogoros S, Griffith KA, Kuenker M, Liu JR (2007) Genistein-induced apoptosis and autophagocytosis in ovarian cancer cells. Gynecol Oncol 105:23–30

Guerra F, Bucci C (2016) Multiple roles of the small GTPase Rab7. Cell 5:34

He L, Zhang J, Zhao J, Ma N, Kim SW, Qiao S, Ma X (2018) Autophagy: the last defense against cellular nutritional stress. Adv Nutr 9:493–504

Hurley JH, Young LN (2017) Mechanisms of autophagy initiation. Annu Rev Biochem 86:225–244

Islam MA, Sooro MA, Zhang P (2018) Autophagic regulation of p62 is critical for cancer therapy. Int J Mol Sci 19:1405

Itakura E, Kishi-Itakura C, Mizushima N (2012) The hairpin-type tail-anchored SNARE syntaxin 17 targets to autophagosomes for fusion with endosomes/lysosomes. Cell 151:1256–1269

Jiang K, Wang W, Jin X, Wang Z, Ji Z, Meng G (2015) Silibinin, a natural flavonoid, induces autophagy via ROS-dependent mitochondrial dysfunction and loss of ATP involving BNIP3 in human MCF7 breast cancer cells. Oncol Rep 33:2711–2718

Kabeya Y, Mizushima N, Yamamoto A, Oshitani-Okamoto S, Ohsumi Y, Yoshimori T (2004) LC3, GABARAP and GATE16 localize to autophagosomal membrane depending on form-II formation. J Cell Sci 117:2805–2812

Kinsey C, Balakrishnan V, O'Dell Michael R, Huang Jing L, Newman L, Whitney-Miller Christa L, Hezel Aram F, Land H (2014) Plac8 links oncogenic mutations to regulation of autophagy and is critical to pancreatic cancer progression. Cell Rep 7:1143–1155

Klionsky DJ, Eskelinen EL, Deretic V (2014) Autophagosomes, phagosomes, autolysosomes, phagolysosomes, autophagolysosomes... wait, I'm confused. Autophagy 10:549–551

Lao Y, Wan G, Liu Z, Wang X, Ruan P, Xu W, Xu D, Xie W, Zhang Y, Xu H, Xu N (2014) The natural compound oblongifolin C inhibits autophagic flux and enhances antitumor efficacy of nutrient deprivation. Autophagy 10:736–749

Lee H-W, Jang KSB, Choi HJ, Jo A, Cheong J-H, Chun K-H (2014) Celastrol inhibits gastric cancer growth by induction of apoptosis and autophagy. BMB Rep 47:697–702

Levine B (2007) Autophagy and cancer. Nature 446:745–747

Li L, Chen H, Gao Y, Wang YW, Zhang GQ, Pan SH, Ji L, Kong R, Wang G, Jia YH, Bai XW, Sun B (2016) Long noncoding RNA MALAT1 promotes aggressive pancreatic cancer proliferation and metastasis via the stimulation of autophagy. Mol Cancer Ther 15:2232–2243

Liang C, Jung JU (2010) Autophagy genes as tumor suppressors. Curr Opin Cell Biol 22:226–233

Liu X, Klionsky DJ (2018) Regulation of autophagic lysosome reformation by kinesin 1, clathrin and phosphatidylinositol-4,5-bisphosphate. Autophagy 14:1–2

Lőrincz P, Juhász G (2020) Autophagosome-lysosome fusion. J Mol Biol 432:2462–2482

Lu Y, Dong S, Hao B, Li C, Zhu K, Guo W, Wang Q, Cheung K-H, Wong CWM, Wu W-T, Markus H, Yue J (2014) Vacuolin-1 potently and reversibly inhibits autophagosome-lysosome fusion by activating RAB5A. Autophagy 10:1895–1905

Machihara K, Namba T (2019) BAP31 inhibits cell adaptation to ER stress conditions, negatively regulating autophagy induction by interaction with STX17. Cell 8:1350

Mahapatra KK, Panigrahi DP, Praharaj PP, Bhol CS, Patra S, Mishra SR, Behera BP, Bhutia SK (2019) Molecular interplay of autophagy and endocytosis in human health and diseases. Biol Rev Camb Philos Soc 94:1576–1590

Mauthe M, Orhon I, Rocchi C, Zhou X, Luhr M, Hijlkema KJ, Coppes RP, Engedal N, Mari M, Reggiori F (2018) Chloroquine inhibits autophagic flux by decreasing autophagosome-lysosome fusion. Autophagy 14:1435–1455

Munck C, Gumpert HK, Wallin AIN, Wang HH, Sommer MOA (2014) Prediction of resistance development against drug combinations by collateral responses to component drugs. Sci Transl Med 6:262ra156

Muniraj N, Siddharth S, Nagalingam A, Walker A, Woo J, Győrffy B, Gabrielson E, Saxena NK, Sharma D (2019) Withaferin A inhibits lysosomal activity to block autophagic flux and induces apoptosis via energetic impairment in breast cancer cells. Carcinogenesis 40:1110–1120

Nagelkerke A, Sieuwerts AM, Bussink J, Sweep FC, Look MP, Foekens JA, Martens JW, Span PN (2014) LAMP3 is involved in tamoxifen resistance in breast cancer cells through the modulation of autophagy. Endocr Relat Cancer 21:101–112

Paha J, Kanjanasirirat P, Munyoo B, Tuchinda P, Suvannang N, Nantasenamat C, Boonyarattanakalin K, Kittakoop P, Srikor S, Kongklad G, Rangkasenee N, Hongeng S, Utaisincharoen P, Borwornpinyo S, Ponpuak M (2019) A novel potent autophagy inhibitor ECDD-S27 targets vacuolar ATPase and inhibits cancer cell survival. Sci Rep 9:9177

Panda PK, Mukhopadhyay S, Das DN, Sinha N, Naik PP, Bhutia SK (2015) Mechanism of autophagic regulation in carcinogenesis and cancer therapeutics. Semin Cell Dev Biol 39:43–55

Parzych KR, Klionsky DJ (2014) An overview of autophagy: morphology, mechanism, and regulation. Antioxid Redox Signal 20:460–473

Qi P, He Z, Zhang L, Fan Y, Wang Z (2016) Rottlerin-induced autophagy leads to apoptosis in bladder cancer cells. Oncol Lett 12:4577–4583

Qiu Y, Hofmann K, Coats JE, Schulman BA, Kaiser SE (2013) Binding to E1 and E3 is mutually exclusive for the human autophagy E2 Atg3. Protein Sci 22:1691–1697

Qiu G, Li X, Che X, Wei C, He S, Lu J, Jia Z, Pang K, Fan L (2015) SIRT1 is a regulator of autophagy: Implications in gastric cancer progression and treatment. FEBS Lett 589:2034–2042

Rong Y, Liu M, Ma L, Du W, Zhang H, Tian Y, Cao Z, Li Y, Ren H, Zhang C, Li L, Chen S, Xi J, Yu L (2012) Clathrin and phosphatidylinositol-4,5-bisphosphate regulate autophagic lysosome reformation. Nat Cell Biol 14:924–934

Sanchez-Wandelmer J, Reggiori F (2013) Amphisomes: out of the autophagosome shadow? EMBO J 32:3116–3118

Santana-Codina N, Mancias JD, Kimmelman AC (2017) The role of autophagy in cancer. Annu Rev Cancer Biol 1:19–39

Singh SS, Vats S, Chia AY, Tan TZ (2018) Dual role of autophagy in hallmarks of cancer. Oncogene 37:1142–1158

Spagnuolo C, Russo GL, Orhan IE, Habtemariam S, Daglia M, Sureda A, Nabavi SF, Devi KP, Loizzo MR, Tundis R, Nabavi SM (2015) Genistein and cancer: current status, challenges, and future directions. Adv Nutr 6:408–419

Vasan N, Baselga J, Hyman DM (2019) A view on drug resistance in cancer. Nature 575:299–309

Wang N, Feng Y, Zhu M, Tsang C-M, Man K, Tong Y, Tsao S-W (2010) Berberine induces autophagic cell death and mitochondrial apoptosis in liver cancer cells: the cellular mechanism. J Cell Biochem 111:1426–1436

Wang K, Tu Y, Wan J-B, Chen M, He C (2019) Synergistic anti-breast cancer effect of pulsatilla saponin D and camptothecin through interrupting autophagic–lysosomal function and promoting p62-mediated ubiquitinated protein aggregation. Carcinogenesis. https://doi.org/10.1093/carcin/bgz140

Weidberg H, Shpilka T, Shvets E, Abada A, Shimron F, Elazar Z (2011) LC3 and GATE-16 N termini mediate membrane fusion processes required for autophagosome biogenesis. Dev Cell 20:444–454

Wong VKW, Li T, Law BYK, Ma EDL, Yip NC, Michelangeli F, Law CKM, Zhang MM, Lam KYC, Chan PL, Liu L (2013) Saikosaponin-d, a novel SERCA inhibitor, induces autophagic cell death in apoptosis-defective cells. Cell Death Dis 4:e720

Yun CW, Lee SH (2018) The roles of autophagy in cancer. Int J Mol Sci 19:3466

Zhang M, Chen L, Wang S, Wang T (2009) Rab7: roles in membrane trafficking and disease. Biosci Rep 29:193–209

Zhang X, Chen L-X, Ouyang L, Cheng Y, Liu B (2012) Plant natural compounds: targeting pathways of autophagy as anti-cancer therapeutic agents. Cell Prolif 45:466–476

Zhang D-M, Liu J-S, Deng L-J, Chen M-F, Yiu A, Cao H-H, Tian H-Y, Fung K-P, Kurihara H, Pan J-X, Ye W-C (2013) Arenobufagin, a natural bufadienolide from toad venom, induces apoptosis and autophagy in human hepatocellular carcinoma cells through inhibition of PI3K/Akt/mTOR pathway. Carcinogenesis 34:1331–1342

Zhang J, Zhou W, Lin J, Wei P, Zhang Y, Jin P, Chen M, Man N, Wen L (2016) Autophagic lysosomal reformation depends on mTOR reactivation in H2O2-induced autophagy. Int J Biochem Cell Biol 70:76–81

Zheng K, Li Y, Wang S, Wang X, Liao C, Hu X, Fan L, Kang Q, Zeng Y, Wu X, Wu H, Zhang J, Wang Y, He Z (2016) Inhibition of autophagosome-lysosome fusion by ginsenoside Ro via the ESR2-NCF1-ROS pathway sensitizes esophageal cancer cells to 5-fluorouracil-induced cell death via the CHEK1-mediated DNA damage checkpoint. Autophagy 12:1593–1613

Zhou J, Hu SE, Tan SH, Cao R, Chen Y, Xia D, Zhu X, Yang XF, Ong CN, Shen HM (2012) Andrographolide sensitizes cisplatin-induced apoptosis via suppression of autophagosome-lysosome fusion in human cancer cells. Autophagy 8:338–349

CPSIA information can be obtained
at www.ICGtesting.com
Printed in the USA
LVHW010900011121
702103LV00004B/115